Embryonic Stem Cells

Practical Approach Series

For full details of the Practical Approach titles currently available, please go to www.oup.com/pas. The following titles may be of particular interest:

Zebrafish (No 261)
Edited by Christiane Nusslein-Volhard

This book not only provides a complete set of instruction that will allow researchers to establish the zebrafish in their laboratory, it also gives a broad overview of commonly used methods and a comprehensive collection of protocols describing the most powerful techniques.

June 2002 0-19-963808-X (Pbk)

The Internet for Molecular Biologists (No 269)
Edited by Clare E. Sansom

This book helps the molecular biologist who is more at home at the laboratory bench than in front of the computer, to use the internet more effectively.

November 2003 0-19-963888-8 (Pbk)

Embryonic Stem Cells

A Practical Approach

Edited by

Elena Notarianni

University of Newcastle upon Tyne,
Comparative Biology Centre,
Medical School,
Framlington Place,
Newcastle upon Tyne, NE2 4HH, UK

Martin J. Evans

Cardiff School of Biosciences,
College of Medicine, Biology,
Health and Life Sciences,
Cardiff University,
Cardiff, CF10 3US, UK

OXFORD
UNIVERSITY PRESS

Great Clarendon Street, Oxford OX2 6DP

Oxford University Press is a department of the University of Oxford.
It furthers the University's objective of excellence in research, scholarship,
and education by publishing worldwide in

Oxford New York

Auckland Cape Town Dar es Salaam Hong Kong Karachi
Kuala Lumpur Madrid Melbourne Mexico City Nairobi
New Delhi Shanghai Taipei Toronto

With offices in

Argentina Austria Brazil Chile Czech Republic France Greece
Guatemala Hungary Italy Japan Poland Portugal Singapore
South Korea Switzerland Thailand Turkey Ukraine Vietnam

Oxford is a registered trade mark of Oxford University Press
in the UK and in certain other countries

Published in the United States
by Oxford University Press Inc., New York

© Oxford University Press 2006

The moral rights of the authors have been asserted
Database right Oxford University Press (maker)

First published 2006

All rights reserved. No part of this publication may be reproduced,
stored in a retrieval system, or transmitted, in any form or by any means,
without the prior permission in writing of Oxford University Press,
or as expressly permitted by law, or under terms agreed with the appropriate
reprographics rights organization. Enquiries concerning reproduction
outside the scope of the above should be sent to the Rights Department,
Oxford University Press, at the address above

You must not circulate this book in any other binding or cover
and you must impose the same condition on any acquirer

British Library Cataloguing in Publication Data

Data available

Library of Congress Cataloging in Publication Data
Embryonic stem cells : a practical approach / edited by Elena Notarianni, Martin J. Evans.
  p. ; cm. — (Practical approach series)
  Includes bibliographical references and index.
  ISBN-13: 978-0-19-855001-3 (alk. paper)
  ISBN-10: 0-19-855001-4 (alk. paper)
  ISBN-13: 978-0-19-855000-6 (alk. paper)
  ISBN-10: 0-19-855000-6 (alk. paper)
1. Embryonic stem cells.
  [DNLM: 1. Stem Cells—physiology. 2. Embryo—physiology. 3. Mice. 4. Models, Animal.
QU 325 E5436 2006] I. Notarianni, Elena. II. Evans, Martin J., Prof. III. Series.
  QP277.E43 2006
  612.6'4—dc22   2006000423

Typeset by Newgen Imaging Systems (P) Ltd., Chennai, India
Printed in Great Britain
on acid-free paper by
Antony Rowe Ltd, Chippenham, Wiltshire

ISBN 0-19-855000-6   978-0-19-855000-6
ISBN 0-19-855001-4 (Pbk.)   978-0-19-855001-3 (Pbk.)

10 9 8 7 6 5 4 3 2 1

Preface

The groundbreaking development of embryonic stem (ES) cells from the mouse in the early 1980s generated a sustained expansion of research worldwide into their exploitation, leading to the genetic engineering of the mouse to an unprecedented level. This ES cell-based methodology has contributed enormously to the understanding of biological processes in the context of the whole animal. With the advent of ES cells from other species including the human, ES cell biology will remain for the foreseeable future a crucial and growing area of research, with far-reaching implications for developmental and comparative biology, and for human health.

An early, landmark volume, *Teratocarcinomas and Embryonic Stem Cells: a Practical Approach*, edited by E.J. Robertson and published by Oxford University Press in 1987, provided the foundation for the nascent technology in terms of protocols for deriving mouse ES cells, and for their use in the generation of chimeric and transgenic mice – crucial advancements for what has become routine genetic manipulation. That volume also contained chapters devoted to embryonal carcinoma (EC) cells of the mouse (their generation, manipulation, and differentiation), and to the isolation of human EC cells from germ cell tumours, reflecting the fact that the successful establishment of ES cells in culture was based largely on antecedent work on EC cells. In the intervening period much progress has been made in the use of mouse ES cells for research into developmental biology, and for the provision of mouse models of human disease syndromes. Progress has been made also in the derivation of ES cells from non-murine species – especially for human and non-human primates where pluripotency has been amply demonstrated. For current projections for ES cell-based clinical therapies to be realized, the challenge is to control ES cell growth and differentiation, most specifically in the human system.

This volume is in the tradition of *Teratocarcinomas and Embryonic Stem Cells*, providing an introduction to the field of ES cell derivation and maintenance as well as up-to-date techniques for experimental manipulation. It also incorporates methods for non-murine species so that common, underlying themes – as well as species-specific differences – may be brought into relief. Topics include: embryoid body formation; cardiac, haematopoietic, and neural differentiation of ES cells; FACS analysis of ES cells and their differentiated derivatives; and ES cell transfection techniques.

We thank our authors for their enthusiasm and efforts in making this volume possible and the editing a pleasure. We are indebted to Ian Sherman, Kerstin Demata, Dave Hames, Melissa Dixon, Abbie Headon, Anita Petrie, Elizabeth Paul and other staff at OUP involved in the production, for patiently facilitating every stage. EN would like especially to thank Rus Hoelzel for his encouragement and advice, Paul Flecknell for all his kind and generous support, and Richard Gardner for his excellent suggestions at the outset of this project. The support of Herbie Newell and Nicola Curtin also is gratefully acknowledged here.

Elena Notarianni and Martin J. Evans
September 2005

Contents

Protocol list xiii
Abbreviations xvii
Contributors xxiii

1 **Introduction**   1
   Martin J. Evans
   1 Aims and scope of the volume   1
   2 Control of differentiation   4
   3 Future perspectives   5

2 **Procedures for deriving ES cell lines from the mouse**   7
   Frances A. Brook
   1 Introduction   7
   2 Derivation of mES cell lines   7
   3 General principles of mES cell culture   11
      3.1 Culture media and supplements   11
      3.2 Preparation of MEFs   11
      3.3 Preparation of feeder layers   13
      3.4 Fetal-calf serum (FCS)   15
      3.5 Routine culture of mES cells   18
   4 Methods for the derivation of mES cell lines   22
      4.1 Derivation of mES cells from intact blastocysts   22
      4.2 Derivation of mES cells from isolated epiblast   23
   5 Validation of new mES cell lines   33
      5.1 Karyotype   33
      5.2 Pluripotency   33
   6 Female mES cell lines   35

3 **Production of ES cell-derived mice**   41
   Yoko Kato and Yukio Tsunoda
   1 Introduction   41
   2 Maintenance of mES cell lines and embryo culture   42

3 Reconstituting tetraploid embryos with mES cells   45
   3.1 Tetraploid embryos   45
   3.2 Mouse ES cell lines for tetraploid-embryo reconstitution   45
   3.3 Methods for reconstituting tetraploid embryos with mES cells   46

4 Hyperthermic treatment of diploid host blastocysts   52

5 Nuclear transfer using mES cells as karyoplasts   53
   5.1 Mouse strains for mES-cell NT   55
   5.2 Cell-cycle synchronization in mES cells   56
   5.3 Enucleation of MII-phase oocytes   58
   5.4 Activation of nuclear-transfer embryos   66

6 Analysis of distribution of mES-cell derivatives in chimeras   68

7 Perinatal mortality and developmental abnormalities in mES cell-derived mice   69

4 Generating animal models of human mitochondrial genetic disease using mouse ES cells   72

Grant R. MacGregor, Wei Wei Fan, Katrina G. Waymire, and Douglas C. Wallace

1 Background   72
   1.1 Mitochondrial genetics and biochemistry   74
   1.2 Deleterious mtDNA mutations in hereditary disease   78
   1.3 Generation, identification, and recovery of mtDNA mutations from mice   78

2 Introduction of mtDNA mutations into the female mouse germline   81
   2.1 Screening for mutations in mtDNA   81
   2.2 Preparation of ρ^0 cells for use as recipient cells for mtDNA transfer   83
   2.3 Preparation of cytoplasts from cell lines harbouring a mtDNA polymorphism   88
   2.4 Generation of brain synaptosomes from mice harbouring mtDNA polymorphisms   88
   2.5 Recovery of mtDNA mutations by electrofusion of cytoplasts or synaptosomes with ρ^0 or mES cells   90
   2.6 Culture of mES cells   93
   2.7 Depletion of mitochondria in mES cells by treatment with R6G   94
   2.8 Culture and selection of *trans*-mitochondrial mES cybrids   94
   2.9 PCR-based allele genotyping of the nuclear genome of cybrids   98
   2.10 Use of X-linked GFP transgenic mice to select female blastocysts for microinjection with *trans*-mitochondrial mES cells   99
   2.11 PCR-based allele genotyping of mice   100

3 Phenotypic analysis of *trans*-mitochondrial, chimeric mice   102

4 Concluding comments   104

5 Controlling the differentiation of mouse ES cells *in vitro* 112
Michael V. Wiles and Gabriele Proetzel

1 Introduction 112

2 Experimental rationale 116

3 Methods 117
    3.1 General considerations for controlled mES-cell differentiation 117
    3.2 Recommendations for CDM preparation and experimental design 119
    3.3 Neuroectoderm differentiation 124
    3.4 Mesoderm and haematopoietic differentiation 124
    3.5 Mesoderm and neuroectoderm differentiation 125

4 Evaluation of mES-cell differentiation 125

5 RNA isolation 125

6 Conclusions 128

6 *In vitro* differentiation of mouse ES cells into muscle cells 130
Yelena S. Tarasova, Daniel R. Riordon, Kirill V. Tarasov and Kenneth R. Boheler

1 Introduction 130

2 Culture of mES cell lines for differentiation into muscle cells 131

3 Differentiation of mES cells into muscle cell lineages 134
    3.1 Differentiation of mES cells into cardiac muscle cells 134
    3.2 Differentiation of mES cells into skeletal myocytes 135
    3.3 Differentiation of mES cells into smooth-muscle cells 141

4 Isolation techniques for differentiated muscle-cell derivatives 142
    4.1 Genetically modified mES cells for enrichment of muscle-cell derivatives 144

5 Assessing muscle-gene expression during *in vitro* differentiation of mES cells 148
    5.1 RNA isolation from cell extracts 148
    5.2 Reverse-transcription PCR, and quantitative reverse-transcription PCR 148
    5.3 Non-radioactive *in situ* detection of transcripts in EBs 149

6 Assessing expression and function of muscle-specific proteins 156
    6.1 Immunofluorescence 156
    6.2 Functional analyses of muscle derivatives 163
    6.3 Electrophysiological measurements in mES cell-derived cardiac cells 164

7 *In vitro* **differentiation of mouse ES cells into haematopoietic cells** 169
Osamu Ohneda and Masayuki Yamamoto
   1 Introduction 169
      1.1 Haematopoietic development 169
      1.2 Haematopoietic differentiation in mouse ES cells 169
      1.3 The mES/OP9-cell coculture system 170
   2 Routine maintenance of cell lines 171
      2.1 Maintenance of mES cells 171
      2.2 MEF culture and feeder-layer preparation 171
      2.3 Maintenance of OP9 stromal cells, and preparation for coculture 171
      2.4 Fetal-calf serum (FCS) 173
   3 mES/OP9-cell coculture system for haematopoietic differentiation 173
      3.1 Primitive erythroid cell differentiation 175
      3.2 Definitive erythroid cell differentiation 176
      3.3 Definitive haematopoietic progenitors and CFU-OP9 181
   4 Analysis of definitive haematopoietic cells produced using the mES/OP9-cell coculture system 183
   5 Evaluation of the mES/OP9-cell coculture system as a model for haematopoietic differentiation 186

8 **Lineage selection and transplantation of mouse ES cell-derived neural precursors** 189
Tanja Schmandt, Tamara Glaser and Oliver Brüstle
   1 Introduction 189
      1.1 Neural differentiation potential of mES cells *in vitro* 189
   2 Controlled differentiation of mES cells into neural precursors 190
   3 Selection of lineage-restricted cell populations from differentiated mES cells 196
      3.1 Transfection of mES cells with selectable marker genes, for neuronal and oligodendroglial lineage selection 196
      3.2 Neuronal lineage selection 198
      3.3 Oligodendroglial lineage selection 203
   4 Transplantation models for studying functional integration of mES cell-derived neural precursors 203
      4.1 Neural precursor integration in organotypic slice cultures 203
      4.2 *In utero* transplantation of mES cell-derived neural progenitors 206
      4.3 Identification and characterization of transplanted cells using donor-specific markers 211
      4.4 Immunohistochemical detection of transplanted cells with donor-specific antibodies 211
   5 Translation to human ES cell-derived neural precursors 211

9 *In vitro* differentiation of mouse ES cells into pancreatic and hepatic cells 218
Przemyslaw Blyszczuk, Gabriela Kania and Anna M. Wobus

1 Introduction 218
2 Maintenance of undifferentiated mES cells 219
3 *In vitro* differentiation procedures 219
   3.1 Generation of mES cell-derived, heterogeneous lineage progenitor cells 219
   3.2 Induction of lineage-specific differentiation 223
4 Characterization of differentiated, pancreatic and hepatic cell phenotypes 227
   4.1 RT-PCR 228
   4.2 Immunofluorescence staining 228
   4.3 ELISA 232
   4.4 Microscopic analysis and imaging of samples labelled by immunofluorescence 234
5 Animal transplantation models 234
   5.1 Pancreatic islet regeneration models 234
   5.2 Liver reconstitution models 235

10 Isolation and characterization of human ES cells 238
Martin F. Pera, Andrew Laslett, Susan M. Hawes, Irene Tellis, Karen Koh and Lihn Nguyen

1 Introduction 238
2 Isolation of hES cells 239
   2.1 Human embryos 239
   2.2 Initiation of primary cultures 241
3 Human ES cell maintenance: serum-supplemented *versus* serum-free media 243
4 Cryopreservation of human ES cells 244
5 Characterization of hES cells 244
   5.1 Immunochemical characterization of hES cells 249
   5.2 FACS analysis of hES-cell antigen expression 253
   5.3 Immunomagnetic isolation of viable hES cells 253
   5.4 Gene expression in hES cells 253
   5.5 Biological assay of hES-cell pluripotentiality 258

11 Differentiation of human ES cells 260
Sharon Gerecht-Nir and Joseph Itskovitz-Eldor

1 Introduction 260
2 Maintenance of hES cell lines 261
   2.1 Spontaneous differentiation of hES cells 264
3 *In vivo* differentiation of hES cells 265

CONTENTS

4 *In vitro* differentiation of hES cells 267
  4.1 Differentiation via the formation of embryoid bodies 267
  4.2 Vascular morphogenesis within EBs 269
  4.3 Cardiac differentiation in EBs 272
5 Guided differentiation 276
  5.1 Haematopoietic differentiation 277
  5.2 Non-EB-mediated hES-cell differentiation: vascular development 277
6 Scale-up procedures 281
  6.1 Dynamic systems for EB production 284
  6.2 Production of hES cell-derived EBs within three-dimentional matrices 288
7 Epilogue: future clinical applications and limitations 289

12 ES cell lines from the cynomolgus monkey (*Macaca fascicularis*) 294

Hirofumi Suemori, Yoshiki Sasai, Katsutsugu Umeda and Norio Nakatsuji

1 Introduction 294
  1.1 Primate *versus* murine ES cells 295
2 Establishment and maintenance of cES cell lines 296
  2.1 Preparation of feeder layers for cES-cell derivation 297
  2.2 Culture of ICMs from cynomolgus monkey blastocysts 298
  2.3 Expansion and routine maintenance of undifferentiated cES cells 301
3 Characterization of cES cell lines 301
  3.1 Expression of cES-cell markers and karyotype analysis 301
4 Genetic modification of cES cells 304
5 Differentiation of cES cells 305
6 Directed differentiation of cES cells induced by coculture 308
  6.1 Neural differentiation of cES cells induced by coculture with PA6 cells 309
  6.2 Haematopoietic differentiation of cES cells induced by coculture with OP9 cells 316
7 Conclusion 316

List of Suppliers 321

Index 327

Protocol list

Mouse ES cells: derivation and manipulation

2 Procedures for deriving ES cell lines from the mouse

    Culture media and stock solutions   12
    Preparation, maintenance, and cryopreservation of MEFs   14
    Maintenance of mES cells on MEF feeder layers   16
    Serum testing   19
    Detection of mycoplasma   21
    Embryo collection   24
    Derivation of mES cells from intact blastocysts   26
    Solutions required for epiblast isolation   29
    Microsurgical isolation of the epiblast   30
    Karyotype analysis   36

3 Production of ES cell-derived mice

    Preparation of mES cells for embryo reconstitution   44
    Production of tetraploid embryos   47
    Lectin-assisted aggregation of mES cells with zona-free, tetraploid embryos   48
    Coculture of mES cells with zona-free, tetraploid embryos   50
    Aggregation of mES cells with zona-cut, tetraploid embryos   51
    Hyperthermic treatment of mouse blastocysts   53
    M-phase and G_1-phase synchronization of mES cells for nuclear transfer   57
    Collection and enucleation of MII-phase mouse oocytes   59
    Inactivated Sendai virus-induced fusion   62
    Electrofusion   64
    Nuclear transfer by direct injection   64
    Strontium chloride-induced activation   67
    Electrical activation   67

4 Generating animal models of human mitochondrial genetic disease using mouse ES cells

    Characterization of mutations in the mouse mitochondrial genome   82
    Preparation of ρ^0 cells   84
    Preparation and isolation of cytoplasts   85
    Production of brain synaptosomes   89

Electrofusion of synaptosomes or cytoplasts with ρ^0 cells or R6G-treated mES cells *92*
Depletion of mitochondria in mES cells by R6G treatment *95*
Selection and maintenance of mES cybrids *97*
Genotyping the nDNA of cybrids *99*
Selection of female blastocysts for microinjection *100*
PCR-based allele genotyping of mice *103*

Mouse ES cells: *in vitro* differentiation systems, lineage selection and analysis, and transplantation models

5 Controlling the differentiation of mouse ES cells *in vitro*
Preparation of mES cells for differentiation in CDM *120*
Mouse ES-cell differentiation in CDM: suspension culture *122*
Mouse ES-cell differentiation in CDM: hanging-drop culture *122*
Mouse ES-cell differentiation in CDM via EB formation followed by attachment culture *123*
Total RNA isolation from limiting amounts of mES cells/EBs *126*
DNAse I treatment of RNA samples *127*

6 *In vitro* differentiation of mouse ES cells into muscle cells
Preparation of mES cell cultures for *in vitro* differentiation *132*
Hanging-drop technique: production of cardiomyocytes *137*
Mass culture of EBs: production of cardiomyocytes *138*
Hanging-drop technique: production of skeletal muscle cells *140*
Hanging-drop technique: production of vascular smooth-muscle cells *141*
Mechanical dissection and recovery of muscle-cell derivatives *142*
Transformation of mES cells for selective enrichment of EB-derived cardiomyocytes *145*
Reverse transcription *151*
Real-time quantitative PCR (Q-PCR) *154*
Digoxigenin-labelled RNA probes *155*
Whole-mount ISH of EBs *157*
Cryosectioning of stained EBs *159*
Immunofluorescence analysis of EB outgrowths *161*
Rate measurements in EBs *163*

7 *In vitro* differentiation of mouse ES cells into haematopoietic cells
Maintenance of OP9 cells and preparation for coculture with mES cells *172*
Induction of primitive and definitive haematopoietic differentiation in mES cells using the mES/OP9-cell coculture system *178*
Isolation of definitive haematopoietic progenitors: the CFU-E, CFU-GM and CFU-OP9 assays *182*
Analysis of definitive haematopoietic cells *184*

8 Lineage selection and transplantation of mouse ES cell-derived neural precursors
Controlled differentiation of mES cells into neural precursors *193*
Transfection of mES cells by electroporation *197*
FACS®-based neuronal lineage selection *199*
Neuronal lineage selection using immunopanning *201*

Oligodendroglial lineage selection using selectable marker genes 202
Establishment of hippocampal slice cultures and
transplantation procedures 207
Intrauterine transplantation of mES cell-derived neural progenitors 210
Species-specific *in situ* hybridization 212

9 *In vitro* differentiation of mouse ES cells into pancreatic and hepatic cells

Preparation of mES cells for *in vitro* differentiation 220
Generation of EBs by the 'hanging-drop' method 221
Differentiation of EBs on an adhesive substratum 222
Induction of pancreatic differentiation 225
Induction of hepatic differentiation 226
Fixation of cells for immunostaining 230
Immunostaining 231
Sample preparation for insulin ELISA 232
Sample preparation for albumin ELISA 233

Human and non-human primate ES cells: derivation, *in vitro* differentiation systems, lineage selection and analysis

10 Isolation and characterization of human ES cells

Isolation of the ICM from human blastocysts 240
Preparation of MEF feeder layers 242
Subculture of hES cells using mechanical dissection of colonies under microscopic guidance 243
Subculture of hES cells under serum-free conditions, with feeder cell support 246
Vitrification of hES cells and recovery from cryopreservation (freezing and thawing) 247
Immunocytochemical examination of hES cells using indirect immunofluorescence 250
Multiplex FACS® analysis of hES-cell antigen expression 252
Immunomagnetic isolation of viable hES cells using antibody GCTM-2 254
Preparation of cDNA by reverse transcription of mRNA from hES cells 256
Formation of teratomas by implantation of hES cells into the testis capsule of SCID mice 257

11 Differentiation of human ES cells

Maintenance of undifferentiated hES cells 262
Teratoma formation by xenotransplantation of hES cells into immunocompromised mice 266
Production of EBs from hES cells in suspension culture 270
Immunofluorescence analysis of vascular structures in EBs 271
Preparation of contractile EBs for analysis 276
Haematopoietic differentiation of hES cells 278
Vascular differentiation of hES cells in monolayer culture 282
Three-dimensional gel assays for vasculogenesis by mesodermal hES-cell derivatives 283
Dynamic formation of EBs from hES cells using the STLV™ 287

12 ES cell lines from the cynomolgus monkey (*Macaca fascicularis*)

Immunosurgical isolation of ICMs from cynomolgus monkey blastocysts *298*
Primary ICM culture for cES-cell derivation *299*
Routine maintenance of cES cells *303*
Transfection of cES cells by electroporation *306*
Production of EBs from cES cells *308*
Maintenance of PA6 cells *311*
SDIA-induced neural differentiation of cES cells *312*
Immunostaining of cES-cell derivatives for neural markers *316*
Induction of haematopoetic differentiation in cES cells by coculture with OP9 cells *317*

Abbreviations

6TG	6-thioguanine
AC	alternating current
ALAS-E	5-aminolevulinate synthase
AP	action potential
bFGF	basic FGF (or FGF2)
BMP	bone morphogenetic protein
BrdU	5-bromodeoxyuridine
BSA	bovine serum albumin
BSS	balanced salt solution
CAFCs	cobblestone area-forming cells
CAP	chloramphenicol
CAP^R	CAP resistant
CAP^S	CAP sensitive
CDM	chemically defined medium
cDNA	complementary DNA (sequence)
CEM	cES-cell medium
cES	cynomolgus ES (cells)
CFU	colony forming unit
CFU-E	CFU-erythroid
CFU-G	CFU-granulocyte
CFU-GM	CFU-granulocyte and macrophage
CNP	2'3'-cyclic nucleotide 3'-phosphodiesterase
CNS	central nervous system
COX	cytochrome c oxidase
CR	control region
C-rich	cytosine-rich
DAB	diamino-benzidine
DAP	DMSO, acetamide and propylene glycol (freezing medium)
db-cAMP	dibutyryl-cyclic AMP
DC	direct current
DEPC	diethyl pyrocarbonate

ABBREVIATIONS

DIG	digoxigenin
Dil-Ac-LDL	Dil-acetylated-low density lipoprotein
DMEM	Dulbecco's modified Eagle's medium
DMSO	dimethyl sulphoxide
DPX	di-n-butyl phthalate/xylene (mountant)
EB	embryoid body
EC	embryonal carcinoma (cell)
ECs	endothelial cells (Chapter 11)
ECM	extracellular membrane
EDTA	ethylenediaminetetraacetic acid
EGF	epidermal growth factor
EGFP	enhanced green fluorescent protein
ELISA	enzyme-linked immunosorbent assay
Epo	erythropoietin
eryD	definitive erythrocytes
eryP	primitive erythrocytes
ES	embryonic stem (cell)
ESGP	mES cell-derived glial precursor
ESNP	mES cell-derived neural precursor
EtBr	ethidium bromide
F_1	first filial (generation)
F_2	second filial (generation)
FACS®	fluorescence-activated cell sorter
FAH	fumarylacetoacetate hydrolase
FCS	fetal calf serum
FGF	fibroblast growth factor
FHM	flushing holding medium
FITC	fluorescein isothiocyanate
F_w	forward (primer)
γmdrMEF	γ-irradiated multiple-drug-resistant MEF
GAPD	glyceraldehyde-3-phosphate dehydrogenase
G-banding	Giemsa-banding
GFAP	glial fibrillary acidic protein
GFP	green fluorescent protein
GM-CSF	granulocyte-macrophage-colony stimulating factor
GPI	glucose phosphate isomerase
G-rich	guanine-rich
GSC	goosecoid
GUP	glucose, uridine and pyruvate (-supplemented medium)
HAT	hypoxanthine, aminopterin and thymidine (-supplemented medium)
HBSS	Hanks' BSS
hCG	human chorionic gonadotropin
HCM	hepatocyte culture medium
HDM	haematopoietic differentiation medium

ABBREVIATIONS

Hepes	4-2-hydroxyethyl-1-piperazineethanesulphonic acid
hES	human embryonic stem (cell)
HLA	human leukocyte antigen
hph (or *hyg*)	hygromycin B phosphotransferase (gene)
HPLC	High-performance liquid chromatography
HPRT	hypoxanthine phosphoribosyltransferase
HQNO	2-n-heptyl-4-hydroxyquinoline N-oxide
HRP	horseradish peroxidase
HSC	haematopoietic stem cell
hyg	see *hph*
hyg^r	hygromycin resistant
ICM	inner cell mass
ICSI	intracytoplasmic sperm injection
IL	interleukin
IMDM	Iscove's modification of DMEM
ISH	*in situ* hybridization
ITSFn	neural selection medium (Chapter 8)
IVF	*in vitro* fertilization
KRBH	Krebs-Ringer bicarbonate Hepes (buffer)
KSOM	potassium simplex optimized medium
KSPG	keratan sulphate proteoglycan
LIF	leukaemia inhibitory factor
M2	embryo culture medium (Chapter 3)
M16	embryo culture medium with Hepes (Chapter 3)
MACS®	magnetic-activated cell sorting
MC	methylcellulose
MDM	mesodermal differentiation medium
MEA	microelectrode array
MEF	mouse embryonic fibroblast
MEM	modified Eagle's medium
mES	mouse ES (cell)
mES cell-NT	nuclear transfer using mES cells as karyoplasts
MHC	myosin heavy chain (Chapter 6)
MHC	major histocompatibility complex (Chapter 11)
mtDNA	mitochondrial DNA
MTG	monothioglycerol
mtPTP	mitochondrial permeability transition pore
NASA	National Aeronautics and Space Administration
NBT/BCIP	nitroblue tetrazolium/5-bromo-4-chloro-3-indolyl-phosphate
NCAM	neural cell adhesion molecule
NCX1	Na^+/Ca^{2+} exchanger
nDNA	nuclear (chromosomal) DNA
neo	neomycin phosphotransferase or aminoglycoside phosphotransferase (gene)
neo^r	neomycin resistant

ABBREVIATIONS

NGF	nerve growth factor
NOD	non-obese diabetic (mouse)
np	nucleotide pair
nt	nucleotide
NT	nuclear transfer
NTC	no-template control
NZB	New Zealand Black
OXPHOS	oxidative phosphorylation
p	passage (number)
pac	puromycin N-acetyl-transferase (gene)
pac^r	puromycin resistant
PAS	periodic acid–Schiff (reaction)
PBS	phosphate-buffered saline
PBS$^-$	PBS without Ca^{2+} or Mg^{2+} (Chapter 10)
PBS$^+$	PBS with Ca^{2+} and Mg^{2+} (Chapter 10)
PBST	PBS containing Tween-20
p.c.	post coitum
PCR	polymerase chain reaction
PDGF	platelet-derived growth factor
PDM	pancreatic differentiation medium
PHA	phytohaemagglutinin
PMSG	pregnant mare's serum gonadotropin
PNA	peptide nucleic acid
PSA	penicillin, streptomycin and amphotericin B (solution)
PSA-NCAM	polysialylated NCAM
PVA	polyvinyl alcohol
PVP	polyvinyl pyrrolidone
qPCR	(real-time) quantitative PCR
qRT-PCR	quantitative RT-PCR
ρ^0	rho zero (mtDNA-deficient cells, Chapter 4)
RA	all-*trans* retinoic acid
RCCS™	Rotating Cell-Culture System™
R6G	rhodamine 6G
ROS	reactive oxygen species
rRNA	ribosomal RNA
r.t.	room temperature
RTNC	negative control for RT reactions
RT-PCR	reverse-transcription PCR
R_v	reverse (primer)
SCF	stem cell factor
SCID	severe combined immunodeficiency
SDH	succinate dehydrogenase
SDIA	stromal cell-derived inducing activity
SED	sucrose, EDTA and dithiothreitol (solution)
SMA	smooth muscle α actin

SMC	smooth-muscle cell
SM-MHC	smooth-muscle myosin heavy chain (Chapter 6)
SR	serum replacement
SRM	serum-replacement hES-cell medium
Sry	sex-determining region, Y chromosome
SSC	saline sodium citrate
SSEA	stage-specific embryonic antigen
SSM	serum-supplemented hES-cell medium
STLV^{TM}	Slow Turning Lateral Vessel^{TM}
STO	SIM (Sandoz Inbred Mouse) embryo-derived thioguanine- and ouabain-resistant (fibroblast cell line)
STZ	streptozotocin
T	Brachyury
TGF	tumour (or transforming) growth factor
TH	tyrosine hydroxylase
TK	thymidine kinase
Tm	annealing temperature
TO-PRO	TO-PRO®-3-iodide
UV	ultraviolet
VCO_2	pulmonary CO_2 excretion
VDM	vascular differentiation medium
VEGF	vascular endothelial cell growth factor
VO_2	pulmonary O_2 consumption
VSM	vascular smooth muscle (Chapter 6)
v-SMC	vascular smooth muscle cell (Chapter 11)
wt	wild type
X^{GFP}	X-chromosome carrying a *GFP* transgene
X^m	maternally derived X-chromosome
X^p	paternally derived X-chromosome
YS	yolk sac

Contributors

Przemyslaw Blyszczuk
In Vitro Differentiation Group
Institute of Plant Genetics and
 Crop Plant Research (IPK)
D-06466 Gatersleben
Germany

Kenneth R. Boheler
Laboratory of Cardiovascular Science
National Institute on Aging, NIH
5600 Nathan Shock Drive
Baltimore, MD
USA

Frances A. Brook
Department of Zoology
University of Oxford
South Parks Road
Oxford OX1 3PS
United Kingdom

Oliver Brüstle
Institute of Reconstructive Neurobiology,
 Life and Brain Center
University of Bonn and
 Hertie Foundation
Sigmund-Freud-Strasse 25
53105 Bonn
Germany

Martin J. Evans
Cardiff School of Biosciences
Wales College of Medicine, Biology,
 Life and Health Sciences
Cardiff University
Cardiff CF10 3US
United Kingdom

Wei Wei Fan
Center for Molecular and
 Mitochondrial Medicine and Genetics
Hewitt Hall
University of California
Irvine, CA 92697–3940
USA

Sharon Gerecht-Nir
Harvard – M.I.T.
Division of Health Sciences and
 Technology
Massachusetts Institute of Technology
Cambridge, Massachusetts
 02139 USA

Tamara Glaser
Institute of Reconstructive Neurobiology,
 Life and Brain Center
University of Bonn and Hertie
 Foundation
Sigmund-Freud-Strasse 25
53105 Bonn
Germany

Susan M. Hawes
Australian Stem Cell Centre
Monash University
Building 75 STRIP, Wellington Road
Clayton Victoria 3800
Australia

Joseph Itskovitz-Eldor
Department of Obstetrics and
 Gynecology
Rambam Medical Center
Bruce Rappaport Faculty of Medicine

Technion – Israel Institute of Technology
Haifa
Israel

Gabriela Kania
In Vitro Differentiation Group
Institute of Plant Genetics and
  Crop Plant Research (IPK)
D-06466 Gatersleben
Germany

Yoko Kato
Laboratory of Animal Reproduction
College of Agriculture
3327-204 Nakamachi
Nara, 631-8505
Japan

Karen Koh
Australian Stem Cell Centre
Monash University
Building 75 STRIP, Wellington Road
Clayton Victoria 3800
Australia

Andrew Laslett
Australian Stem Cell Centre
Monash University
Building 75 STRIP, Wellington Road
Clayton Victoria 3800
Australia

Grant R. MacGregor
Center for Molecular and
  Mitochondrial Medicine and Genetics
Hewitt Hall
University of California
Irvine, CA 92697-3940
USA

Norio Nakatsuji
Division of Development and
  Differentiation
Institute for Frontier Medical Sciences
Kyoto University
53 Kawaharacho, Shogoin
Sakyo-ku, Kyoto 606-8507
Japan

Lihn Nguyen
Australian Stem Cell Centre
Monash University
Building 75 STRIP, Wellington Road
Clayton Victoria 3800
Australia

Osamu Ohneda
Graduate School of Comprehensive
  Human Science
University of Tsukuba
1-1-1 Tennoudai
Tsukuba 305-8575
Japan

Martin F. Pera
Australian Stem Cell Centre
Monash University
Building 75 STRIP, Wellington Road
Clayton Victoria 3800
Australia

Gabriele Proetzel
Scil Proteins
Heinrich-Damerow-Str. 1
06120 Halle
Germany

Daniel R. Riordon
Laboratory of Cardiovascular Science
National Institute on Aging, NIH
5600 Nathan Shock Drive
Baltimore, MD
USA

Yoshiki Sasai
Neurogenesis and Organogenesis Group
Riken Center for Developmental Biology
2-2-3 Minatojima-Minamimachi
Chuo, Kobe, 650-0047
Japan

Tanja Schmandt
Institute of Reconstructive Neurobiology,
  Life and Brain Center
University of Bonn and Hertie
  Foundation

Sigmund-Freud-Strasse 25
53105 Bonn
Germany

Hirofumi Suemori
Stem Cell Research Center
Institute for Frontier Medical Sciences
Kyoto University
53 Kawaharacho, Shogoin
Sakyo-ku, Kyoto 606–8507
Japan

Kirill V. Tarasov
Laboratory of Cardiovascular Science
National Institute on Aging, NIH
5600 Nathan Shock Drive
Baltimore, MD
USA

Yelena S. Tarasova
Laboratory of Cardiovascular Science
National Institute on Aging, NIH
5600 Nathan Shock Drive
Baltimore, MD
USA

Irene Tellis
Australian Stem Cell Centre
Monash University
Building 75 STRIP, Wellington Road
Clayton Victoria 3800
Australia

Yukio Tsunoda
Laboratory of Animal Reproduction
College of Agriculture
3327-204 Nakamachi
Nara, 631-8505
Japan

Katsutsugu Umeda
Department of Pediatrics
Graduate School of Medicine
Kyoto University
54 Kawahara-cho, Shogoin
Sakyo-ku, Kyoto 606–8507
Japan

Douglas C. Wallace
Center for Molecular and
    Mitochondrial Medicine and Genetics
Hewitt Hall
University of California
Irvine, CA 92697-3940
USA

Katrina G. Waymire
Center for Molecular and
    Mitochondrial Medicine and Genetics
Hewitt Hall
University of California
Irvine, CA 92697-3940
USA

Michael V. Wiles
The Jackson Laboratory
600 Main Street
Bar Harbor
Maine 04609-1500
USA

Anna M. Wobus
In Vitro Differentiation Group
Institute of Plant Genetics and
    Crop Plant Research (IPK)
D-06466 Gatersleben
Germany

Masayuki Yamamoto
Graduate School of Comprehensive
    Human Science
University of Tsukuba
1-1-1 Tennoudai
Tsukuba 305-8575
Japan

Chapter 1

Introduction

Martin J. Evans

Cardiff School of Biosciences, Wales College of Medicine, Biology, Life and Health Sciences, Cardiff University, Cardiff CF10 3US, UK.

1 Aims and scope of the volume

This book, a practical handbook for researchers using embryonic stem (ES) cells, may be thought of as a cookbook. Some cooks reading a cookbook will just slavishly copy a recipe or two whereas others will learn from it, use it as a sourcebook, and provide their own interpretations resulting in novel dishes. It is hoped that you, the scientists using this volume, will emulate the adventurous chefs and use the protocols and ideas presented here as a stimulus for your own original studies and developments, which should take the field forward and fulfil its great potential.

Downing and Battey, who recently reviewed the field (1), provide an interesting analysis of the rapid increase in annual numbers of publications on mouse ES (mES) cells: they found 2700 publications, including 1338 primary papers, up to 2001. A similar search in 2005 gives a gross number of 6933 for 'mouse embryonic stem cells', and 5629 for 'human embryonic stem cells' (of which 1324 are reviews, the total including also numerous short reports on the political and ethical scene).

There are perhaps two main drivers of the increasing interest in ES cells (discussed below). These both depend upon the fundamental property of ES cells of their pluripotency, and their concomitant ability to differentiate into a wide variety of, if not all, cell types either *in vitro* in the cell culture dish or *in vivo* in the context of an embryo. The *in vitro* differentiation, though dramatic, is very often grossly chaotic although well organized on a local basis. It was, for instance, noticed at an early stage that the differentiation seen is typically nonrandom and with tissue-level coordination (2). *In vivo*, however, in the environment of a normal embryo, perfectly organized differentiation takes place (3). Although mES cells have provided the opportunity for the study of developmental processes, their use in this way has been overshadowed by that as a vector for genetic manipulation of the mouse germline (4). Now, with the availability of human ES (hES) cells, where there is really no opportunity for study in the context of a normal embryo, *in vitro* differentiation and fundamental studies of early development become much more important.

The drivers of the research into ES cells have therefore been largely (a) their use for achieving germline chimerism in the mouse, and (b) the opportunity to derive from them a wide range of developmental tissue types in the human. Much of the interest in hES cells is stimulated by the therapeutic scenario of damaged tissue being repaired by appropriate tissue-specific stem and precursor cells. These precursor cells, which are of themselves difficult to isolate and proliferate, may possibly be derived by specific differentiation from hES cell cultures. Moreover, the possibility of obtaining histocompatibility in the therapeutic cells so generated, either by choice from a very large bank of hES cell lines or, even more appropriately, by dedifferentiating somatic cells (self-donated by the transplant patient) to an ES cell, has proved to be so powerful a concept that it is dominating the field. Thus we have moved from a situation where ES cells were providing a very important platform technology for gaining knowledge of human genetics and for drug discovery (via mouse genetics and animal models of human disease) to one where they may be seen to provide another platform technology, for a direct therapeutic opportunity.

There has been considerable anxiety, ethical debate, and regulatory and legislative intervention on both national and international scales in response to the advent of culturable hES cells, and to the far-reaching ideas for their application in future cell-based therapies. Although undoubtedly appropriate and necessary, much of this debate has far outrun both the present and potential scientific and medical realities. Moreover, some of it is based on misconception and misinterpretation. I have elsewhere discussed scientific and medical possibilities, which might lead towards more rational and considered approaches to regulating the derivation and use of hES cells (5).

Perhaps one of the most important things to remember – and one that should cut through much of the angst and hype currently surrounding ES cells – is that they are normal cells; cells in tissue culture. They do, however, have distinct and useful properties. They may be maintained in a nontransformed state of immortality; they represent an early stage in embryonic development; they are pluripotential and their descendants may differentiate into virtually any type of embryonic or adult cell. It is worth emphasizing that this is a particular developmental cell state, and is not the same as being undifferentiated. In earlier years of study of mouse teratocarcinoma stem cells there was some debate about their true origin. Did they represent germ cells or early embryonic cells? This is still a debate today (6). Embryonic stem cells do share many characteristics and cell type markers (but not all (6)) with primordial germ cells, but perhaps the important point is that there is interconvertibility between the two. It has been reported that both sperm and oocytes may be formed by *in vitro* differentiation from ES cell cultures (7–9) (as well as *in vivo*, in the developing embryo), and early embryo cells derived from cleavage stages through to early gastrulation may readily give rise to cultures of ES cells.

Despite the heightened interest there is little very new. Embryonic stem cells have always been a subject for cell developmental analysis, but the new imperative of the therapeutic potential for hES cells brings this into a sharper focus.

With the greatly increased resources now being applied to the field, this volume is a timely collection of practical methods and provides a handbook for future work. Readers are also recommended to browse the two volumes of the *Handbook of Stem Cells* (10) for alternative descriptions.

Chapters 2, 10, and 12 provide methods for isolation of ES cells from mouse, human, and monkey embryos, respectively. It is noticeable that there is very little difference in the fundamental methodology. Excellent cleanliness and growth conditions are essential. It is very likely that one of the worst types of contaminant is laboratory detergent, and that is one of the reasons why these methods work best using only new, disposable plastic materials. The serum or its equivalent replacement is a critical component. It is interesting to note that many modern protocols use immunosurgery to isolate the inner cell mass from the blastocyst, but experience in the mouse showed that this was always an optional procedure. An implantationally delayed blastocyst is optimum in the mouse – but is not available from the primates. Isolation of ES cells from a number of other species has also been reported. It is interesting to note that to this date there has been no successful isolation of ES cells from rat embryos, despite much effort and the fact that development is very similar to that in the mouse, and that these embryos can be put into implantation delay.

Chapter 3 details methods for recreating mice from cultured mES cells. The standard method to transfer the genetic material from an experimental culture to the mouse genome is via a germline chimera, by breeding. Chimeras can be especially useful in this process, as the normal cells of the carrier embryo can provide support for an mES cell that is unable to give rise to all the tissues of the mouse but is still capable of colonizing the germline. Kato and Tsunoda, however, also provide detailed protocols that will allow development of mice that may be wholly derived from the cultured mES cells. This *tour de force* is particularly useful for immediate analyses at the fetal stage. In particular, their method of heat-treating the host blastocyst is a very simple and effective technical refinement that can be invaluable when mES cell lines with the potential for complete chimerism are being used. They conclude with practical, well-described methodology for oocyte nuclear transplantation (or 'nuclear transfer').

Already thousands of designer mutations have been introduced into the mouse chromosomal genome, and many of these have either illuminated an understanding of human disease processes or indeed provided an animal model of a specific human genetic disease. MacGregor *et al.* in Chapter 4 present us with a whole new raft of techniques to allow transfer of variant or mutant mitochondrial DNA into mice via mES cells. This opens up a whole new area of investigation. One technique that they report may also be of considerable general interest: that is, the use of an X-linked *GFP* transgene to identify female blastocysts for microinjection. Their *Figure* 6 illustrates this perfectly.

The remaining Chapters 5 to 9 and 11 all approach the present crucial aim of refining methods for obtaining specific types of differentiated cell. In consideration of the proposed therapeutic protocols, it is important to remember that it is not the ES cells themselves that are needed for therapy but their derived, committed, precursor populations. So an understanding of the modes of differentiation, and development of specific methods for their effective and reproducible manipulation, will be essential. Chapters 5 to 9 concentrate on differentiation of mES cells into specific products. It seems probable that these animal studies are, at the moment, the most useful because there is the possibility of directly testing the tissue-reconstitution potential of the derived cells in the whole animal. The human scenario must be further removed, but the methods being reported by Gerecht-Nir and Itskovitz-Eldor (Chapter 11) for the directed differentiation of hES cells give an indication of the rapid progress in this area. Further, it must not be forgotten that such differentiating cell populations provide an excellent entry for pharmaceutical testing.

Methods for ES cell differentiation *in vitro* are generally relatively empirical. It is possible to arrange the ablation of all cell types apart from the desired product, or to provide growth conditions that favour only the desired product. When differentiation is started from an embryoid body (EB), this system is effectively still a black box. In the very first *in vitro* differentiations seen with mouse embryonal carcinoma (EC) cells the technique was to repeatedly feed and overgrow the culture. There was a colossal cell death and, when the mists cleared, dramatic differentiation was observed. It was soon found that these conditions had allowed EBs to form; and the alternative condition that gave rise to differentiation was to allow a single colony to overgrow. In both cases there was produced a rind of endoderm surrounding an inner core of differentiating cells; and it was realized that the differentiation was not a random process, but was a caricature of the early egg cylinder. Multiple cellular and tissue interactions must have been powering the process. Some of the growth factors involved have now been identified and many of the transcription factors described. The processes involved are still, however, far from being fully understood. In Chapter 5, Wiles and Proetzel give an extremely good introduction to the situation. In the future, understanding will be imperative, but for now some degree of empiricism allows a forward progress.

2 Control of differentiation

Although it would be desirable to have better control over ES cell differentiation, the ability to dedifferentiate cells is particularly sought in the human context. This is because currently the only satisfactory route known to the generation of ES cells is from an early embryo. Immunohistocompatible isogenic hES cells could be tailormade for a particular patient by nuclear transfer of a self-donated somatic cell into an enucleated oocyte, but this raises both practical and regulatory issues.

It is self-evident that at the earliest stages of development there will be dividing populations of cells that have a broad potential developmental fate. Experimental embryology has shown that this fate becomes progressively restricted as cell lineages develop. It is not self-evident however at which, if any, stages along this progression from totipotential to single lineage restriction the cells may cycle

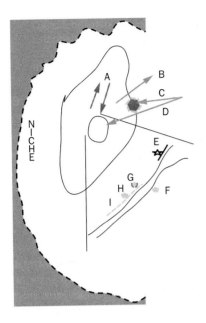

Figure 1 Interactions and factors maintaining and controlling the differentiative state of a cell. Cellular level – A: nucleocytoplasmic interactions, B: maintenance and growth factors produced endogenously, C: external factors interacting with cell surface receptors, D: external factors interacting with nuclear receptors (possible after specific transport). Chromosomal level – E: transcription factors, F: DNA methylation, G: histone acetylation, H: histone methylation, I: chromatin packing. For many cells these interactions take place in populations (homoiogenic induction-community effect) and in specific 'niches'.

mitotically and keep their developmental fate statically intact. Indeed, this question is intimately bound up with our understanding of the stability of cell determination and the maintenance of the differentiated state. A precursor population is not the same as a stem cell population. In addition, stem cells both during development and in adult organisms reside in specific conditions, in niches; and it is not necessarily straightforward to maintain or replicate the niche conditions in isolation, *in vitro*. In order to isolate and maintain stem cells such conditions need to be provided, or their necessity obviated by other means. Another factor to be considered is that in some adult conditions stem cells may represent very slowly growing reserve populations that give rise to transit committed populations, in which the amplification of rapid cell division occurs.

A full understanding of these processes and of the conditions under which they may take place will be desirable in order to have full manipulative control over cellular commitment and differentiation. Indeed, if and when we have this practical manipulative ability to dedifferentiate and, at will, to redifferentiate cells, there will be no need for the isolation of ES cells directly from an embryo. This would remove all the ethical considerations which apply to the use of hES cells (5). There is, however, a long way to go. Some of the areas that need to be explored are set out in the annotated diagram above (*Figure 1*).

3 Future perspectives

The differentiated state of a cell is the result of a series of developmental steps, each of which typically narrows the prospective fate of the cell and its descendants. This is, in most cases, an irreversible restriction although there are some well-documented examples of transdifferentiation. We know, however, that the genetic constitution of most vertebrate differentiated cells remains unchanged, despite their restricted fate. We know furthermore that the differentiated state of an adult cell nucleus can be reversed. (This is most clearly demonstrated by the nuclear transfer experiments where a nucleus from a differentiated cell is transplanted into an enucleated oocyte, leading to reprogramming of that nucleus to the embryonic state.) The metastable state of differentiation is set up by nucleo-cytoplasmic interaction, and maintained by transcription factor networks and secondary modification of chromatin and of the DNA. All these processes are reversible. And it is clear that this reversion of differentiation should be amenable to the same types of experimental intervention as those required directly to manipulate forward differentiation. Not only should it be possible to modify environment, culture conditions, and exogenous growth factors, but also it is noteworthy that most of the actions and interactions should be directly responsive to pharmaceutical intervention.

References

1. Downing, G. J. and Battey, Jr, J. F. (2004). *Stem Cells*, **22**, 1168–80.
2. Evans, M. J. and Martin, G. R. (1975). The differentiation of clonal teratocarcinoma cell culture in vitro. In *Roche Symposium on Teratomas and Differentiation* (ed. D. Solter, and M. Sherman), Academic Press, San Diego.
3. Papaioannou, V.E., McBurney, M.W., Gardner, R.L., and Evans, M.J. (1975). *Nature*, **258**, 70–3.
4. Evans, M.J. (1989). *Mol. Biol. Med.*, **6**, 557–66.
5. Evans, M. (2005). *Nat. Rev. Mol. Cell. Biol.*, **6**, 663–7.
6. Zwaka, T.P. and Thomson, J.A. (2005). *Development*, **132**, 227–33.
7. Geijsen, N., Horoschak, M., Kim, K., Gribnau, J., Eggan, K., and Daley, G.Q. (2004). *Nature*, **427**, 148–54.
8. Hubner, K., Fuhrmann, G., Christenson, L. K., Kehler, J., Reinbold, R., De La Fuente, R., Wood, J., Strauss, J.F. 3rd, Boiani, M., and Scholer, H. R. (2003). *Science*, **300**, 1251–6.
9. Surani, M. A. (2004). *Nature*, **427**, 106–7.
10. Lanza, R., Gearhart, J., Hogan, B., Melton, D., Pedersen, R., Thomson, J., and West, M., eds (2004). Handbook of Stem Cells, Volume 1, Academic Press, San Diego.

Chapter 2
Procedures for deriving ES cell lines from the mouse

Frances A. Brook
Department of Zoology, University of Oxford, South Parks Rd., Oxford OX1 3PS, UK.

1 Introduction

Although germline-competent mouse ES (mES) cell lines are widely available, most are derived from substrains of the 129 mouse. This strain and, to a lesser extent, C57BL/6 and BALB/c have proved to be the predominant ones from which mES cell lines have been derived, not least because the testing for germline competence is a time-consuming and laborious process, and therefore efforts have been focused on deriving lines from preferred strains. There are many situations in which it is desirable to derive novel mES cell lines, and from mice other than 129 and C57BL/6. The 129 mouse is not favoured as a strain in which to develop model systems (1, 2), and so extensive backcrossing usually is required to transfer genetic modifications from a 129 background to the strain of choice. (Even then, the DNA immediately upstream and downstream of a targeted locus would remain of 129-strain origin.) In many areas of mouse genetics, much time and effort has been invested in working with a particular strain (few mutants are maintained on a 129 background), and the availability of the corresponding mES cell lines would be highly advantageous for gene targeting and *in vitro* differentiation studies. Furthermore, emerging techniques for the generation of mice that are entirely mES cell-derived benefit from the use of cell lines from hybrid, rather than inbred 129 or C57BL/6, mice in terms of increased efficiency of production and viability of live-born offspring (Chapter 3).

It is the aim of this chapter to describe procedures for the derivation of mES cell lines that may be applied to all, and not only the more permissive, strains of mice.

2 Derivation of mES cell lines

Most extant mES cell lines, from 129 and other strains, have been isolated using variations of the principal procedure of Evans and Kaufman (3), detailed by Robertson (4) and modified by Abbondanzo *et al.* (5). (A version of this method is provided in Section 4 of this Chapter). Briefly, blastocysts are explanted onto a feeder layer of mitotically inactivated, mouse embryonic fibroblasts (MEF). After 2–5 d of culture the inner cell mass (ICM) is picked, dissociated, and seeded onto fresh feeder layers. Resulting mES-cell colonies are either picked and dissociated individually or trypsinized *en masse*, and passaged onto progressively larger feeder layers until a stable culture is safely established. The feeder cells serve as a source of leukaemia inhibitory factor (LIF) and probably other factors, but in most protocols exogenous murine LIF is added to the medium at 400–5000 U/ml. Moreover, mES cell lines have been obtained in the absence of feeders but with the addition of LIF to the medium (6).

With the 129 strain and using standard techniques, 20–25% of explanted blastocysts may be expected to give rise to mES cell lines. However, the success rate with other strains is much lower: for example, the rate of derivation of lines from DBA, BALB/c, and MRL blastocysts was, respectively, 1.2 %, 2.5%, and 7% (7–9), whilst ICR (or CD-1) blastocysts yielded no lines at all (10). Although mES cell lines from outbred (as well as hybrid and diverse inbred) strains were first produced in the 1980s, early in the development of the technology (11), there is a general perception that outbred strains are more refractory to the derivation of mES cell lines than are inbred strains.

Modifications to the culture procedures have improved success rates of mES-cell derivation with some of the less permissive strains. The nature of the feeder cells is of importance (8, 12), with MEFs and selected substrains of the STO cell line proving the most effective for mES-cell derivation and maintenance. The use of medium conditioned by '5637' human bladder carcinoma cells, which are a source of human LIF, together with the use of these cells as feeders during the initial stages of derivation, increased the success rate with C57BL/6 blastocysts from 20.5% to 62.5% (13), and permitted the establishment of mES cell lines from 18.5% of BALB/c embryos (14). Schoonjans *et al.* (15) reported that culture medium conditioned by an immortalized, rabbit fibroblast cell line that was transformed with a rabbit LIF construct increased the efficiency of derivation of mES cell lines from C57BL/6 embryos from 34% to 61%; and enabled the derivation of lines from mouse strains that were previously considered refractory, such as CBACa, C3HheN, and FVB/N. Improved generation of C57BL/6 mES cell lines has also been reported using a serum-free, defined medium (16).

Alternative strategies to improve the efficiency of mES-cell derivation involve modification of the starting material. For example, the use of delayed-implanting blastocysts, produced by either ovariectomy or treatment of donors with tamoxifen and Depo-Provera (17), can enhance the rate at which mES cell lines are obtained (4, 8, 12). Theoretically, this may be due to an increase during the period of delay in the total number of ICM cells, although the observed increase is modest (3). Alternatively, it has been suggested that dependency on gp130 signalling is activated during delayed implantation, thus priming the embryonic cells for proliferation when cultured in medium containing LIF (18).

To circumvent the nonpermissiveness of certain mouse strains, one method that has been used successfully is to derive mES cell lines from hybrid embryos produced by crosses between the strain of choice and either 129 or C57BL/6. Yagi et al. (19) obtained mES cell lines from {C57BL/6 × CBA}F_1 embryos, whilst 88% of {129/Ola × NOD}F_1 blastocysts gave rise to mES cell lines–notwithstanding that the NOD strain *per se* has proved to be completely refractory to mES-cell derivation (20). For the DDK strain, Kress et al. (21) found that a single backcross to generate a 75% 129/Sv and 25% DDK background was necessary before several mES cell lines could be derived.

Another strategy that has been successfully employed is to use blastocysts that are transgenic for the bacterial neomycin-resistance gene, *neo*, controlled by regulatory sequences of the *Oct4* gene. Expression of this transgene (and concomitant G418-resistance) is restricted to undifferentiated cells, and G418 selection against (differentiated) cells in which expression of *Oct4* is down-regulated should therefore enrich for the undifferentiated precursors of mES cells. To this end, mice carrying the (*pOctneo 1*) transgene were produced by pronuclear injection into {C57BL/6 × CBA}F_1 embryos, and the founder animals were intercrossed and bred with CBA mice, to generate males that were homozygous for the transgene on a 75% CBA background. These were mated with CBA females to produce blastocysts whose genetic background was 87.5% CBA and 12.5% C57BL/6; and when explanted into culture with G418 in the medium, these subsequently gave rise to mES cell lines with an efficiency of 10.5% (22).

However, a problem associated with the hybrid and transgenic strategies described above is that a significant proportion of the mES-cell genotype is of 129 or C57BL/6 origin: the goal remains to obtain 'pure' mES cell lines from the more refractory strains of mice. A great improvement in the efficiency of obtaining mES cell lines can be achieved by microsurgically separating the tissues of the implanting blastocyst and culturing the embryonic portion of the ICM, the epiblast, in isolation on feeder cells, free from contamination by extraembryonic cell types (trophectoderm and primitive endoderm). This procedure, which is described in detail in this Chapter (Section 4), is based on theoretical considerations of the origin of mES cells, the starting premise being that the *in vivo* progenitors of the mES cells reside in the primitive ectoderm, or epiblast (see Chapter 1). By this technique the success rate with 129 blastocysts was increased from 25% to 52%. However, a success rate of 100% was achieved when delayed-implanting 129 blastocysts were used. Further, this technique was used to produce germline-competent mES cell lines from 56% of CBA blastocysts, and has been successfully applied to other inbred strains. Of the two outbred strains that have been attempted with this technique, mES cell lines were obtained from PO (Pathology, Oxford), but not from ICR, mice (12). Further studies will show whether or not this approach has applicability to refractory strains generally, and in particular to nonstandard strains.

The relative merits and efficiencies of procedures for mES-cell derivation described in this Chapter are summarized in *Table 1*.

Table 1 Comparison of methods for the derivation of mES cell lines

Method	Starting material	Advantages	Disadvantages	Efficiencies of mES cell line derivation for mouse strains
A. Blastocyst culture (4,5) and *Protocol 7*	Day 3.5 p.c. blastocyst[a]	Established procedure for 129 strain No micromanipulation required	Less effective with non-129 strains	*Inbred strains* ~25% with strain 129; 1.2% with DBA (7); 2.5% with BALB/c (8), 7% with MRL (9); other strains are nonpermissive *Hybrid strains* {C57BL/6 × CBA}F$_1$ (19) {75% 129/Sv × 25% DDK} (21) *Modified culture conditions* ^{c} >60% with C57BL/6 (13), and 18.5% with BALB/c (14) ^{d} >60% for C57BL/6, and derivation with CBACa, DBA/2, DBA/1, C3HheN, BALB/c and FVB/N (15) ^{e} Improved derivation with C57BL/6 (16) ^{f} 10.5% with transgenic (87.5% CBA × 12.5% C57BL/6) (22)
B. Epiblast culture (12) and *Protocol 9*	Day 4.5 p.c. blastocyst	Improved success rate over Method A	Requires micromanipulation	~50% with strain 129; ineffective for conventionally refractory strains
C. Epiblast culture (12) and *Protocol 9*	Delayed-implanting blastocyst[b]	Improved success rate over Methods A and B. Enables derivation from otherwise nonpermissive strains. May prove effective for otherwise nonpermissive strains	Requires micromanipulation	100% for strain 129 56% of CBA blastocysts (12) 88% with {129/Ola × NOD}F$_1$ (20) Outbred strain P0, but not with ICR (12)

[a] Use of delayed-implanting blastocysts improves efficiency of derivation, for example to >30% with strain 129 (4).
[b] The animal surgery involved can be obviated using the method described in ref. (17).
[c] Medium conditioned by 5637 human bladder carcinoma cells, and use of these cells as feeders.
[d] Medium conditioned by a transformed rabbit fibroblast cell line.
[e] Serum-free, defined medium.
[f] Use of p*Octneo 1* transgenic blastocysts and G418 selection.

3 General principles of mES cell culture
3.1 Culture media and supplements

The culture of mES cells, whilst not difficult, is a labour-intensive process that requires considerable care and attention to detail. Although mES cells may continue to proliferate under suboptimal conditions, they readily lose their ability to contribute to chimeras and be transmitted through the germline. Hence it is vitally important to maintain optimal culture conditions at all times.

All reagents should be of tissue-culture grade, and solutions made up with AnalaR® water. Milli-Q® water is acceptable provided that the purification system is well maintained. Wherever possible, use of disposable plastics is highly recommended. If this is unfeasible, all glassware for cell culture should be reserved exclusively for this purpose, and washed and stored separately from other, possibly contaminated, glassware. Rigorous aseptic technique is required throughout. It is good practice to carry out routine tissue culture without the addition of antibiotics to the medium. However, as the derivation of mES cell lines from mouse blastocysts necessitates some initial procedures to be undertaken on the open bench, the mES cell culture medium described in *Protocol 1* contains antibiotics, which can be omitted once the cell line is established. Where solutions are sterilized by filtration, membranes with a pore size of 0.22 μm should be used.

In many laboratories the plastic Petri dishes for mES cell culture are precoated with gelatin; but in the author's experience this is unnecessary. Similarly, the addition of nucleosides to mES-cell medium is recommended in some protocols (4) (see Chapter 3, *Protocol 1*), and left out of others with no deleterious effects. Therefore the use of these reagents for derivation and maintenance of *de novo* mES cell lines is at the discretion of the researcher; but where a cell line is imported, it should be cultured according to the recommendations of the laboratory of origin.

All cell cultures are maintained in a humidified incubator, gassed with 5% CO_2 in air, at 37 °C.

3.2 Preparation of MEFs

Mouse ES cells are routinely cultured on a feeder layer of mitotically inactivated mouse fibroblasts. These may be low-passage cultures derived from mouse embryos (i.e. MEFs, *Protocol 2*) or continuously proliferating STO cells (a thioguanine- and ouabain-resistant line of SIM mouse fibroblasts; ATCC No. CRL 1503). The latter cells have the advantage of being easy to grow, but the disadvantage that sublines vary in their effectiveness as feeder cells; and continuous culture predisposes towards the selection of faster growing, less effective variants. MEFs have been shown to be more effective than STO cells when establishing new mES cell lines (12), and are therefore recommended. However, as MEFs proliferate for only a limited period in culture, new stocks must be prepared regularly. Aliquots may be cryopreserved, and so fresh stocks need only be prepared every few months. The effectiveness of MEFs is independent of the

Protocol 1
Culture media and stock solutions

Reagents and equipment

- Dulbecco's modified Eagle's medium (DMEM), powdered: formulation for 4500 mg/ml glucose, without pyruvate or sodium bicarbonate (Cat. No. D5648, Sigma)
- $NaHCO_3$, crystalline
- Glutamine, 200 mM (Cat. No. G7513, Sigma)
- MEM nonessential amino acids, 100 × solution (Cat. No. M7145, Sigma)
- β-mercaptoethanol
- Penicillin (Cat. No. P4687, Sigma)
- Streptomycin (Cat. No. S1277, Sigma)

- Phosphate-buffered saline (PBS) tablets, without Ca^{2+} or Mg^{2+} (Cat. No. BR14, Oxoid Ltd)
- AnalaR® or Milli-Q® water
- Fetal calf serum (FCS)
- Murine recombinant LIF (sold as ESGRO®; Cat. No. ESG1106, Chemicon International)
- Glass rod
- Vacuum filtration system with 0.2–0.22 µm pore size
- Disposable syringe filters with 0.2 µm pore size

Methods

DMEM stock:

1. Measure out 90% of the final volume (usually 1 L) of water.
2. Add to the water the entire contents of a package of powdered DMEM (sufficient for 1 L); rinse out the package and add the residue to the solution.[a]
3. Add 2.2 g/L $NaHCO_3$.
4. Make up to the final volume with water, and stir with a glass rod until dissolved.
5. Immediately sterilize the medium by filtration, and store at 4 °C.

PBS:[b] Dissolve 1 PBS tablet in 100 ml water, sterilize by filtration or autoclaving, and store at 4 °C.

β-mercaptoethanol stock, 100 × solution:[c] Add 7 µl of pure β-mercaptoethanol to 10 ml PBS, filter sterilize and store at 4 °C for up to 1 week.

Antibiotic stock, 100 × solution: Add 5×10^5 U penicillin and 500 mg streptomycin to 100 ml PBS, filter sterilize, aliquot under sterile conditions (e.g. 2 ml is a convenient volume), and store at −20 °C.

Glutamine, 200 mM stock solution: Aliquot the supplied stock (e.g. 2 ml) under sterile conditions, and store at −20 °C.

Nonessential amino acids solution, 100 × stock: Aliquot the supplied stock (e.g. 10 ml) under sterile conditions, and store at 4 °C.

> **Protocol 1 continued**
>
> *Fetal-calf serum (FCS)*: This must be tested for optimal ability to support mES cell culture (Protocol 4), although batches used for MEF culture need not be of such high quality. Aliquot into 20 ml and 15 ml for MEF and mES cell culture, respectively, and store at $-20\,°C$.
>
> *LIF*: Murine recombinant LIF, or ESGRO®, is supplied as 10^6 or 10^7 U/ml in PBS. Dilute to 10^5 U/ml in sterile DMEM, aliquot and store at 4 °C. Use 10 µl/ml medium, equivalent to a final concentration of 10^3 U/ml.
>
> *MEF culture medium*: For ~200 ml, combine the following constituents.
>
> DMEM 175 ml
>
> FCS 20 ml
>
> Glutamine (200 mM stock) 2 ml
>
> β-mercaptoethanol stock 2 ml
>
> *Mouse ES-cell culture medium*: For ~100 ml, combine the following constituents.
>
> DMEM 80 ml
>
> FCS 15 ml
>
> Glutamine, 200 mM stock 1 ml
>
> β-mercaptoethanol stock 1 ml
>
> Nonessential amino acids stock 1 ml
>
> Antibiotic stock 1 ml
>
> [a] It is inadvisable to keep opened packages, as powdered media are hygroscopic and liable to contamination.
>
> [b] The PBS used throughout is without calcium or magnesium ions.
>
> [c] Pure, liquid β-mercaptoethanol should be replaced every 6–12 months.

strain of mouse from which they are derived, and it is therefore advantageous to choose a vigorous strain with a large litter size for MEF preparation, for example MF1.

3.3 Preparation of feeder layers

Feeder cells must be mitotically inactivated to prevent their proliferation and consequent overgrowth of the mES cell culture. This can be achieved using either γ-irradiation or mitomycin-C treatment. Gamma-irradiation is quicker and avoids the risk of carrying over mitomycin C into the mES cell culture. However, as γ-sources are not standard equipment, mitomycin-C treatment is employed more commonly and a detailed method is provided here (Protocol 3).

Protocol 2

Preparation, maintenance, and cryopreservation of MEFs

Equipment and reagents

- Pregnant mouse at day 13.5–14.5 p.c. (where the morning of finding the copulation plug is designated day 0.5 p.c.)
- Forceps and scissors for dissection, plus two pairs of Watchmaker's forceps: these are sterilized by autoclaving
- 70% alcohol in a wash bottle
- Spirit lamp to flame-sterilize instruments during dissections
- 60 mm bacteriological-grade Petri dishes
- Scalpel or razor blade
- Dissecting microscope
- 5 ml transfer pipette
- Sterile, 50 ml conical tubes
- Water bath set at 37 °C
- 75 cm^2 and 150 cm^2 tissue-culture flasks
- PBS (*Protocol 1*)
- 0.125% trypsin/0.01% EDTA solution: prepare by diluting a 0.25% trypsin/0.02% EDTA solution (Cat. No. T4049, Sigma) in its own volume of PBS
- MEF culture medium with the addition of 1 ml antibiotic stock (*Protocol 1*) per 100 ml medium
- Haemocytometer

Methods

Preparation of MEFs:

1. Sacrifice the mouse humanely and swab its abdomen with 70% alcohol. Using sterile forceps and scissors, cut open the abdominal cavity, dissect both uterine horns of the reproductive tract and place them in a Petri dish.

2. Trim away excess fat and mesentery from the uterine horns, rinse them briefly with PBS and transfer to a fresh dish containing PBS.

3. Under a dissecting microscope, tear open the horns carefully with Watchmaker's forceps, release the conceptuses one at a time and transfer them into a fresh dish containing PBS.

4. Gently tear open the embryonic membranes, and remove them and the placenta from each embryo. Using forceps, pinch off the head, open the abdomen and remove the internal organs. Transfer the remaining carcase into a fresh dish containing PBS.

5. Working in a sterile hood, transfer the carcases into a dish containing 5 ml 0.125% trypsin/0.01% EDTA solution. Mince the tissue as finely as possible using sterile forceps and a sterile scalpel or razor blade. Draw the suspension up and down a 5 ml pipette to dissociate the tissue as much as possible, and transfer into a 50 ml conical tube. Rinse the dish with a further 5–10 ml of 0.125% trypsin/0.01% EDTA solution, and pool with the existing tissue.

Protocol 2 continued

6. Incubate the tissue suspension in a water bath for 5 min at 37 °C. Meanwhile, dispense 15 ml MEF medium into each of four 75 cm² flasks.

7. Retrieve the tube from the water bath and shake vigorously to dissociate the tissue. Allow large debris to settle.

8. Remove as much as possible of the supernatant and transfer this into a second 50 ml tube containing 5 ml MEF medium, to halt trypsinization.

9. Add 5 ml fresh trypsin solution to the remaining tissue debris, shake, and incubate again for 5 min at 37 °C.

10. Repeat from Step 7 until three trypsin extractions have been carried out in total. Pool the supernatants, distribute equally between the four flasks and place in an incubator for culture. Change medium the following day to remove floating debris.

11. When these flasks of primary MEFs are confluent, normally after 2–3 d, passage each into a 150 cm² flask, as follows. Use (undiluted) 0.25% trypsin/0.02% EDTA solution for this (first) and subsequent passages of MEFs.

12. Aspirate the medium from each 75 cm² flask, wash with 4 ml PBS, and add 2 ml trypsin solution. Replace in the incubator for 2–3 min. Rock the flask until the cells detach, add 8 ml MEF medium and pipette up and down to give a single cell suspension, and transfer into a 150 cm² flask in 35 ml MEF medium in total.

13. When these flasks also are confluent, after a further 2–3 d, harvest and freeze MEFs in 1 ml aliquots at about 10^7 cells/ml (using the freezing procedure given in *Protocol 3*), and store the aliquots under liquid nitrogen.

Maintenance of MEFs:

14. Rapidly thaw a frozen aliquot of MEFs by holding the vial momentarily in a water bath set at 37 °C; thoroughly swab the outside of the vial with 70% alcohol, and transfer the contents to a centrifuge tube containing 10 ml MEF medium.

15. Pellet the cells in a benchtop centrifuge at 1000 rpm for 5 min. Aspirate the supernatant, and gently resuspend the cell pellet in ~2 ml MEF medium. Add the cell suspension to 25 ml MEF medium in a 75 cm² flask, and incubate.

16. Change medium the following day only–there is no need for further medium changes between passages. The flask should reach confluence after 4–5 d of incubation, when the cells can be either used to make feeder monolayers or passaged further. MEFs can be subcultured three or four times from this stage (i.e. to passage 5 or 6) at a split ratio of 1:3, following which they usually cease to proliferate. A confluent 75 cm² flask of MEFs contains $1.0-1.5 \times 10^7$ cells.

3.4 Fetal-calf serum (FCS)

The quality of the FCS in the culture medium is a critical factor determining the success of mES-cell derivation. Sera vary considerably between batches in their

Protocol 3

Maintenance of mES cells on MEF feeder layers

Reagents and equipment

- One or more confluent 75 cm^2 flasks of MEFs (*Protocol 2*)
- Mitomycin C stock solution, 1 mg/ml: to a 2 mg vial of mitomycin C (Cat. No. M4287, Sigma) add 2 ml PBS to dissolve, protect the solution from light, and store at 4 °C for short-term use (or at −20 °C for longer term)
- PBS (*Protocol 1*)
- MEF medium (*Protocol 1*)
- mES-cell medium (*Protocol 1*)
- Trypsin solution: 0.25% trypsin/0.02% EDTA solution (Cat. No. T4049, Sigma)
- 2 × freezing mixture: combine 20% dimethyl sulphoxide, 20% FCS, and 60% DMEM stock (*Protocol 1*)
- Sterile, 15 ml or 50 ml conical tube
- Sterile Pasteur pipette
- Haemocytometer
- Inverted phase-contrast microscope
- 25 cm^2 tissue culture flasks, 35 mm tissue-culture dishes, or 4-well tissue-culture plates, depending on the type of feeder layers required
- Cryovials

Methods

Preparation of feeder layers using mitomycin C:

1. From a confluent flask of MEFs aspirate the medium and replace with 5 ml MEF medium to which has been added 50 μl of 1 mg/ml mitomycin C stock solution (i.e. to give a final concentration of 10 μg/ml). Resume incubation of MEFs for 2–3 h.

2. Aspirate the mitomycin C-containing medium (disposing of it as hazardous waste), and wash the cell monolayer at least three times with PBS.[a] Trypsinize the now mitotically inactive feeder cells and resuspend in MEF medium, as described in Steps 11–12, *Protocol 2* for MEFs.

3. Transfer the suspension into a 15 ml tube (or a 50 ml tube if harvesting more than one flask), centrifuge at 1000 rpm for 5 min to pellet the cells, and resuspend in MEF medium.

4. Count the cells using a haemocytometer, and seed the flasks/dishes/wells at an appropriate density of MEFs to give a confluent monolayer, as follows:[b]
   (a) 25 cm^2 flask, 1.5×10^6 cells in 5 ml medium
   (b) 35 mm dish, 6×10^5 cells in 2 ml medium
   (c) 4-well plate, 2×10^5 cells/well in 1 ml medium

5. Incubate feeder layers overnight, and for up to a week after preparation, before coculture with mES cells. Immediately prior to use, aspirate MEF medium from the feeder layers and replace with mES-cell medium.

Protocol 3 continued

Alternative preparation of feeder layers, using γ-irradiation:

1. Trypsinize a confluent flask of MEFs and resuspend in MEF medium as described in Steps 11–12, Protocol 2. Transfer into a sterile 15 ml tube.^{c}
2. Place the tube inside a ^{60}Co source, and γ-irradiate the cell suspension with 3000 rads.
3. Count the cells, and plate as for mitomycin C-inactivated feeder cells, above.

Passaging mES cells on feeder layers:

1. To passage mES cells maintained in a 25 cm^2 flask, aspirate medium from the flask and wash cells with 3–4 ml PBS.
2. Add 0.5 ml of trypsin solution, and incubate the flask for 5 min at 37 °C; rock the flask from side to side to detach the cells, and resume incubation for a further 5 min.
3. Halt trypsinization by addition of 0.5 ml mES-cell medium. It is essential at this point to dissociate colonies into single mES cells, which is achieved by pipetting the cell suspension vigorously with a Pasteur pipette.
4. Monitor the progress of dissociation under a microscope and once a single cell suspension is obtained, transfer this into a sterile tube, add medium to a convenient volume (usually 10 ml), and count the mES cells.^{d}
5. Seed fresh 25 cm^2 feeder flasks with 10^6 mES cells from the suspension, in 5 ml mES-cell medium.

Cryopreservation and thawing of mES cells:

1. Obtain a single-cell suspension of mES cells in mES-cell medium (Protocol 3).
2. Count the total number of mES cells in suspension, pellet the cells (at 1000 rpm for 5 min), and resuspend in mES cell-medium, in *half* the final volume that will contain $7-9 \times 10^6$ cells/ml (for freezing).
3. Make the suspension up to the final volume with 2× freezing mixture, added dropwise.
4. Transfer 0.5 ml aliquots of cell suspension into cryovials, and freeze slowly in a purchased freezing container (e.g. Nalgene's 'Mr Frosty') according to the manufacturer's instructions, or in a polystyrene box placed at -70 °C, before storing under liquid nitrogen.^{e}
5. After thawing as follows, one 0.5 ml aliquot of mES cells is sufficient to seed a 25 cm^2 feeder flask. Place the vial in a 37 °C water bath to thaw quickly, swab with 70% alcohol, and transfer the contents directly into the 25 cm^2 feeder flask containing 10 ml of mES-cell medium, and incubate.^{f}
6. Replace medium after 3–5 h with 5 ml fresh mES-cell medium supplemented with LIF (where appropriate). Normally recovery is good, and the cells should be ready to passage after incubation for 3 d.

> **Protocol 3 continued**
>
> ᵃ This step is to remove any residual traces of mitomycin C (4).
>
> ᵇ Feeder cells surplus to immediate requirements may be cryopreserved for future use, as detailed for MEFs (*Protocol 2*, Step 13). A 1 ml aliquot of feeder cells when thawed will seed a confluent feeder layer in three 25 cm² flasks – it is important to achieve the correct seeding density, which is much higher than when using fresh feeder cells.
>
> ᶜ Alternatively, MEFs may be irradiated *in situ* and trypsinized afterwards, for convenience.
>
> ᵈ The contribution of feeder cells to this suspension is low enough to be regarded as negligible.
>
> ᵉ Mouse ES cells may be stored at – 70 °C for several months, but longer-term storage should be under liquid nitrogen.
>
> ᶠ Cells are not pelleted (to remove the freezing mixture) here, in order to avoid their fragmentation.

ability to support the proliferation of pluripotent mES cells in the undifferentiated state, and it is therefore crucial to test a range of samples and select only the best for mES cell culture. It is advisable to screen as many samples as can be handled at once; these should be requested from several suppliers (e.g. HyClone, Sigma), and the required number of bottles reserved until completion of the test. Once a batch is chosen, purchase on reserve from the supplier sufficient bottles to ensure that this laborious procedure need not occur more often than annually. Serum can be stored frozen at – 20 °C for at least 1 year, but if kept much longer (e.g. for 2 years) its ability to support mES cell culture may deteriorate. Pretested serum, sold as 'mES cell-qualified serum', is available at a higher price; but in the author's experience this may not be optimal for culturing nonstandard mES cell lines.

The batch testing of FCS involves plating single-cell suspensions of mES cells at low density onto feeder layers, and comparing the number of colonies produced with the different sera in the medium at a concentration of 15%. The morphology of the colonies is also taken into account: sera producing large, undifferentiated colonies and a plating efficiency of 20% or higher are optimal. Serum toxicity, which is assessed by culturing the cells with 30% serum in the medium, may be due at least in part to high levels of complement; hence FCS commonly is heated at 56 °C for 30 min, for complement inactivation. However, other (beneficial) components of the serum also may be destroyed, and therefore many laboratories choose not to perform this treatment. The decision whether or not to heat-inactivate FCS must be taken prior to batch testing.

3.5 Routine culture of mES cells

3.5.1 Passaging

Established mES cells divide rapidly and are maintained at a relatively high density, and therefore require regular maintenance: the medium should be changed daily to prevent its acidification, and the cells passaged approximately

Protocol 4
Serum testing

This test is for nine samples of FCS and one control, which is usually the serum currently in use, or one of proven ability to support mES cell proliferation. The test is organized to include five replicate dishes per FCS sample at a concentration of 15% in the medium, and one dish at 30%.

Equipment and reagents

- At least six confluent 75 cm^2 flasks of MEFs: a total of 9×10^7 cells is required for a test with 10 samples, therefore it is advisable to have one or two extra flasks in reserve to ensure enough cells
- One 25 cm^2 flask of mES cells, growing exponentially (i.e. at subconfluence)[a]
- Sixty 60 mm tissue-culture plates
- One sterile, 150 cm^2 tissue-culture flask
- Benchtop centrifuge
- Haemocytometer
- Sterile Pasteur pipette

- Nine serum samples plus one control serum
- PBS (Protocol 1)
- Trypsin solution (Protocol 3)
- MEF medium (Protocol 1)
- FCS-supplemented mES-cell medium (Protocol 1)
- Serum-free mES-cell medium (Protocol 1)
- Methylene blue stain: 2% aqueous solution, filtered before use

Method

1. Prepare a suspension of feeder cells from the confluent flasks of MEFs either by treatment with mitomycin C or by γ-irradiation (according to Protocol 3); and having trypsinized the mitotically inactivated cells, and added MEF medium to neutralize the trypsin solution, pellet (at 1000 rpm for 5 min) and resuspend the feeder cells in *serum-free* mES-cell medium. Count the cells.

2. Prepare a single-cell suspension of mES cells from the subconflent flask (according to Protocol 3); and having trypsinized the cells, added mES-cell medium, and obtained a single-cell suspension by vigorous pipetting, pellet (at 1000 rpm for 5 min) and resuspend the mES cells also in *serum-free* mES-cell medium. Count the cells.[b]

3. Dispense 260 ml *serum-free* mES-cell medium into a 150 cm^2 flask. Add 9×10^7 feeder cells together with a defined number of mES cells (between 5 and 10×10^4). Mix well. Dispense uniform aliquots of 4.25 ml into each of 60 tissue-culture dishes (60 mm diameter).

4. Arrange the dishes containing the suspension of feeder and mES cells into ten rows of six plates, and label each row with an appropriate FCS sample number.

5. For each FCS sample add 0.75 ml of that serum to each of five dishes, and 1.75 ml to the remaining, sixth, dish.

> **Protocol 4** continued
>
> 6 Incubate the dishes for 7–10 d. Periodically observe the morphology of the growing mES-cell colonies, and note any serum samples that induce a high level of differentiation.
>
> 7 After this time the colonies should be visible to the naked eye. Aspirate medium, wash the cells with PBS, and stain them with methylene blue solution for 2–5 min. Rinse dishes with distilled water and allow to air dry.
>
> 8 Invert each dish and count the total number of colonies. The percentage of plated mES cells that form colonies (i.e. 'efficiency of plating') usually varies between 5% and 30%. Select the serum sample giving the highest average number of colonies at a concentration of 15%, equal to or better than the control, and which is nontoxic at 30%.[c]
>
> [a] Any established mES cell line can be used for this purpose.
> [b] It is crucial that a single-cell suspension is achieved at this stage.
> [c] Toxicity is indicated by a reduction in the number and/or size of colonies.

every 3 d (see *Protocol 3*). Many laboratories choose not to add LIF to the medium for established mES cell lines growing on feeder layers, with good results. However, the author considers that more efficient germline transmission is achieved when targeted cells are maintained in medium containing LIF, in addition to being cocultured with feeder cells.

3.5.2 Mycoplasma contamination

Contamination with mycoplasma can be highly problematic for mES cell culture, being readily apparent neither macroscopically nor microscopically, and having calamitous consequences for the pluripotency of mES cells. Crucially, blastocyst injection of mycoplasma-contaminated mES cells results in embryo lethality, and a dramatic reduction in the number of chimeras obtained as well as the level of mES-cell contribution to their tissues (23). Elimination of mycoplasma contamination is impractical, treatment with antimycoplasma agents being detrimental to mES cells. Therefore once mycoplasma is detected in either mES cells or feeder cells, the best course of action is to discard the cultures. The most common sources of mycoplasma infection are other contaminated cultures, sera, or laboratory personnel using poor aseptic technique.

The routine screening of mES cell lines and MEF cultures therefore is essential; and all incoming cultures should be tested for mycoplasma, and quarantined until proven clean. The most commonly used tests to reveal mycoplasma involve DNA-staining with a fluorescent dye such as Hoechst 33258 (*Protocol 5*), enzyme-linked immunosorbent assay (ELISA), or polymerase chain reaction (PCR). Kits are available commercially for ELISA- and PCR-based tests (e.g. from Roche or Stratagene). Each method has its relative merits, and the choice of which to use is largely one of convenience. For example, Hoechst staining is less sensitive but

Protocol 5

Detection of mycoplasma

(Adapted, courtesy of the Cell Culture Bank, Sir William Dunn School of Pathology, University of Oxford.)

Reagents and equipment

- Mouse ES cells or MEFs to be tested: cells should be cultured without antibiotics for at least two passages
- Single-cell suspension of (mycoplasma-free) detector cells at 5×10^5 cells/ml: it is usually convenient to use mitotically inactivated feeder cells prepared from an uncontaminated batch of MEFs (*Protocol 3*)
- Bis-benzimide flourochrome stain: prepare a stock concentrate by dissolving 5 mg Hoechst 33258 (Cat. No. B 1155, Sigma) in 100 ml PBS, dispense 20 μl aliquots into small Eppendorf tubes, and store in the dark at −20 °C; and for the working dilution add the contents of one tube to 20 ml PBS, shake well, and protect from light
- Mounting fluid (McIvain's buffer): 22 ml of 0.1 M citric acid, 28 ml of 0.2 M di-sodium hydrogen phosphate, and 50 ml glycerol
- Fixative: prepare a 3:1 mixture of methanol and glacial acetic acid, fresh before use
- Plastic, 35 mm and 90 mm bacteriological-grade Petri dishes
- Sterile forceps
- Glass coverslips, 22×22 mm: degrease in acetone for 10 min, wash in running water for 1–2 h, rinse in Milli-Q® water and drain; place on a disc of filter paper in a 90 mm glass Petri dish, and sterilize in an oven
- Glass slides for microscopy
- Fluorescence microscope fitted with $\times 100$ objective and appropriate filters for Hoechst 33258, which absorbs at a wavelength of 350 nm and emits at 460 nm

Method

1. Under sterile conditions and using forceps, place a glass coverslip into each 35 mm Petri dish. Use one dish for each sample to be tested plus one for the negative control.
2. Dispense 2.5 ml mES-cell medium into each test dish, and 3.5 ml into the control. Ensure that there are no air bubbles under the coverslips.
3. Add 1–2 drops of detector-cell suspension to each dish, including the control: the number of cells is not critical, but must be enough to form a subconfluent layer on the coverslip.
4. To each test dish add 1 ml of medium conditioned by the cultures to be tested. Label the dishes clearly.
5. Place each dish into a 90 mm Petri dish, and incubate for 3 d.[a]
6. Remove the dishes from the incubator and discard the larger Petri dishes. Pre-fix cells by adding 1–2 drops of fixative to the medium, and leave the dishes on the bench for 5 min.

Protocol 5 continued

7. Aspirate the medium and add 2 ml fixative to cover the coverslips. Leave for 5 min.

8. Aspirate the fixative, rinse dishes with PBS, and add sufficient stain to cover the coverslips. Leave for 7 min, covered with foil to protect the stain from light.

9. Remove the stain and rinse the dishes with water. Lift the coverslips from the dishes with forceps, and leave propped up on filter paper on the bench to dry, still protected from light.

10. Add 1 drop of mounting fluid to a slide and place a coverslip, cell-side down, onto the slide.

11. Observe samples under UV-epifluorescence using ×100 objective. The presence of mycoplasma is revealed by the fluorescence of small cocci or filaments. This appears as graininess in the cytoplasm, especially around the nucleus. A clean preparation should have stained nuclei only, and clear cytoplasm.

[a] This arrangement is to prevent the potential cross-contamination of cultures.

---

relatively rapid and straightforward (Protocol 5). PCR analysis, in comparison, is highly sensitive but requires rigorous controls to exclude false positives. ELISA- or PCR-based methods may be preferred in laboratories where those techniques are routine.

4 Methods for the derivation of mES cell lines

The classical and standard method for deriving mES cell lines is to culture intact day 3.5 p.c blastocysts on feeder layers for 2-5 d (during which time the blastocysts hatch and attach to the substratum), and then to collect, dissociate, and subculture the cells of the explanted ICM (Protocol 7, Method A). With blastocysts from strain 129 or C57BL/6 mice, lines should be obtained efficiently by this technique without recourse to other methods. Otherwise, improved efficiencies may be achieved by culturing the isolated epiblast after microsurgical removal of the primitive endoderm and trophectoderm. This procedure may be carried out on either implanting day 4.5 p.c. blastocysts (Method B), or on delayed implanting blastocysts (Method C). As outlined in Table 1, each method offers practical advantages depending on the mouse strain of choice; but it should be emphasized that the microsurgical techniques require considerable effort to master, and a substantial investment in equipment.

4.1 Derivation of mES cells from intact blastocysts (Method A)

Blastocysts are collected from day 3.5 p.c. pregnant mice by Protocol 6, explanted intact into culture, and mES cells derived by Protocol 7, which is based on references (4), (5), (24) and (25). Stages in the derivation are shown in *Figure 1*, and a schedule for manipulations and culture is given in *Figure 2*.

4.2 Derivation of mES cells from isolated epiblast

By day 4.5 p.c., the ICM of the mouse embryo develops into the epiblast and the primitive endoderm: the epiblast is invested on the blastocoelic surface by the layer of primitive endoderm, and is overlaid externally by trophectoderm. The aim of the microsurgical technique is to separate the epiblast cleanly from endoderm and trophectoderm, prior to culture. The procedure is the same for both day 4.5 p.c. and delayed blastocysts, but the subsequent schedule for epiblast culture differs slightly (*Figure 2*). Delayed blastocysts are produced by ovariectomy of mice on day 2.5 p.c. followed by the administration of a progestagen, and are recovered 5–7 d later. Protocols for ovariectomy may be found elsewhere (24). An alternative and effective procedure that avoids the need for this major

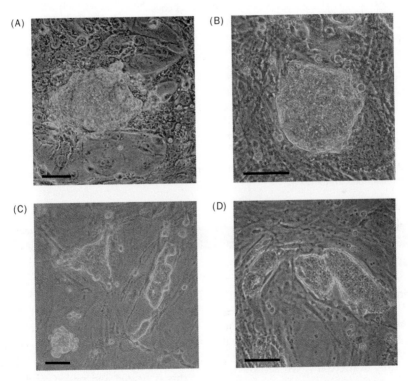

Figure 1 Stages in the derivation of mES cells from an intact, explanted blastocyst. (A) Appearance of the blastocyst after 4 d of culture on feeder cells: the ICM has formed a central, rounded mass of proliferating, undifferentiated cells surrounded by outgrowths of trophoblast giant cells. (B) One mES-cell colony 5 d after the dissociation of the cultured ICM: this colony shows the classic multilayering and clearly demarcated boundary with the feeder cells. (C) Several mES-cell colonies 3 d after the dissociation of the colony shown in (B). (D) The resultant culture of mES cells at passage 1: it is apparent that the colonies are composed of small cells with a high nuclear-to-cytoplasmic ratio. Scale bars: 100 μm.

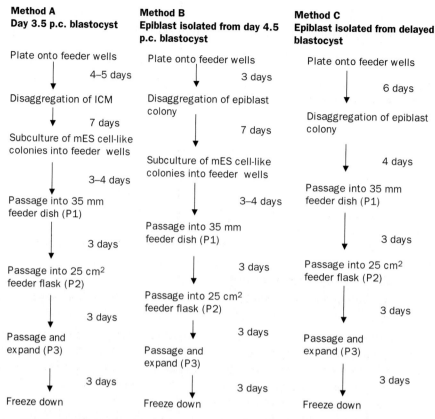

Figure 2 Schedules of manipulations and culture for the derivation of mES cell lines using the methods described in the text.

Protocol 6

Embryo collection

Equipment and reagents

- Pregnant mouse at day 3.5 p.c. for Method A; *or* at day 4.5 p.c. for Method B; *or* at day 7.5–9.5 p.c., where implantation has been delayed by ovariectomy (see Section 4.2), for Method C
- Sterile forceps and scissors for dissection, and a sterile pair of Watchmaker's forceps
- 70% alcohol in wash bottle
- Spirit lamp to flame-sterilize instruments
- 60 mm bacteriological-grade Petri dish
- 35 mm bacteriological-grade Petri dishes
- Pasteur pipette for collecting blastocysts: this is heated and pulled over the spirit lamp, and broken to give a diameter a little larger than that of a blastocyst (~200 μm internal diameter)
- Aspirator tube assembly for mouth-controlled pipetting (Cat. No. A5177, Sigma)

Protocol 6 continued

- Flushing instrument: either a Pasteur pipette (pulled over the spirit burner, broken at a diameter of ~0.5 mm, and flame polished at the tip) or a 25 Gauge hypodermic needle attached to a 1 or 2 ml syringe
- KSOM-Hepes medium for flushing (*Protocol 8*)
- Dissecting microscope with *circa* ×20 objective

Method

1. Sacrifice the mouse humanely, swab its abdomen with 70% alcohol and, using flame-sterilized forceps and scissors, cut open the abdominal cavity.

2. Move aside the intestines and locate the uterus. With the Watchmaker's forceps grasp the uterus at the cervical end, and cut through the cervix. Lift the uterine horns upwards and cut off the mesometrium, towards the utero-tubal junction. Cut through – on the uterine side – this junction, which acts as a valve and, if left in place, makes subsequent flushing exceedingly difficult.

3. Transfer the uterus into a 60 mm Petri dish, and rinse with KSOM-Hepes medium. Trim away any fat and mesentery left attached.

4. Fill the flushing instrument, either a mouth-controlled Pasteur pipette or an hypodermic syringe, with about 0.4 ml KSOM-Hepes medium. Place the uterus into a 35 mm dish. With the Watchmaker's forceps secure the uterus near the cervix, and insert the tip of the flushing instrument into the uterine lumen through the cut cervix. Flush each horn with ~0.2 ml medium.[a,b]

5. Place a drop of medium (~0.4 ml) into a fresh 35 mm dish. Under a dissecting microscope at magnification of about ×20, draw the blastocysts two or three at a time into the pulled Pasteur pipette using mouth control. Expel them gently into the drop of clean medium without introducing air bubbles.

6. Proceed immediately to explantation (*Protocol 7*).

[a] The flushing procedure is easier if performed under a dissecting microscope.

[b] Day 4.5 p.c. blastocysts will have hatched from the zona pellucida and be adherent, and therefore more difficult to flush. To aid blastocyst recovery, ensure that the tip of the flushing instrument fits tightly into the cervical end of the uterus, and clamp the oviductal end with Watchmaker's forceps; expel medium to gently inflate the uterus, and then release the clamp to flush out the contents. Additional embryos may be recovered by repeating this procedure in the opposite direction. However, the total number of blastocysts recovered will almost always be lower than for day 3.5 p.c. blastocysts.

Protocol 7

Method A: Derivation of mES cells from intact blastocysts

Equipment and reagents

- Blastocysts isolated at day 3.5 p.c. (*Protocol 6*)
- 4-well feeder plates prepared 1 d previously (*Protocol 3*)
- LIF-supplemented mES-cell medium (*Protocol 1*)
- 60 mm bacteriological-grade Petri dishes
- Pasteur pipettes drawn over a spirit burner and broken to give a fine capillary: one should be heat-sealed to give a blunt, closed tip
- Aspirator tube assembly (*Protocol 6*)
- PBS (*Protocol 1*)
- Trypsin solution (*Protocol 3*)
- Light mineral oil (Cat. No. M8410, Sigma)

Methods

Explantation of blastocysts into culture:

1. Aspirate MEF medium from feeder layers in 4-well plates, and replace with 0.6 ml/well of LIF-supplemented mES-cell medium.
2. Transfer blastocysts singly onto feeder layers in wells, using a pulled Pasteur pipette under mouth control; and incubate for 4–5 d.[a]
3. Change medium after 3–4 d, or when it starts to become acidic.

Dissociation of cultured ICMs:

After 4–5 d of culture the ICM is ready to be picked and dissociated (*Figure 1A*). The cultured ICMs will vary in size and appearance, with some having grown or differentiated more than others; and it is impossible to predict precisely from their morphological appearance at this stage which are more likely to give rise to mES cell lines. However, those cultured ICMs that have a dense core of material surrounded by a 'rind' of endoderm-like cells, or which have elongated into an egg cylinder-like structure, are not likely to produce undifferentiated mES cell lines.

4. Dispense (~30 μl) microdrops of trypsin solution in rows in a 60 mm Petri dish. Carefully pipette mineral oil down the side of the dish to just cover the drops.
5. Aspirate medium from the blastocyst cultures, and replace with 0.5 ml PBS.
6. Under a dissecting microscope, use the blunt pipette to dislodge the ICM from the underlying trophoblast. Transfer the ICM into a microdrop of trypsin solution, using a pulled pipette under mouth control.
7. Leave the ICM in trypsin solution for 5 min at r.t.; introduce a small amount of mES-cell medium to neutralize the trypsin; and using a pipette which has been pulled and broken at a diameter about half that of the ICM, draw the ICM up and down several times to dissociate it into a mixture of small clumps and single cells.

Protocol 7 continued

8   Transfer the disaggregated ICM into a fresh feeder well, and resume incubation.

Identification and harvesting of mES-cell colonies

9   After 2–3 d of incubation of the disaggregated ICM, discrete colonies of cells may be observed under an inverted microscope with either × 10 or × 20 objectives.[b]

10 By 4–7 d of culture, definitive mES-cell colonies should be discernible in a proportion of dissociated ICM cultures (*Figure 1B*). These colonies, which are multilayered and have a sharp, clear boundary, are composed of characteristic, small cells with a high nuclear-to-cytoplasmic ratio, and which are difficult to distinguish individually by light microscopy. Definitive mES-cell colonies may be picked and dissociated individually following the method used for ICMs (above).[c, d, e]

11 Transfer each disaggregated colony into a fresh feeder well containing LIF-supplemented mES-cell medium for expansion.

Expansion of mES cells:

After 2 d of culture small, secondary colonies of mES cells will be visible in the majority of wells or dishes. Discard cultures consisting only of differentiated cells. By 3–4 d of culture, the undifferentiated colonies will have reached a size suitable for expansion (*Figure 1C*), which should carried out even if few colonies are present.[f]

12 Aspirate medium, wash cells with 1 ml PBS, add 0.2 ml trypsin solution, and incubate for 5 min. Add 0.5 ml mES-cell medium to neutralize the trypsin, and using a Pasteur pipette draw the cell suspension up and down several times to break up the colonies into single cells.[g]

13 When a single-cell suspension is achieved, transfer this onto a feeder layer in a 35 mm dish containing LIF-supplemented mES-cell medium. Contents of wells originating from the same blastocyst may be combined at this stage, and the mES cell culture is designated as 'passage 1' (*Figure 1D*).

14 After a further 3–4 d of culture, trypsinize the dish and transfer the contents into a 25 cm^2 feeder flask (passage 2). Cells are now cultured as described in *Protocol 3*, and aliquots should be frozen down as soon as sufficient cells are available (usually by passage 3).

[a] The blastocysts hatch from the zona pellucida and attach to the dish during overnight culture. Thereafter, trophoblast cells grow outwards over the surface of the dish whilst the cells of the ICM form a clump in the middle of the outgrowth.

[b] At this stage most colonies will consist of differentiated cells, and may include trophoblast, endoderm and epithelial-like cells. Illustrations of these are given in (4), (5), and (24). Before becoming giant cells, trophoblast colonies in particular may be mistaken for mES-cell colonies; and so it is important to monitor periodically the development of putative mES-cell colonies.

[c] If the dissociated ICM cells have been blown hard *via* the mouthpiece into the well, mES-cell colonies become located near the edge of the well where they are difficult to see; and so it is advisable to search the periphery particularly carefully.

> ^d These primary mES-cell colonies dissociate more readily than cultured ICMs, and form a single cell suspension with a minimum of pipetting.
>
> ^e Alternatively, one may trypsinize the entire contents of a well *in situ* and transfer the whole onto another feeder well (5, 26) or a larger feeder layer (if there are many mES-cell colonies).
>
> ^f When allowed to become too large, these secondary mES-cell colonies fail to dissociate into single cells on trypsinization.
>
> ^g Care should be taken to minimize formation of air bubbles, which make it difficult to monitor the progress of dissociation under a dissecting microscope.

surgical intervention has been described (17). Briefly, this uses tamoxifen as a pharmacological block to oestrogen action, thus obviating the need for ovariectomy; and progesterone activity is maintained as in the ovariectomy method by the coinjection of Depo-Provera.

Microsurgical dissection is performed with glass needles and requires a micromanipulator assembly, consisting of a pair of micromanipulators attached to a base-plate and a fixed stage microscope with image-erected optics (*Figure 3*). The reader is referred to primary sources for comprehensive descriptions of the preparation of apparatus, and for rudiments in micromanipulation (23, 24).

4.2.1 Culture of epiblast from day 4.5 p.c. blastocysts (Method B)

Epiblasts isolated from day 4.5 p.c. blastocysts are cultured (on feeder layers, with LIF-supplemented mES-cell medium) for 3 d, by which time a single colony of undifferentiated cells should be present in each well (*Figure 6A*). If contaminating primitive endoderm was transferred with the epiblast there may be several additional, small colonies of differentiated cells (*Figure 6B*). After 3 d, the epiblast colonies are picked and dissociated as described in *Protocol 7* for cultured ICMs. However, detaching the epiblast colony from the dish is not as straightforward as with ICM clumps, but can usually be achieved by cutting through the feeder layer around the colony with the blunt-tipped pipette, and then pushing the colony with attached feeder cells away from the surface of the dish. The epiblast colonies should dissociate in trypsin solution more easily than ICMs to give mostly single cells, which are transferred into a fresh feeder well and cultured for 7 d in LIF-supplemented mES-cell medium. The medium is changed after 4 d.

Colonies of cells usually can be found after ~3 d. As occurs following ICM dissociations, many of these colonies are of differentiated cells, often endoderm-like. In contrast to ICM dissociations, epiblast dissociations give rise to fewer colonies in total, and trophectoderm and giant cells are rarely encountered. This facilitates the identification of individual mES-cell colonies, which appear from ~4 d of culture and may vary in number from one to 30 per well. After 7 d, mES-cell colonies are harvested and dissociated as in *Protocol 7*, and seeded into fresh feeder wells. Numerous mES-cell colonies should be apparent 2 d after this second dissociation. After 3-4 d (*Figure 6D*), the well is trypsinized

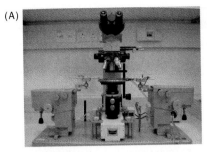 

Figure 3 Micromanipulator assembly for microsurgical dissection of epiblasts. (A) Here, Leitz micromanipulators are attached to a baseplate, and a microscope with erect optics is set between them. (B) Close-up to show a Puliv chamber on the microscope stage, with glass needles mounted in each micromanipulator.

Protocol 8
Solutions required for epiblast isolation

Ca^{2+} – and Mg^{2+} – free Tyrode's saline:
For 1 L, combine the following reagents:

NaCl 8.00 g KH_2PO_4 0.025 g

KCl 0.3 g $NaHCO_3$ 1.0 g

$NaH_2PO_4.5H_2O$ 0.093 g   Glucose 2.0 g

(or $NaH_2PO_4.2H_2O$ 0.05 g)

Make up in AnalaR® water, filter sterilize, aliquot (e.g. 5 ml), and store at 4 °C.

Hepes-buffered KSOM, or KSOM-Hepes (27):
For 1 L, combine the following reagents:

NaCl 5.5518 g $CaCl_2.2H_2O$ 0.2514 g

KCl 0.1868 g   EDTA 0.0038 g

KH_2PO_4 0.0476 g   Hepes 5.206 g

$MgSO_4.7H_2O$ 0.0493 g   BSA 4.0 g

Pyruvate (sodium salt) 0.022 g   Penicillin 0.060 g

Glucose 1.002 g   Streptomycin 0.050 g

$NaHCO_3$ 0.4201 g   Phenol Red 0.01 g

Lactate (Na salt) syrup, 60% (w/w) 0.8623 ml   Glutamine (200 mM) 5.0 ml

Make up in AnalaR® water, pH to 7.2, filter sterilize, aliquot (e.g. 10 ml), and store at 4 °C.

Protocol 9
Microsurgical isolation of the epiblast

(Based on methods described by Gardner and Davies (28).)

Reagents and equipment

- Day 4.5 p.c. blastocysts, *or* delayed blastocysts collected at day 7.5–9.5 p.c. (*Protocol 6*)
- 4-well feeder plates prepared 1 d previously (*Protocol 3*)
- LIF-supplemented mES-cell medium (*Protocol 1*)
- Trypsin powder (Cat. No. 215240, Difco)
- Pancreatin powder (Cat. No. P 3292, Sigma)
- Ca^{2+}- and Mg^{2+}-free Tyrode's saline (*Protocol 8*)
- KSOM-Hepes (*Protocol 8*)
- 0.22 µm syringe filter
- Puliv chamber, which may be manufactured in a workshop: it requires a sterile, siliconized coverslip (32 × 22 mm), KSOM-Hepes medium, Vaseline® or other high vacuum grease, heavy-grade liquid paraffin (may be obtained from a pharmacy), and a Pasteur pipette with the tip broken at a diameter of about 1.5 mm
- Benchtop centrifuge
- Light mineral oil (*Protocol 7*)
- A pair of glass needles for microdissection: their preparation requires thick-walled glass tubing (e.g. Leitz capillary, 1 mm outer diameter and 0.85 mm inner diameter), a microburner, and an electrode puller (e.g. from Sutter Instrument)
- Micromanipulator assembly (e.g. from Leitz)
- Pulled Pasteur pipette and aspirator tube assembly (*Protocol 6*)
- 35 mm bacteriological-grade Petri dish

Method

Embryo collection:

1. Before recovering blastocysts, freshly prepare a trypsin/pancreatin solution as follows: add 25 mg trypsin and 125 mg pancreatin to 5 ml Ca^{2+}- and Mg^{2+}-free Tyrode's saline (*Protocol 8*) and mix thoroughly; spin at 2500 rpm for 25 min, and sterilize the supernatant using a 0.22 µm syringe filter.

2. Place a large drop of trypsin/pancreatin solution in a 35 mm bacteriological-grade Petri dish, and cover with light mineral oil. Refrigerate the dish at 4 °C.

3. Collect either day 4.5 p.c. or delayed implanting blastocysts, as described in *Protocol 6*, in KSOM-Hepes medium.

Chamber preparation (Figure 4):

4. To set up the Puliv chamber, first smear Vaseline® or other high vacuum grease along the top of the walls (*Figure 4A*). Take a sterile coverslip and place a row of microdrops (about 1 µl) of KSOM-Hepes down the centre. Working quickly (because such small drops rapidly evaporate), invert the coverslip and press into place on top of the walls. Fill the chamber completely with paraffin using the wide-bore Pasteur pipette, by bringing the end of the pipette against the underside of the coverslip and

Protocol 9 continued

away from the drops, and expelling the oil whilst taking care not to introduce air bubbles (*Figure 4B* and *C*).

5 Transfer the blastocysts to the chamber for microdissection, introducing one blastocyst per hanging drop of KSOM-Hepes using a pulled Pasteur pipette under mouth control.

Needles for microdissection:

6 Prepare a pair of solid, sharp-tipped glass needles by rotating a piece of thick-walled glass tubing in a microburner flame until the glass solidifies. The solidified region is then placed within the heating filament of an electrode puller, and pulled to give a needle with a gradual taper and a sharp tip. Mount the needles in each micromanipulator with the instrument holder tilted slightly upward, so that the tips of the needles can make firm contact with the underside of the chamber coverslip.

Microsurgical dissection of the blastocyst (Figure 5):

7 Stab a blastocyst with both needles near the abembryonic pole, well clear of the ICM. (Hereafter the term 'ICM' refers to the epiblast plus primitive endoderm). Raise the needles until the blastocyst caught on the tip is in contact with the undersurface of the coverslip (*Figure 5A*).

8 Working against the coverslip, move the two needles apart to tear open the trophectoderm, keeping clear of the ICM (*Figure 5B*).

9 When several blastocysts have been opened in this way remove them from the hanging drops, pool and transfer them into the precooled trypsin/pancreatin solution, and incubate for 22–24 min at 4 °C.[a]

10 Rinse the embryos briefly in KSOM-Hepes, and return them individually to the chamber drops.

11 Recommence with an embryo: push one needle into the trophectoderm, away from the ICM, and again raise the blastocyst against the coverslip. Insert the tip of the second needle into the trophectoderm, and move the two needles apart so as to spread the trophectoderm out as a sheet along the undersurface of the coverslip. The ICM should now hang downwards, the whole structure resembling the yolk and white of a fried egg (*Figure 5C*).

12 Holding the trophectoderm in place with one needle, move the second until it catches the side of the ICM. Gently lower this needle whilst moving it laterally to scrape the primitive endoderm away from the surface of the epiblast (*Figure 5D*). Care must be taken with the level of the needle as, if it is not lowered sufficiently, the entire ICM may become detached. Once the endoderm begins to peel away, the bipartite structure of the ICM becomes obvious and the boundary between the

> **Protocol 9 continued**
>
> endoderm and epiblast can usually be seen. Continue until the endoderm is completely removed. (*Figure 5E, F*).
>
> 13 If necessary, clear the needle of debris by withdrawing it from the microdrop into the paraffin, and back into the microdrop. Then move the tip of the needle to one edge of the epiblast, raise the tip slightly and scrape the epiblast away from the overlying trophectoderm (*Figure 5F*).[b]
>
> 14 Once isolated, transfer each epiblast directly into a well of a 4-well feeder plate in which MEF medium has been replaced with 0.6 ml/well of LIF-supplemented mES-cell medium, and incubate (Sections 4.2.1 and 4.2.2).
>
> [a] This enzyme treatment loosens contact between the epiblast and both trophectoderm and endoderm, and facilitates the next stages. But if blastocysts are left in the trypsin/pancreatin solution for too long, the trophectoderm cannot be opened out as a sheet without fragmentation.
>
> [b] It is important to conduct this stage of the operation at a high enough magnification (about × 500) to give a limited depth of focus, and to monitor progress carefully to ensure that the overlying polar trophectoderm remains intact as the epiblast is scraped away from it.

and the contents transferred into a 35 mm feeder dish, and labelled 'passage 1'. After a further 3 d of culture the cells are passaged into a 25 cm^2 feeder flask, and subsequently the line is expanded, and frozen down.

4.2.2 Culture of epiblast from delayed-implanting blastocysts (Method C)

Epiblasts isolated from delayed-implanting blastocysts are cultured (on feeder layers with LIF-supplemented mES-cell medium) for 6 d with a medium change after 4 d, or earlier if it becomes acidified. Colonies – usually one per well – first become apparent by ~ 4 d as a small cluster of undifferentiated cells. By 5 d the undifferentiated colonies are larger and have a well-defined boundary, resembling a single large mES-cell colony in appearance (*Figure 6C*). After a further day in culture, the colonies begin to flatten and loose their sharp boundaries. At this stage they should be picked and dissociated as described in *Protocol 7*, and any colonies consisting solely of differentiated cells discarded. Undifferentiated colonies should dissociate very easily in trypsin solution into single cells; if not, they have probably already differentiated and are therefore unlikely to give rise to mES cell lines.

The dissociated cells are plated onto a fresh feeder well in LIF-supplemented mES-cell medium. Numerous small mES-cell colonies should be visible after 1–2 d of culture with almost no differentiated colonies present. By 3–4 d the colonies will be large enough to passage into a 35 mm feeder dish ('passage 1'). Thereafter the culture is passaged into a 25 cm^2 feeder flask, expanded, and frozen down.

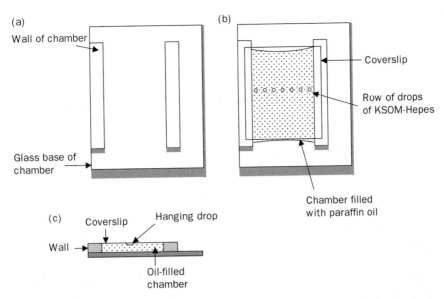

Figure 4 Preparation of a Puliv chamber for micromanipulation. (A) An empty Puliv chamber. This could be made in a workshop. (B) The walls of the chamber are smeared with Vaseline® or high-vacuum grease and a sterile, siliconized coverslip carrying a row of microdrops of KSOM–Hepes is inverted and pressed into place on the walls. The chamber is then filled with heavy paraffin oil. (c) A cross-section of the filled chamber shows a microdrop of medium hanging downwards from the coverslip, surrounded by oil in the chamber.

5 Validation of new mES cell lines

5.1 Karyotype

The first stage in validating a new mES cell line is to determine its sex and ensure that it has a normal karyotype. Much time and effort can be wasted in pursuing work with a line that is subsequently shown to be karyotypically abnormal. Loss of an X chromosome, and trisomies 8 and 11, are common abnormalities. Hence it is vital to analyse the exact chromosome constitution; a simple count cannot, for instance, distinguish between a normal karyotype and one that is XO with trisomy 8. We discard any line in which more than 10% of cells are trisomic, as in our experience the proportion of such cells may increase rapidly with successive passages.

5.2 Pluripotency

To demonstrate unequivocally that a newly derived mES cell line is pluripotent requires the generation of chimeras among whose offspring are individuals sharing the cell line's genetic background: that is, the mES cell line is germline competent. There are two principal methods for producing chimeras using mES cells, by blastocyst injection or morula aggregation, for which various protocols are available (24, 25, 31, 32). To be useful for genetic modification, a novel mES

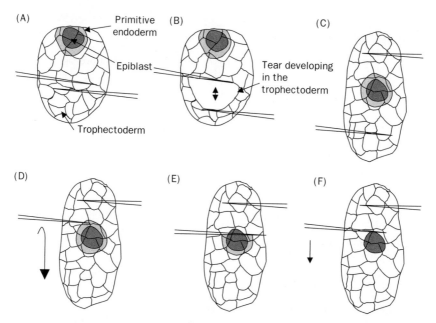

Figure 5 Stages of blastocyst dissection (modified with the authors' and publisher's permission from (28)). All views are as seen through the microscope, looking vertically down onto the blastocyst through the coverslip. (A) The blastocyst is stabbed with the needles, away from the ICM, and raised until it is in contact with the coverslip. The blastocyst is positioned with the ICM to one side, away from where the tear will be made. (B) The two needles are moved apart, as indicated by the arrow, to tear the trophectoderm open. The blastocyst is then incubated in trypsin/pancreatin solution. (C) Once returned to the chamber drop, one needle is pushed into the trophectoderm and the blastocyst is raised up against the coverslip. The second needle is inserted and the two moved apart, to spread the trophectoderm against the coverslip as a horizontal sheet of cells. (D) One needle is left in position holding the blastocyst whilst the second is moved to the same side of the ICM, lowered slightly, and used to scrape away the endoderm. (E) Once this manoeuvre is started, the endoderm should peel away readily. (F) The second needle should be returned to the starting position as in (D), raised slightly, and used to scrape away the epiblast from the overlying polar trophectoderm.

cell line should give chimeras at a rate of at least 50% of the total number of pups born, and most of the male chimeras should transmit through the germline. For purposes that do not require the generation of chimeric mice *per se*, such as *in vitro* differentiation experiments, a lower rate of chimera production may be acceptable.

Mouse ES cells are characterized by markers including the stage-specific embryonic antigen, SSEA-1 (33), alkaline phosphatase activity, and expression of the genes *Oct4* (34–36) and *Nanog* (37–38). Expression of these markers is consistent with – but not evidence of – pluripotency. However, if the pluripotency of a novel cell line is not an issue, it may be sufficient to use the presence of such markers to demonstrate mES cell-like characteristics.

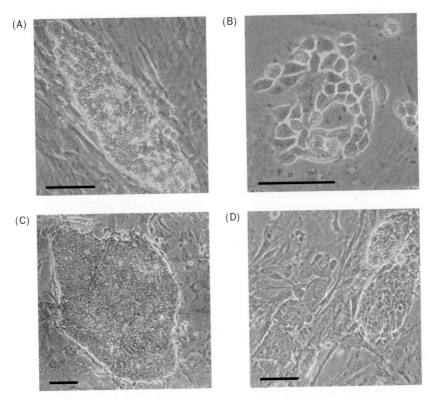

Figure 6 Stages in the derivation of mES cells from cultured epiblasts. (A) Epiblast isolated from day 4.5 p.c. blastocyst and cultured on feeder cells for 3 d. The colony is small with clearly defined boundaries, and individual cells are not distinguishable. (B) Contaminating colony of extraembryonic endoderm cells in epiblast culture. The individual cells are clearly distinguishable, with highly refractile cell borders. (C) Epiblast isolated from delayed-implanting blastocyst and cultured on feeder cells for 5 d. The colony has a clearly defined boundary and resembles in morphology a single, large mES-cell colony. (D) Cultured mES-cell colonies on feeder cells, 4 d after the second dissociation of isolated epiblast. Scale bars: 100 μm.

6 Female mES cell lines

Almost all the mES cell lines currently in use are male, as the XY genotype confers significant advantages where germline transmission is required: when injected into XX blastocysts, XY mES cells may colonize the gonads in sufficient numbers to convert the host embryo to a phenotypic male; and as none of the host (female) germ cells can develop in a male gonad, all the offspring are necessarily of the mES cell's genotype. In addition, male chimeras produce many more offspring than females over their lifetime. Nevertheless, there are occasions when XX mES cells are specifically required, e.g. for the study of X-linked hemizygous lethal mutations, or of mitochondrial disease (see Chapter 4).

Female lines are produced when deriving novel mES cells, although not with the 50% frequency one might expect. In only four out of 12 studies in which the

Protocol 10
Karyotype analysis

(The following methods have been provided by Dr E. P. Evans.)

A. Harvesting cells and preparing slides

Equipment and reagents

- 25 cm² flask of mES cells, 2 or 3 d after passaging
- 25 cm² feeder flask containing mES-cell medium
- 25 Gauge hypodermic needle connected to a 1 ml syringe
- Cell scraper
- Pasteur pipettes, one pulled to a smaller diameter in the flame of a spirit burner
- 15 ml plastic centrifuge tubes
- Benchtop centrifuge
- Microscope slides, cleaned by soaking overnight in a 1:1 mixture of absolute alcohol to concentrated hydrochloric acid: the slides are washed in running tap-water, rinsed in deionized water, and stored in a 1:1 mixture of absolute alcohol to diethyl ether; and before use, the slides are dried using grease-free tissues, taking care to avoid finger-marks
- 0.25% Trypsin/0.02% EDTA solution (Cat. No. T4049, Sigma)
- PBS (*Protocol 1*)
- Mouse ES-cell medium (*Protocol 1*)
- Colcemid solution: 10 µg/ml in HBSS (Cat. No. D1925, Sigma)
- Hypotonic solution: 0.56% aqueous solution of KCl (w/v)
- Fixative: a freshly prepared, 3:1 mixture of methanol and glacial acetic acid

Method

1. On the day before harvesting, trypsinize the flask of mES cells as in *Protocol 3*, and add half the cell suspension to a fresh feeder flask containing 8–10 ml of mES-cell medium.[a]

2. On the following day, add one drop of Colcemid stock solution to the culture to arrest mitosis, using a 25 Gauge hypodermic needle and syringe, and replace in the incubator for 20–30 min.

3. Remove all the cells from the flask, including feeders, with a cell scraper. Using a Pasteur pipette gently aspirate and transfer the cell suspension into a tube, and centrifuge at 1000 rpm for 5 min (this speed and time is used throughout, with braking).

4. Remove the supernatant, resuspend the pellet in ~10 ml of hypotonic solution, and incubate at r.t. for 6 min exactly.[b]

5. Centrifuge the suspension. Remove the supernatant and briefly (over 10 s) fix the pellet in methanol/acetic acid. Remove the fixative, flick the pellet to disperse the cells, and quickly resuspend in fresh fixative. Recentrifuge the suspension, remove the fixative, and resuspend the pellet finally in 0.5–1.0 ml fixative for slide making.[c]

Protocol 10 continued

6. Leave the tube for 30 min to allow the feeder cells to settle, thereby enriching the supernatant with mES cells.

7. To make chromosome spreads, release one drop of the cell suspension onto a clean slide using a pulled Pasteur pipette. Allow the drop to spread to its maximum diameter, and then rapidly dry it by blowing over it. Cell density can be monitored under a phase contrast microscope, and further drops added if necessary.[d]

[a] To obtain satisfactory mitotic preparations, the cells need to be in an active growth phase; if the medium is acidic they will have progressed too far and fewer dividing cells will be obtained.

[b] Mouse ES cells are very sensitive to hypotonicity, and any overrun of this time will result in excessive mitotic-cell breakage.

[c] Preparations are best made within 3 h of the final fixation.

[d] Although slides can be processed immediately, the quality of G-banding improves if slides are first 'aged' by being kept in closed boxes at r.t. for 1-2 weeks.

B. G-banding

Although a number of different methods can be used to differentiate chromosomes for analysis, a method for Giemsa- or G-banding is described here as it requires minimal preparation and uses conventional light microscopy.

Equipment and reagents

- Chromosome spreads fixed on glass slides (see above)
- Lidded staining jars (Coplin-style) for five slides: some glass jars tend to crack if subjected to a temperature of 65 °C, therefore test before use
- Duck-billed forceps for handling slides
- Pasteur pipettes.
- Water bath set at 60-65 °C
- Light microscope equipped with $circa \times 20$ (dry) and $\times 100$ (oil immersion) objectives

- $2 \times$ SSC solution: dissolve 4.4 g trisodium citrate and 8.8 g sodium chloride in 500 ml distilled water
- Normal saline solution: reconstituted from tablets (Cat. No. BR0053, Oxoid Ltd)
- Buffer solution, pH 6.8: reconstituted from Gurr's tablets (Cat. No. 331932D, VWR Int.)
- Trypsin powder (e.g. Cat. No. 215240, Difco)
- Giemsa stain: commercial stains are available specifically for chromosome banding (e.g. Cat. No. GS500, Sigma)

Method

1. Place slides in $2 \times$ SSC in a lidded Coplin jar, in a water bath at 60-65 °C for 1.5-2 h. Meanwhile, prepare a solution of 0.025% trypsin in normal saline.

2. Cool the jar and contents by placing under running tap water, and replace the $2 \times$ SSC with normal saline.

> **Protocol 10** continued
>
> 3. Remove one slide, drain by touching onto absorbent paper, and flood with 0.025% trypsin solution from a Pasteur pipette. Leave for a trial period of, e.g. 2 min. Stop the tryptic activity by returning the slide to the saline.
>
> 4. Rinse the slide in pH 6.8 buffer solution, and stain in Giemsa which has been diluted to 1 ml stain in 50 ml buffer solution, pH 6.8.
>
> 5. After 10 min remove the slide, wipe the back dry with tissue paper, and monitor the quality of staining/banding under the microscope using a *circa* × 20 objective. Assess the treatment of further slides by reference to the following guidelines:[a]
>
>   (a) Overall dark staining, with poor/no visible banding: increase duration of trypsin treatment.
>
>   (b) Chromosomes appear 'bloated' with pale staining: decrease duration of trypsin treatment.
>
>   (c) Lack of staining: increase the concentration of Giemsa.
>
> 6. When satisfactory staining/banding has been achieved, rinse the slide in pH 6.8 buffer solution, and quickly blow dry with an air current from a 'puffer'.
>
> 7. Analyse chromosome spreads microscopically.[b,c]
>
> [a] Following a period of trial and error, optimal conditions can be determined for the laboratory, and these will usually remain unchanged.
>
> [b] Slides do not necessarily need mounting under a coverglass, but can be directly analysed under immersion oil. Some oils, however, dissolve out the stain whereas others (e.g. from Lamb) do not and the slides will remain usable for years. If in doubt, make the staining permanent with a mountant such as DPX.
>
> [c] With experience, mouse chromosomes can be analysed directly under the microscope without recourse to photography. A number of karyotypes have been published and the standard ideogram is available (29, 30).

sex of newly derived mES cell lines was reported did the frequency of male to female lines approximate 50% (6, 9, 12, 39). In the remaining eight studies the majority of the lines were XY; and these included lines of 129, C57BL/6, BALB/c, and DBA genotypes (7, 8, 14, 20, 21, 40-42). A further complication with female mES cell lines is the facility with which they lose an X chromosome to become largely 39XO. In our experience, over 25% of newly derived female mES cell lines are already substantially XO by passage 4-6; and these lines show continuous loss of an X chromosome during culture so that by passage 15, more than half the cells are XO. This is often accompanied by a rapid increase in the proportion of cells with trisomy 8 (F.A. Brook and E.P. Evans, unpublished observations). Nevertheless, it should be emphasized that loss of an X chromosome, whilst clearly an issue when studying an X-linked hemizygous-lethal mutation, does not preclude transmission of the mES cell genotype through the germline (see Chapter 4).

References

1. Simpson, E.M., Sargent, E.E., Davisson, M.T., Mobraaten, L.E., and Sharp, J.J. (1997). *Nat. Genet.*, **16**, 19-27.
2. Threadgill, D.W., Yee, D., Matin, A., Nadeau, J.H., and Magnuson, T. (1997). *Mamm. Genome*, **8**, 390-3.
3. Evans, M.J. and Kaufman, M.H. (1981). *Nature*, **292**, 154-6.
4. Robertson, E.J. (1987). Embryo-derived stem cell lines. In *Teratocarcinomas and Embryonic Stem Cells: A Practical Approach*, (ed. E.J. Robertson), pp. 71-112. IRL Press, Oxford.
5. Abbondanzo, S.J., Gadi, I., and Stewart, C.L. (1993). Derivation of embryonic stem cell lines. In *Guide to Techniques in Mouse Development*, Methods in Enzymology, Volume 225 (eds P.M. Wasserrman and M.L. DePamphilis), pp 803-23. Academic Press, San Diego.
6. Nichols, J., Evans, E.P., and Smith, A.G. (1990). *Development*, **110**, 1341-8.
7. Roach, M.L., Stock, J.L., Byrum, R., Koller, B.H., and McNeish, J.D. (1995). *Exp. Cell Res.*, **221**, 520-5.
8. Kawase, E., Suemori, H., Takahashi, N., Okazaki, K., Hashimoto, K., and Nakatsuji, N. (1994). *Int. J. Dev. Biol.*, **38**, 385-90.
9. Goulet, J.L., Wang, C.Y., and Koller, B.H. (1997). *J. Immunol.*, **159**, 4376-81.
10. Suzuki, O., Matsuda, J., Takano, K., Yamamoto, Y., Asano, T., Naiki, M., and Kusanagi, M. (1999). *Exp. Anim.*, **48**, 213-6.
11. Robertson, E.J., Kaufman, M.H., Bradley, A., and Evans, M.J. (1983). Isolation, properties and karyotypic analysis of pluripotential (EK) cell lines from normal and parthenogenetic embryos. In *Teratocarcinoma stem cells*, (eds L.M. Silver, G.R. Martin, and S. Strickland), Cold Spring Harbor Conferences on Cell Proliferation, Vol. 10, pp. 647-63. Cold Spring Harbor Press, New York.
12. Brook, F.A. and Gardner, R.L. (1997). *Proc. Natl. Acad. Sci. USA*, **94**, 5709-12.
13. Ledermann, B. and Burki, K. (1991). *Exp. Cell Res.*, **197**, 254-8.
14. Noben-Trauth, N., Kohler, G., Burki, K., and Ledermann, B. (1996). *Transgenic Res.*, **5**, 487-91.
15. Schoonjans, L., Kreemers, V., Danloy, S., Moreadith, R.W., Laroche, Y., and Collen, D. (2003). *Stem Cells*, **21**, 90-7.
16. Cheng, J., Dutra, A., Takesono, A., Garrett-Beal, L., and Schwartzberg, P.L. (2004). *Genesis*, **39**, 100-4.
17. MacLean Hunter, S. and Evans, M. (1999). *Mol. Reprod. Dev.*, **52**, 29-32.
18. Nichols, J., Chambers, I., Taga, T., and Smith, A. (2001). *Development*, **128**, 2333-9.
19. Yagi, T., Tokunaga, T., Furuta, Y., Nada, S., Yoshida, M., Tsukada, T., Saga, Y., Takeda, N., Ikawa, Y., Aizawa, S., and Yagi, T. (1993). *Anal. Biochem.*, **214**, 70-6.
20. Brook, F.A., Evans, E.P., Lord, C.J., Lyons, P.A., Rainbow, D.B., Howlett, S.K., Wicker, L.S., Todd, J.A., and Gardner, R.L. (2003). *Diabetes*, **52**, 205-8.
21. Kress, C., Vandormael-Pournin, S., Baldacci, P., Cohen-Tannoudji, M., and Babinet, C. (1998). *Mamm. Genom.*, **9**, 998-1001.
22. McWhir, J., Schnieke, A.E., Ansell, R., Wallace, H., Colman, A., Scott, A.R., and Kind, A.J. (1996). *Nat. Genet.*, **14**, 223-6.
23. Bradley, A. (1987). Production and analysis of chimeric mice. In *Teratocarcinomas and Embryonic Stem Cells: A Practical Approach*, (ed. E.J. Robertson), pp. 113-51, IRL Press, Oxford.
24. Nagy, A., Vintersten, K., Gertsenstein, M., and Behringer, R., eds (2003). *Manipulating the Mouse Embryo*, Cold Spring Harbor Laboratory Press, New York.
25. Papaioannou, V. and Johnson, R. (2000). Production of chimeras by blastocyst and morula injection of targeted ES cells. In *Gene Targeting: a Practical Approach*, (ed. A.L. Joyner), pp. 133-75, Oxford University Press, Oxford.

26. Matise, M., Auerbach, W., and Joyner, A.L. (2000). Production of targeted embryonic stem cell clones. In *Gene Targeting: a Practical Approach* (ed. A.L. Joyner), pp. 101-32, Oxford University Press, Oxford.
27. Summers, M.C., Bhatnagar, P.R., Lawitts, J.A., and Biggers, J.D. (1995). *Biol. Reprod.*, **53**, 431-7.
28. Gardner, R.L. and Davies, T.J. (2000). *Methods Mol. Biol.*, **135**, 397-424.
29. Lyon, M.F., Rastan, S., and Brown, S.D.M., eds. (1996). *Genetic Variants and Strains of the Laboratory Mouse*, pp. 1446-9, Oxford University Press, Oxford.
30. Mammalian Genetics Unit, Harwell, UK. *Introduction to Anomaly Data.* http://www.mgu.har.mrc.ac.uk/anomaly/anomaly-intro.html.
31. Stewart, C.L. (1993). Production of chimaeras between embryonic stem cells and embryos. In *Guide to Techniques in Mouse Development*, Methods in Enzymology, Vol. 225 (eds P.M. Wasserman and M.L. DePamphilis), pp 823-55. Academic Press, San Diego.
32. Nagy, A. and Rossant, J. (2000). Production and analysis of ES cell aggregation chimeras. In *Gene Targeting: a Practical Approach* (ed. A.L. Joyner), pp. 176-206, Oxford University Press, Oxford.
33. Solter, D. and Knowles, B.B. (1978). *Proc. Natl. Acad. Sci. USA*, **75**, 5565-9.
34. Okamoto, K., Okazawa, H., Okuda, A., Sakai, M., Muramatsu, M., and Hamada, H. (1990). *Cell*, **60**, 461-72.
35. Rosner, M.H., Vigano, M.A., Ozato, K., Timmons, P.M., Poirier, F., Rigby, P.W., and Staudt, L.M. (1990). *Nature*, **345**, 686-92.
36. Scholer, H.R., Ruppert, S., Suzuki, N., Chowdhury, K., and Gruss, P. (1990). *Nature*, **344**, 435-9.
37. Chambers, I., Colby, D., Robertson, M., Nichols, J., Lee, S., Tweedie, S., and Smith, A. (2003). *Cell*, **113**, 643-55.
38. Mitsui, K., Tokuzawa, Y., Itoh, H., Segawa, K., Murakami, M., Takahashi, K., Maruyama, M., Maeda, M., and Yamanaka, S. (2003). *Cell*, **113**, 631-42.
39. Voss, A.K., Thomas, T., and Gruss, P. (1997). *Exp. Cell Res.*, **230**, 45-9.
40. Auerbach, W., Dunmore, J.H., Fairchild-Huntress, V., Fang, Q., Auerbach, A.B., Huszar, D., and Joyner, A.L. (2000). *Biotechniques*, **29**, 1024-8, 1030, 1032.
41. Brown, D.G., Willington, M.A., Findlay, I., and Muggleton-Harris, A.L. (1992). *In Vitro Cell Dev. Biol.*, **28A**, 773-8.
42. Sukoyan, M.A., Kerkis, A.Y., Mello, M.R., Kerkis, I.E., Visintin, J.A., and Pereira, L.V. (2002). *Braz. J. Med. Biol. Res.*, **35**, 535-42.

Chapter 3
Production of ES cell-derived mice

Yoko Kato and Yukio Tsunoda
Laboratory of Animal Reproduction, College of Agriculture,
3327-204 Nakamachi, Nara, 631-8505, Japan.

1 Introduction

The reintroduction of mouse ES (mES) cells into an embryonic environment allows expression of their pluripotency, as defined by the capacity to contribute to all tissue and cell types, including the germ cells (1, 2). This most powerful feature of mES cells has led to their exploitation in transgenic research, providing a direct route for the insertion of genetic modifications into the germline of fertile animals. Techniques for generating chimeric mice using mES cells – specifically, blastocyst injection and morula aggregation – are now standard worldwide (3); and recent refinements to those techniques have increased the efficiency of production of mice that are derived entirely – rather than partially – from the introduced mES cells, that is 'mES cell-derived mice'. These refinements include the use of host embryos that have been rendered tetraploid (4), and which in consequence are incompetent to undergo fetal development; and the hyperthermic treatment of host embryos selectively to kill their inner cell masses (ICMs) (5). To compromise the viability of the host embryo's ICM by either of these means favours its reconstitution by exogenously added mES cells, and consequently the production of mES cell-derived fetuses and offspring. In addition, mES cell-derived mice may be cloned by the technique of nuclear transfer using mES cells as the donor cells, or 'karyoplasts' (6). In this chapter we provide our current methods for producing mES cell-derived mice by embryo reconstitution (morula aggregation and blastocyst injection) using tetraploid and hyperthermically treated host embryos, and by nuclear transfer. Owing to space limitations, the reader is referred to primary texts for comprehensive coverage of supplementary methods in mouse husbandry, embryo micromanipulation, and surgical transfer (7).

Methods for the production of mES cell-derived mice are as yet too inefficient to replace conventional embryo-reconstitution techniques (i.e. using normal, diploid host embryos) for the production of chimeric mice (see *Table 1*). Furthermore, neonatal mortality and developmental abnormalities accompany the production of mES cell-derived mice, possible reasons for which are discussed in Section 7. Nevertheless, these emergent techniques offer significant advantages arising from the reproducible, and relatively rapid, production of fetal or adult material that is entirely mES-cell derived:

(a) The requirements for intermediate generations of chimeric founder animals, and lengthy breeding programmes to produce mutant animals, are circumvented.

(b) For loss-of-function studies, development in mES cell-derived fetuses may progress beyond stages observed with heterozygous intercrosses (8).

(c) Where a genetic modification results in lethality or loss of fertility in heterozygous mice, these alternative techniques may allow the production of homozygous-mutant tissues.

(d) Systems for the production of mES cell-derived mice provide optimum models for investigating epigenetic instability in mES cells, which phenomenon is manifest phenotypically by developmental abnormalities. Epigenetic instability is currently an important problem in the nascent ES cell-based technologies.

Once the problem of inefficiency inherent to these methods is overcome, and causes of neonatal mortality are elucidated, the great experimental potential of mES cell-derived mice will be realized.

2 Maintenance of mES cell lines and embryo culture

To preserve the *in vivo* developmental capacity and germline competence of mES cells, for the production of either mES cell-derived mice or chimeras, the same critical factors apply: namely, the specific mES cell line used, its period in culture (i.e. passage number), and a high standard of routine maintenance. Regarding the suitability of cell lines for the production of mES cell-derived mice, considerations apply depending on the particular method, as discussed in the respective sections below. However, it is generally valid for all embryo reconstitution experiments that low-passage cultures of mES cell lines are preferable, so that the developmental capacity is maximized; and recommended passage numbers include p5–11 (9), <p14 (10), and <p20 (11). It should be noted that for lines that have reached higher passage numbers, sublines may be cloned that retain high developmental capacity (10); and from this it is inferred that high-passage cultures are heterogeneous cell populations with respect to retention of pluripotency.

Methods for the derivation, maintenance, cryopreservation, and karyotypic analysis of mES cell lines are described in detail in Chapter 2. We routinely

Table 1 Collated efficiencies of production of mES cell-derived mice by various techniques

Technique	Status of host embryo	Offspring (%)[a]	Chimeras (%)[b]	ES cell-derived mice (%)[b]	Surviving ES cell-derived mice (%)[b]	References
Blastocyst injection	Diploid	36	49	0	0	(5)
Blastocyst injection	Tetraploid	4	9	40	0	(19, 37)
Blastocyst injection	Diploid, hyperthermically treated	3	36	64	14	(19, 37)
Morula aggregation	Diploid	8–47	22–67	0	0	Kato and Tsunoda, unpublished
Morula aggregation	Tetraploid	3–16	–	12	10	(8)
Mouse ES cell-NT	–	1.0–24.6	0	100	0.7–12.3	(6, 20, 25, 27, 31, 37)

[a] Percentage of embryos transferred.
[b] Percentage of total offspring.

passage mES cells on MEF feeder layers in gelatinized dishes (see Chapter 6, *Protocol 1*), using mES-cell medium that is supplemented with 1000 U/ml leukaemia inhibitory factor (LIF). The proportion of cells in a line having a diploid karyotype should be at least 75% for the successful production of mES cell-derived mice. Mycoplasma testing must be performed on mES cell lines on a regular basis (see Chapter 2). In *Protocol 1* we describe a simple and convenient method for obtaining a suspension of several hundred mES cells, sufficient for embryo-reconstitution experiments and nuclear transfer. Unless otherwise stated, the incubation and culture of cells and embryos are at 37 °C, in a humidified incubator gassed with 5% CO_2 in air.

Embryo culture throughout is in microdrops (5–10 μl) of medium under oil, contained in bacteriological-grade Petri dishes, unless otherwise indicated. We recommend: either M16 or modified potassium simplex optimized medium (KSOM) for the culture of both diploid and tetraploid embryos; and the Hepes-buffered counterparts, M2 and FHM, respectively, for embryo manipulations and incubations in atmospheric air.

Embryos produced and manipulated according to these protocols, and having reached the compacted morula or blastocyst stages, may be transferred into pseudopregnant recipients via either the oviduct (day 0.5 p.c., pseudopregnant) or the uterus (day 2.5–3.5 p.c., pseudopregnant). Typically, 10 embryos are transferred per oviduct or uterine horn.

Protocol 1
Preparation of mES cells for embryo reconstitution

Immediately prior to experimentation (whether involving morula aggregation, blastocyst injection, or nuclear transfer), small-scale cultures of mES cells are harvested mechanically and disaggregated into a single-cell suspension of several hundred cells, sufficient for embryo reconstitution or as a source of karyoplasts.

Reagents and equipment

- Mouse ES cell culture on a feeder layer, in a multiwell or tissue-culture dish
- Mouse ES cell medium (for components see 'mES-cell medium' in Chapter 2, *Protocol 1*) supplemented with 10^3 U/ml LIF (or ESGRO®; Cat. No. ESG1106, Chemicon International)[a, b]
- Ca^{2+}- and Mg^{2+}-free PBS (Chapter 2, *Protocol 1*)
- Trypsin/EDTA solution: 0.25% trypsin and 0.02% EDTA in HBSS (Cat. No. T4049, Sigma)
- Hepes-buffered embryo-culture medium: M2 (Cat. No. M7167, Sigma), FHM (MR-122-D, Specialty Media) or Hepes-buffered KSOM (Chapter 2, *Protocol 8*)
- Mouse ES cell-qualified fetal-calf serum (FCS; see Chapter 2) or KnockOut™ SR[a]
- 35 mm tissue-culture dishes (Cat. No. 100–035, Iwaki)
- Sterile, Pasteur pipettes, one finely drawn and connected to an aspirator tube assembly (consisting of mouthpiece, tubing, and pipette holder; Cat. No. A-5177, Sigma)
- Stereo dissecting microscope

Method

1. Aspirate medium from the culture of mES cells, and wash cells once with PBS.
2. Add trypsin/EDTA solution to cover the base of the dish or well, and incubate for 1–2 min.
3. Tilt the dish gently from side to side, and as soon as mES-cell colonies start detaching from the substratum transfer the dish to the stage of a dissecting microscope. Select, by their morphological appearance, several undifferentiated colonies, harvest them using a Pasteur pipette, and transfer them into 1 ml trypsin/EDTA solution in a dish or well. Disaggregate the colonies into a single-cell suspension by repeatedly drawing them up and down a finely drawn Pasteur pipette under mouth control.[c]
4. Transfer the single-cell suspension into a dish containing 2–3 ml fresh Hepes-buffered embryo-culture medium supplemented with either 10–15% FCS or 15% KnockOut™ SR as appropriate, for immediate use.

[a] As an alternative to the powdered DMEM for mES-cell medium described in Chapter 2, *Protocol 1*, for convenience DMEM may be purchased in liquid form (high-glucose formulation; Cat. No. 10829–018, Invitrogen); and 15% KnockOut™ Serum Replacement (KnockOut™ SR; Cat. No. 10828–028, Invitrogen) may be used in place of 10–15% FCS.

> **Protocol 1** continued
>
> [b] We recommend the additional supplementation of mES-cell medium with nucleosides. To prepare a 100 × stock solution, dissolve 80 mg adenosine, 85 mg guanosine, 73 mg cytidine, 73 mg uridine, and 24 mg thymidine in 100 ml double-distilled water at 37 °C, filter sterilize, and store at 4 °C. Warm the stock solution to dissolve precipitates before use.
>
> [c] Mouse ES cells in suspension are readily discriminated (by their smaller size, rounded shape, and refractile borders) from contaminating feeder cells (which are large and irregular in shape). Furthermore, as feeder cells settle faster than mES cells, a feeder-depleted suspension of mES cells may be obtained by incubating the dish for ∼20 min (during which time feeder cells adhere to the substratum) and then carefully harvesting the supernatant.

3 Reconstituting tetraploid embryos with mES cells

3.1 Tetraploid embryos

When mouse embryos are induced experimentally to become tetraploid at early cleavage stages, their capacity for fetal development is compromised; and when a tetraploid and a diploid morula are aggregated, the tetraploid cells contribute solely to the extraembryonic lineages, and the diploid cells to the fetus. These distinct propensities of tetraploid and diploid embryonic cells were exploited by Nagy *et al.* (4) in devising a method for the aggregation of (diploid) mES cells with tetraploid morulae: in consequence, the relative incorporation of mES cells into the host embryos was enhanced to the effect that viable and fertile mES cell-derived mice were produced (10). A subsequent modification has been to use tetraploid-blastocyst injection (8) to generate mES cell-derived mice, in this case carrying genetic modifications. (Tetraploid blastocysts have fewer numbers of total cells and of ICM cells than diploid blastocysts and, theoretically, the introduced mES cells are at an additional, numerical advantage for colonizing the ICM.)

Tetraploid embryos are produced most commonly by electrofusion of blastomeres at the two-cell embryonic stage, as described in *Protocol 2* (and see references 12, 13; and the Nagy Laboratory's website, http://www.mshri.on.ca/nagy). The resultant 'pseudo-one-cell' embryos may be cultured *in vitro* to either the morula or blastocyst stages for embryo reconstitution with mES cells, which is achieved by conventional methods as used for diploid host embryos without substantial modification. For tetraploid embryos, as for diploid embryos, morula compaction and blastocoele formation occur at 96 h and 112 h post-hCG injection (*Protocol 2*), respectively.

3.2 Mouse ES cell lines for tetraploid-embryo reconstitution

Eggan and coworkers (12) have established that mES cell lines from mice with an F_1 hybrid genotype (e.g. C57BL/6 × 129/Sv, 129/Sv × FVB) are considerably superior to lines from inbred strains for the production of viable and fertile mES cell-derived mice by tetraploid-morula aggregation, and are therefore recommended here. When considering appropriate mouse strains to act as donors for the production of tetraploid embryos, it is essential to select a genetic background that is distinct from that of the mES cell line used, and which further allows rapid discrimination of offspring by coat colour or by isozyme analysis of tissues. When the mES cells are derived from hybrid mice, it is preferable to use host tetraploid embryos from outbred rather than (less vigorous) inbred strains, to maintain an overall balance of viability within reconstituted embryos. In particular, we recommend the outbred strain, CD-1 (or ICR), and this should carry GPI isozymes (Gpi-1a/a or Gpi-1b/b) that are distinguishable from those of the mES cell line in use.

3.3 Methods for reconstituting tetraploid embryos with mES cells

Mouse ES cells may be introduced into tetraploid embryos at either the morula or blastocyst stages. Accordingly, pseudo-one-cell, tetraploid embryos are incubated *in vitro* until the desired stage is reached.

Three techniques are available for introducing mES cells into tetraploid morulae, as with diploid morulae:

(a) aggregation of mES cells with zona pellucida-free (or 'zona-free') morulae;

(b) simple coculture of mES cells with zone-free morulae; and

(c) microinjection of mES cells sub-zonally, that is into the perivitelline space of zona-intact morulae.

Microinjection is more technically demanding, and for handling larger numbers of embryos the zona-free methods are recommended. Aggregation techniques require neither special equipment nor expertise in micromanipulation, and give efficiencies of production of germ-line chimeras comparable to those obtained by blastocyst injection. Zonae pellucidae of morulae may be removed *en masse* either chemically using acid Tyrode's solution, or enzymatically using pronase solution; but for smaller numbers of embryos, mechanical removal (or cutting) by micromanipulation is also practical and effective. In this section we highlight our particular aggregation methods for zona-free and zona-cut embryos, and a modified method for coculture based on Wood *et al.* (14). (Our methods can also be employed for aggregations using diploid host embryos.) Other suitable methods may be found elsewhere (7, 12). Tetraploid blastocysts may be injected with mES cells using essentially the same techniques as described for diploid blastocysts (7).

Protocol 2
Production of tetraploid embryos

Reagents and equipment

(Standard reagents are from Sigma, unless otherwise stated.)

- Female donor and male stud mice, e.g. CD-1 (or ICR) strain, carrying GPI isozymes Gpi-1a/a or Gpi-1b/b as appropriate (available from Charles River Laboratories, or CLEA Japan, Inc.): females should be at least 6 weeks old
- Pregnant mare's serum gonadotropin (PMSG; supplied as 1000 IU Serotropin from Aska Pharmaceutical Co. Ltd): resuspend at 100 IU/ml in 0.6% NaCl
- Human chrorionic gonadotropin (hCG; supplied as 1500 IU Puberogen from Sankyo LifeTech): resuspend at 100 IU/ml in 0.6% NaCl
- Embryo-culture medium: M16 (Cat. No. M7292, Sigma) or KSOM (Cat. No. MR-106, Specialty Media)
- Hepes-buffered embryo-culture medium: M2 or FHM (Protocol 1)
- Sterile instruments for dissection
- 27 Gauge hypodermic needle with a square-cut tip, flattened by grinding, and attached to a 1 ml syringe
- Aspirator tube assembly (Protocol 1)
- Culture dish or glass cavity block
- Stereo dissecting microscope
- Embryo-tested light mineral oil (Cat. No. M8410, Sigma) or mES cell-qualified silicon oil (Cat. No. ES-004-C, Specialty Media)
- Pulse generator or electrofusion apparatus, e.g. BTX ECM2001 (BTX Technologies Inc.) or CF-100 (Biochemical Laboratory Service)
- Electrofusion chamber consisting of two platinum electrodes set 1 mm apart on a glass slide, or the manufactured chamber of an electrofusion apparatus (GSS-500, Biochemical Laboratory Service)
- Non-electrolyte buffer (or electrofusion medium), either: mannitol solution, consisting of 0.3 M mannitol with 0.5% polyvinyl pyrrolidone (PVP-0930, Sigma) in double-distilled water, pH 7.2; or Zimmerman's fusion medium, freshly prepared, consisting of 9.584 g sucrose, 0.011 g magnesium acetate tetrahydrate (($CH_3CO_2)_2$ $Mg.4H_2O$), 0.002 g calcium acetate monohydrate (($CH_3CO_2)_2Ca.H_2O$,), 0.017 g dipotassium phosphate, and 0.003 g glutathione dissolved in ~80 ml double-distilled water, adjusted to pH 7.0 with 1 M HCl and to 100 ml, supplemented with 1 mg BSA (Cat. No. A4378, Sigma) and filtered

Method

1 At ~18:00 hours, induce superovulation in female mice by intraperitoneal (i.p.) injection of 5 IU PMSG, followed 40–48 h later (typically at ~17:00 hours) by i.p. injection of 5 IU hCG; and immediately cage the females with stud males of the same strain, using one female per male.

2 Next morning, check for vaginal plugs to confirm mating (day 1 p.c.).[a]

3 Sacrifice mated females 45–46 h after hCG injection (on day 1.5 p.c.), and collect the oviducts by dissection. Harvest two-cell embryos by carefully flushing oviducts with M2 or FHM into a dish or glass cavity block, using a flat-tipped needle connected to a 1 ml syringe.[b, c]

Protocol 2 continued

4 Wash embryos free of debris by passing them through drops of M2 or FHM, and culture in microdrops of M16 or KSOM, prior to electrofusion.

5 Place the electrofusion chamber on the base of a suitably sized Petri dish, and the Petri dish onto the stage of a dissecting microscope. Place a large, flat drop (~500 μl) of non-electrolyte buffer centrally, across both electrodes.

6 Collect and transfer batches of up to 10 embryos into a drop of electrofusion medium in a culture dish. Wait for the embryos to sink through the medium, to equilibrate; and transfer them into the drop on the electrofusion chamber. Either align the embryos manually using an aspirator tube assembly so that the individual blastomeres are parallel to the electrodes, or allow alignment to occur automatically during the initial pulse of AC described below.

7 Induce electrofusion of blastomeres using the following series of pulses and conditions: alternating current (AC) at 0.5 MHz and 15 V/mm for 15 s; direct current (DC) at 100 V/mm for 100 μs; AC at 0.5 MHz, 15 V/mm for 10 s; and DC at 100 V/mm for 100 μs. Blastomeres can be observed to fuse over 30 min (usually with over 90% efficiency) to form pseudo-one-cell, tetraploid embryos. *Immediately after electrofusion transfer the embryos into M2 or FHM.*

8 Wash embryos by passing them through drops of M2 or FHM, transfer them into microdrops of M16 or KSOM under oil, and incubate.[d]

9 Culture the tetraploid embryos for ~1 d (or until 72 h post-hCG injection) to reach the two-cell stage, ~1.5 d for the precompacted morula stage, or up to 3 d (or until 112 h post-hCG) for the blastocyst stage.

[a] For reference, diploid embryos at various stages may by obtained by harvesting one-cell, fertilized embryos at 24 h post-hCG, and culturing them *in vitro* until approximately: 48 h post-hCG, for the two-cell stage; 72 h post-hCG, for the 4–8 cell stage; 96 h, for compacted morulae; and 112 h post-hCG, for blastocysts.

[b] Alternatively, transfer the entire oviducts into the dish containing M2, and tear them carefully to release the embryos.

[c] From a superovulated, pregnant CD-1 female, 20–30 two-cell embryos should be harvested.

[d] Tetraploid embryos may be obtained at various stages by culturing pseudo-one-cell embryos *in vitro* until approximately: 72 h post-hCG, for the two-cell stage; 96 h post-hCG, for the 4–8 cell stage and compacted morulae; and 112 h post-hCG, for blastocysts.

Protocol 3

Lectin-assisted aggregation of mES cells with zona-free, tetraploid embryos

This particular method for aggregation features the use of a plant lectin, phytohaemagglutinin (PHA), to rapidly enhance the process of adhesion of mES cells to the tetraploid embryos.

Protocol 3 continued

Reagents and equipment

(Standard reagents are from Sigma, unless otherwise stated.)

- Single-cell suspension of mES cells in M2 or FHM (Protocol 1)
- Precompacted, tetraploid morulae following 1.5 d of incubation postelectrofusion (Protocol 2)
- Acid Tyrode's solution: dissolve 0.8 g NaCl, 0.02 g KCl, 0.024 g $CaCl_2.2H_2O$, 0.01 g $MgCl_2.6H_2O$, 0.2 g glucose, and 0.4 g PVP (Protocol 2) in ~80 ml double-distilled water; adjust to pH 2.5 (with c. HCl) and to 100 ml final volume
- Pronase solution: dissolve 0.5% (w/v) pronase (Actinase E; Kaken Pharmaceutical Co. Ltd) in PBS, and incubate at 37 °C for 2 h; dialyse against PBS at 4 °C overnight, to remove low-molecular-weight toxins
- Hepes-buffered embryo-culture medium: M2 or FHM (Protocol 1)
- PHA-supplemented medium: dissolve 5 µg/ml phytohaemagglutinin (PHA; Cat. No. L-9017, Sigma) in M2 or FHM
- Bacteriological-grade Petri dish
- Embryo-transfer capillary: finely drawn glass capillary, 80–100 µm diameter, connected to a holder and mouth-piece
- Cell-transfer capillary: finely drawn glass capillary, 20–50 µm diameter, connected to a holder and mouth-piece
- Aggregation dishes: these consist of 60 mm BD Falcon™ No. 353002 tissue-culture dishes (Cat. No. 08-772B, Fisher Scientific), containing 5–10 microdrops of PHA-supplemented medium overlaid with light mineral oil (Protocol 2); and into each microdrop 4–5 depressions are made using a sterile darning needle or similar metal implement (15, 16, 17)
- 60 mm BD Falcon™ tissue-culture dishes containing microdrops of M16 or KSOM supplemented with 5–15% FCS (Protocol 1) under oil, into each of which 4–5 depressions are made (as above)
- Stereo dissecting microscope

Method

1. Remove zonae pellucidae from tetraploid morulae by incubation in 10–20 µl microdrops of either acid Tyrode's solution (at r.t., for less than a minute) or pronase solution (at 37 °C, for several minutes) under oil in a bacteriological-grade Petri dish, constantly monitoring the procedure using a dissecting microscope.[a]

2. As soon as the zonae pellicidae become barely perceptible, immediately transfer the embryos into drops of M2 or FHM.

3. Wash the zona-free embryos by passing through several microdrops of M2 or FHM, and transfer them singly into the depressions in microdrops of PHA-supplemented medium, in aggregation dishes.

4. Using either a cell-transfer or embryo-transfer capillary under mouth control, collect 10–15 mES cells from the cell suspension and place them in the depression, in close contact with a zona-free embryo – between blastomeres if possible. (If there are small clumps of adhering mES cells in the suspension, these are preferentially used.) Alternatively, sandwich the mES cells between *two* zona-free embryos. The cells will

> **Protocol 3** continued
>
> be seen immediately to adhere to juxtaposed blastomeres. Manipulations should be performed as quickly as possible, ideally within 5–10 min to sustain the viability of the embryos in PHA-containing medium.
>
> 5 Wash the successfully aggregated embryos through several microdrops of M2 or FHM, and transfer to fresh culture dishes, into depressions in microdrops of FCS-supplemented M16 or KSOM, and culture overnight.
>
> 6 Next morning, select good-quality, expanded blastocysts for transfer to recipient females (see Section 2).
>
> [a]Alternatively, zonae pellucidae may be removed by micromanipulation (*Protocol 5*).

> **Protocol 4**
>
> ## Coculture of mES cells with zona-free, tetraploid embryos
>
> ### Equipment and reagents
>
> - Single-cell suspension of mES cells at 0.5×10^6–1.0×10^6 cells/ml in mES-cell medium
> - Zona-free, tetraploid morulae (*Protocol 3*)
> - Aggregation dishes: these consist of 60 mm dishes containing microdrops of mES cell suspension overlaid with mineral oil, and into each of which 4–5 depressions are made (*Protocol 3*)
> - 60 mm dishes containing microdrops of M16 or KSOM overlaid with mineral oil, and into each of which 4–5 depressions are made (*Protocol 3*)
> - Embryo-culture medium: M16 or KSOM (*Protocol 2*)
> - Stereo dissecting microscope
>
> ### Method
>
> 1 Place zona-free, tetraploid morulae singly into the depressions in microdrops of mES-cell suspension, in the aggregation dishes, and incubate for 3–4 h.
>
> 2 Transfer those embryos to which mES cells have attached to the depressions in microdrops of M16 or KSOM in 60 mm dishes, and culture overnight.
>
> 3 Next morning, select the good-quality, expanded blastocysts for transfer to recipient females (see Section 2).

Protocol 5
Aggregation of mES cells with zona-cut, tetraploid embryos[a]

Reagents and equipment

- Mouse ES cell suspension in either M2 or FHM (Protocol 1)
- Tetraploid host embryos, obtained by electrofusion of diploid embryos at the two-cell stage (Protocol 2) and cultured for 1.5 d postelectrofusion in microdrops of M16 or KSOM, to reach the precompacted morula stage
- Hepes-buffered embryo-culture medium, M2 or FHM (Protocol 2)
- Mineral oil (Protocol 2)
- Culture dishes containing microdrops of M16 or KSOM, overlaid with mineral oil
- Micromanipulator assembly, consisting of an inverted fixed-stage microscope (or upright fixed-stage microscope with image-erected optics), micromanipulators fitted with a holding pipette and a microinjection pipette, and a micromanipulation chamber or glass depression slide

Method

1. Set up microdrops of M2 or FHM in the microinjection chamber, under oil.
2. Secure an embryo to the holding pipette by applying negative pressure to the zona pellucida.
3. Push the fine needle through the zona pellucida and into the perivitelline space of the embryo, and out again through the zona pellucida (Figure 1A).
4. Release the embryo from the holding pipette whilst it remains impaled on the needle (Figure 1B).
5. Press the needle repeatedly against the wall of the holding pipette to produce a slit through the zona pellucida, along 10–20% of the length of the embryo (Figure 1C): be careful not to press too hard against the holding pipette, else the needle will break.[b]
6. Reorientate the embryo by micromanipulation with the holding pipette, so that the holding pipette is opposite to the slit.
7. Replace the needle with a microinjection pipette, and inject 5–10 mES cells through the slit in the zona pellucida and into the perivitelline space of the embryo (Figure 1D).
8. Pool and culture manipulated and injected embryos in microdrops of M16 or KSOM.
9. After 1–2 d of culture, transfer those embryos having reached the compacted morula or blastocyst stages to recipient females (see Section 2).

[a] Alternatively, mES cells may be injected into tetraploid blastocysts, which are obtained by electrofusion at the two-cell stage (Protocol 2) followed by culture for 3 d in microdrops of M16 or KSOM. Typically, 5–10 mES cells are injected per blastocyst.

[b] By extending the slit to 50% of the length of the embryo, this manipulation may conveniently be used to remove entirely the zonae pellucidae from small numbers of embryos (see Protocol 3).

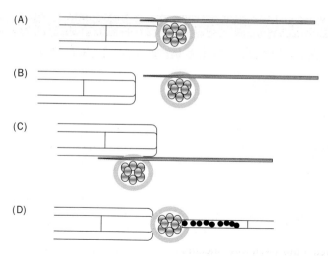

Figure 1 Diagram illustrating the technique for cutting the zona pellucida of a tetraploid morula, and microinjection of mES cells. Stages (A) to (D) of the procedure are explained in *Protocol 5*.

4 Hyperthermic treatment of diploid host blastocysts

For many mammalian species, it has been established that exposure of pregnant females to high environmental temperatures results in embryo mortality. This phenomenon is attributable to a direct effect of hyperthermia on the preimplantation embryo, based on evidence from *in vitro* systems. For example incubation of rabbit embryos for 6 h at 40 °C prior to transfer to recipients significantly reduces the proportion of conceptuses containing live fetuses on day 12 p.c., whilst the implantation rate remains unchanged (18). In the blastocyst-stage mouse embryo, the ICM is relatively more sensitive than trophectoderm to hyperthermia, as *in vitro* incubation for 15–20 min at 45 °C inhibits proliferation of ICM, but not trophectoderm, cells (5). Here, also, the implantation rate is unaffected by the hyperthermic treatment, whilst the production of live fetuses is abrogated. This phenomenon of the selective killing of ICMs with unimpaired implantation forms the basis of our strategy for hyperthermic treatment of host blastocysts, as an alternative to inducing a state of tetraploidy, for embryo reconstitution with mES cells.

We have devised a relatively straightforward and rapid method for hyperthermic treatment of mouse blastocysts that effectively compromises the developmental capacity of the ICM. Such hyperthermically treated blastocysts not only are easy to produce but, crucially, function well as host embryos for the production of mES cell-derived mice (5, 19).

Protocol 6
Hyperthermic treatment of mouse blastocysts

Reagents and equipment

- Mouse (diploid) embryos from super-ovulated (*Protocol 2*), pregnant females of strain CD-1 (Gpi-1 a/a or b/b, as appropriate): these embryos are recovered either as expanded blastocysts at day 3.5 p.c, or as two-cell embryos at 1.5 d p.c. (or 45–46 h post-hCG injection) and cultured in M16 or KSOM to the blastocyst stage (at 110–114 h post-hCG)
- Hepes-buffered embryo-culture medium: either M2 or FHM (*Protocol 1*)
- Sterile, 10 ml glass test tubes with caps or rubber bungs
- Water bath set at 42 °C
- Dissecting microscope with a graticule fitted to an eyepiece
- Wide-bore transfer pipette, or a Pasteur pipette cut and polished to have a wide-bore, smooth-edged tip

Method

1. Using a dissecting microscope and graticule, select and pool those blastocysts having a diameter of 100–120 µm (i.e. those that are fully expanded), and transfer them into glass tubes (with 3–8 blastocysts per tube) containing 1 ml M2 or FHM. Close the tubes.
2. Incubate blastocysts in the tubes for 20 min at 45 °C.[a]
3. Recover the heat-treated blastocysts from the tubes using a wide-bore transfer or Pasteur pipette, and pass them several times through 1–2 ml of fresh M2 or FHM in dishes or glass cavity blocks at r.t. Proceed immediately to blastocyst microinjection.[b]

[a] The stated blastocyst diameter of 100–120 µm, and incubation period for blastocysts of 20 min at 42 °C, have been ascertained as optimal parameters for hyperthermic treatment (5).
[b] The microinjection procedure for heat-treated blastocysts is the same as for normal blastocysts. Usually, 5–10 mES cells are injected into the blastocoel cavity, and embryos are cultured for several hours to overnight, until the blastocoele cavity re-expands. Around 10 expanded, injected blastocysts are then transferred per oviduct or uterine horn of recipient females (see Section 2).

5 Nuclear transfer using mES cells as karyoplasts

Recently, mES cell-derived mice have been produced by the technique of cloning by nuclear transfer using mES cells as karyoplasts, or 'mES cell-NT' (6, 20). Post-implantational development of nuclear-transfer mouse embryos is more successful when mES cells are used as the source of karyoplasts rather than somatic cells (6, 9, 21, 22), with the percentage of blastocysts surviving to adulthood

Table 2 Effect of culturing line TT2 mES cells under feeder-free conditions on the developmental potential of mES cell-NT embryos to the blastocyst stage

Days of mES cell culture without feeder cells	Oocytes fused (%)	NT embryos developing to blastocyst (%)
0	165/200 (83)	116/163 (71)[a]
3	180/224 (80)	99/180 (55)[b]
5	206/261 (79)	89/206 (43)[c]
7	95/106 (90)	35/95 (37)[c]

[a,b,c] Values with different superscripts are significantly different.

increasing by a factor of 10–20 fold (9, 23); this indicates that either the requisite nuclear reprogramming is less extensive for (undifferentiated) mES cells than for differentiated cells, or that the epigenetic status of mES cells is more conducive to embryonic development. The former deduction is supported by the reduced capacity of mES cell-NT embryos for blastocyst formation with increasing time of culture of the mES cells in the absence feeder cells, when spontaneous differentiation is induced (Table 2); thus, undifferentiated mES cells are more amenable to nuclear reprogramming than differentiating populations. Although the technique of mES cell-NT has not yet been fully established, its potential and practical significance are great. Here we describe our current protocols for mES cell-NT, by which the efficiency of producing nuclear-transfer blastocysts is 34–88%, and of live-born mES cell-derived mice, 1–5% (Table 1).

For mES cells, as for somatic cells, there are two basic methods for conducting nuclear transfer:

(a) *'Membrane fusion' or 'cell fusion'*: Here, fusion is induced between the plasma membranes of the cytoplast (i.e. recipient, enucleated oocyte) and juxtaposed karyoplast. Cell fusion can be readily achieved using inactivated Sendai virus (i.e. the haemagglutinating virus of Japan, or HVJ) as a fusogen (24), or alternatively can be induced by delivering electrical pulses (25). Of the two methods, Sendai viral infection results in consistently higher rates of cell fusion (typically >80%, compared with 60–70% on average for electrofusion; see *Table 3*). However, electrofusion is a flexible and highly reproducible method, whilst Sendai virus-induced fusion requires virological expertise and resources; and so electrofusion has come to be more widely adopted (25). The potential of successfully fused, nuclear-transfer embryos subsequently to develop into blastocysts and to term is broadly similar using either cell fusion method, but particular care must be taken with electrofusion to ascertain appropriate parameters for the electrical pulses (*Table 3*).

(b) *Direct injection*: The direct-injection method for nuclear transfer is a relatively recent innovation, having been adapted from techniques for intracytoplasmic sperm injection (26). It exploits the small size of mES cells, which are injected directly into the enucleated oocytes using a Piezo-micropipette driving unit (6, 21). There are two corollaries of this method: (i) the plasma membrane

Table 3 Comparison of cloning efficiencies by mES cell-NT using different activation methods and parameters

Activation method	Oocytes fused (%)	Oocytes developed to blastocyst (%)	Blastocysts transferred	Dead fetuses (%)[a]	Live young (%)[a]	Placentas only(%)[a]	Surviving mice
Strontium	258/309 (83)	179/249 (72)	140	4 (3)	1 (0.7)	9 (6)	0
Electrical							
(a) 120 V/mm, 20 µs	88/131 (67)	65/84 (77)	42	0	2 (5)	3 (7)	0
(b) 180 V/mm, 9 µs	243/418 (58)	113/229 (49)	76	0	2 (3)	3 (4)	0
(c) 150 V/mm, 50 µs; and 50 V/mm, 50 µs	84/84 (100)	40/77 (52)	40	1 (3)	1 (3)	0	1

[a] Percentage of blastocysts transferred. (a) to (c) show strength and duration of DC pulses (see *Protocol 13*).

of the mES cell is deliberately broken during the process of injection, and so the total volume of karyoplast cytoplasm that is introduced may be less than by other methods, whilst (ii) the karyoplast nucleus may be exposed transiently to the medium. Perhaps in consequence, the potential of nuclear-transfer embryos to develop into blastocysts is lower when the direct-injection method is used, compared with cell-fusion methods (27); but the rate of production of live offspring from transferred blastocysts is higher by direct injection, especially when mES cell lines from a hybrid background are used as karyoplasts (9).

Successful mES cell-NT requires the appropriate tools and skills for micromanipulation. Considerable care is needed to ensure that the microinjection pipettes are of the correct diameter. For example, for Sendai virus-induced fusion, if the diameter is too large the virus suspension becomes diluted by medium in the micropipette and consequently the fusion rate may decrease. And for direct injection, the microinjection pipette's diameter must be small enough (10–20 µm) that the karyoplast's cell membrane is destroyed during micropipetting whilst the ooplasm membrane is relatively less damaged. As the manipulations involved in successful nuclear transfer are complex and technically demanding, the authors strongly advise the novice who wishes to acquire them to visit an established laboratory for practical guidance. Further details of the manipulations and procedures involved in nuclear transfer are provided elsewhere (25).

5.1 Mouse strains for mES-cell NT

As general guidelines for cloning by mES-cell NT, reproducibly higher efficiencies of production of live-born mES-cell-derived mice are achieved using mES cell lines having hybrid genetic backgrounds. Such suitable lines have been derived from mating F_1 hybrid mice (e.g. C57BL/6 × 129/Sv, 129/Sv × C57BL/6, 129/Sv × *M. castaneus*, 129/Sv × FVB, BALB/c × 129/Sv, and 129/Sv × 129/Sv-Cp). In contrast, when mES cell lines from inbred strains are used, neonatal death occurs (with 129/Sv or C57BL/6 cell lines in particular) or the survival rate to adulthood is very low (with BALB/c and 129/Ola cell lines). Consequently, for karyoplasts we routinely employ low passage numbers of karyotypically male mES cell lines derived by mating F_1 hybrid (C57BL/6 × C3H, Gpi-1 b/b) females with males of the same genetic background (20, 27, 28), or other F_1 hybrid females and males having a genetic component from the 129/Sv strain. For cytoplasts, we use unfertilized oocytes from F_1 females, which are produced by mating inbred C57BL/6 females with CBA, DBA (DBA/2 or DBA/10), or C3H males: oocytes from females with a C57BL/6 component in the hybrid genotype are particularly suitable for the *in vitro* manipulations involved, as they are not prone to the two-cell block.

Other factors affecting the success of mES cell-NT, with respect to embryo development, include the degree of confluence of the mES cell cultures used as sources of karyoplasts – when cells are permitted to reach 80–90% confluence, the rate of formation of blastocysts and the efficiency of production of live young are significantly higher than obtained from cells at 60–70% confluence (11). As for the possible effects of mES-cell culture medium on the success of NT, in our laboratory three different commercial preparations (DMEM, KnockOut™-DMEM, and DMEM/F12 mixture, from Invitrogen) have proved equally effective in terms of rates of blastocyst formation and numbers of offspring produced.

5.2 Cell-cycle synchronization in mES cells

Extensive investigations have highlighted the importance of cell-cycle synchronization in the donor nucleus (karyoplast) and recipient oocyte (cytoplast) for successful nuclear transfer to occur. Such synchronization is necessary to prevent an abnormal DNA complement from arising in the nuclear-transfer embryo. For the mouse, recommended cell-cycle combinations are: either G_0/G_1-phase or M-phase for karyoplasts; and the second metaphase (MII-phase) for cytoplasts (i.e. enucleated, unfertilized oocytes). It is difficult, however, to synchronize mES cells at the G_0/G_1 transition as standard methods (such as serum starvation or contact inhibition) prove ineffective for that cell type. However, mES cells may readily be synchronized at M-phase by nocodazole treatment, without detriment to viability or developmental competence; and on withdrawal of nocodazole the mES cells resume cell division (29, 30). There is an added advantage in that M-phase mES cells are characteristically spherical, and therefore easily discerned in monolayer culture. When mES-cells that are synchronized in M-phase in this way are harvested and used as karyoplasts, nuclear-transfer embryos develop to the blastocyst stage at a high rate, and some 1–10% of these develop to term (6, 27, 28, 31). Furthermore, M phase-synchronized mES cells may subsequently be resynchronized at the G_1 phase using aphidicoline treatment (32).

Another simpler, morphological approach to estimating the cell-cycle stage of individual mES cells is based on their size (6); the smaller mES cells in a population are more likely to be in the G_1 phase. Furthermore, randomly selected mES cells are most likely to be in the G_1 phase.

Protocol 7
M-phase and G_1-phase synchronization of mES cells for nuclear transfer

Reagents and equipment

- Mouse ES cells cultured in multi-well or tissue-culture dishes, e.g. lines NR2 and TT2 established in our laboratory (5)
- Mouse ES-cell medium (*Protocol 1*)
- Nocodazole-supplemented media, for M- and G_1-phase synchronization: stock solution of 1 mg/ml nocodazole (Cat. No. M-1404, Sigma) in DMSO (Cat. No. D-2650, Sigma), diluted to 1 μg/ml in mES-cell medium and in either M2 or FHM
- Aphidicoline-supplemented medium, for G_1-phase synchronization: stock solution of 1 mg/ml aphidicoline (Cat. No. A-0781, Sigma) in DMSO, diluted to 5 μg/ml in mES-cell medium
- Embryo-culture media: either FHM and KSOM, or M2 and M16 (*Protocols 1 and 2*)
- 35 mm tissue-culture dishes
- Cell-transfer capillary connected to a holder and mouth-piece *(Protocol 3)*
- Micromanipulator assembly *(Protocol 5)*

Methods

M-phase synchronization:

1. Aspirate conditioned medium from the mES cells, replace with nocodazole-supplemented mES-cell medium, and incubate cells for 3–5 h.
2. Harvest individual mES cells in M-phase using a cell-transfer capillary under mouth control: M-phase (i.e. mitotic) mES cells are spherical, and easily visualized by light microscopy (*Figure 2*).
3. Pool and transfer M-phase mES cells into a droplet of nocodazole-supplemented M2 or FHM under oil, on the micromanipulation chamber, in preparation for nuclear transfer.

G_1-phase synchronization (32):[a]

1. Prepare microdrops (5–10 μl) of aphidicoline-supplemented mES-cell medium under oil, in a culture dish; and of aphidicoline-supplemented M2 or FHM under oil, on the microinjection chamber.
2. Manually harvest M-phase mES cells (see steps 1 and 2, above), and wash through several 1 ml drops of nocodazole-free, aphidicoline-supplemented mES-cell medium.
3. Pool and transfer the M-phase mES cells into a final microdrop of aphidocoline-supplemented mES-cell medium, and culture for 50–60 min for cell division to become complete.

> **Protocol 7** continued
>
> 4 Collect the cleaved daughter cells and transfer them into a microdrop of aphidicoline-supplemented M2 or FHM on the microinjection chamber, in preparation for nuclear transfer.
>
> [a] An alternative, simpler method of harvesting G_1-phase mES cells is simply to prepare a single-cell suspension using trypsin/EDTA solution (*Protocol 1*), transfer cells into mES-cell medium, and select the smallest cells (which may represent the G_1-phase fraction) using a cell-transfer capillary and phase-contrast microscopy.

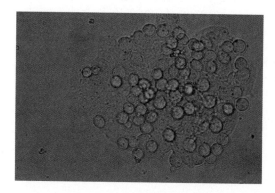

Figure 2 M-phase mES cells. Cells were treated with 3 μg/ml nocodazole for 5 h. The rounded cells in this colony are blocked at metaphase.

Experiments to ascertain the best cell-cycle combinations and method for mES cell-NT are still ongoing, and the heterogeneous nature of mES cells in culture (23) signifies that the precise efficiency of producing mES cell-derived mice may be largely dependent upon the particular cell line used and the conditions of its maintenance. However, it has been reported that the rate of development into blastocysts was considerably higher when M-phase mES cells were used as karyoplasts, compared with G_1-phase mES cells: 56.8–70.0% *vs* 5.9–14.1% by *Zhou et al.* (31); and 30–67% *vs* 2–5% by Yabuuchi *et al.* (27). Following transfer of such blastocysts into recipient females, full-term development was achieved at similar rates for both cell-cycle stages in the karyoplast (1.5–3.1% by *Zhou et al.* (31)), or was obtained only using M-phase karyoplasts (3–4%) (27).

5.3 Enucleation of MII-phase oocytes

Recipient oocytes for nuclear transfer are collected at MII-phase from superovulated, F_1 hybrid female mice, for example B6C3F$_1$ (C57BL/6 × C3H) or BDF$_1$ (C57BL/6 × DBA). When cell-fusion methods are to be employed for nuclear transfer, the zona pellucida should be cut partially in preparation for enucleation. When the direct-injection method is to be used, the zona-cutting process is unnecessary; instead, the enucleation pipette is placed in contact with the surface of the zona pellucida, which it penetrates when Piezo-pulses are applied.

Protocol 8
Collection and enucleation of MII-phase mouse oocytes

Reagents and equipment

- F_1 hybrid female mice having strain C57BL/6 maternal genotype, e.g. C57BL/6 × C3H ('B6C3.F_1') or C57BL/6 × DBA ('BD.F_1'), at least 5 weeks old, and superovulated as described in *Protocol 2*
- Embryo-culture media: M2 and M16, or FHM and KSOM (*Protocol 2*)
- Hyaluronidase solution: 0.1% hyaluronidase (Cat. No. H4272 Sigma) in M2 or FHM supplemented with PVP (*Protocol 2*)
- Sterile dissection kit
- Stereo dissecting microscope
- Plastic tissue-culture dishes
- Sterile, glass cavity block with covers
- Ungassed incubator, set at 37 °C
- 27 Gauge hypodermic injection needle
- Fine or Watchmaker's forceps
- Pasteur pipette
- Hoechst 33342-supplemented medium: stock solution of 1 mg/ml Hoechst 33342 (Cat. No. 382065, Calbiochem) in double-distilled water, diluted to a working solution of 5–10 µg/ml in M2 or FHM
- Cytochalasin B-supplemented medium: stock solution of 1 mg/ml cytochalasin B (Cat. No. C-6762, Sigma) in DMSO (Cat. No. D-2650, Sigma), diluted to a working concentration of 5 µg/ml in M2 or FHM
- Micromanipulator assembly (*Protocol 5*), with inverted microscope fitted with Nomarski optics and ×10 and ×20 objectives.[a]

Methods

Oocyte collection:

1. Collect oviducts from superovulated mice 15–16 h post-hCG injection, and transfer to hyaluronidase-supplemented medium in tissue-culture dishes.

2. Tear open the oviducts with forceps, collect the cumulus–oophorus complexes using a Pasteur pipette, and transfer them into tissue-culture dishes or a glass cavity block containing hyaluronidase-supplemented medium.

3. Incubate the cumulus–oophorus complexes in hyaluronidase-containing medium for 2–3 min at 37 °C, without gassing with CO_2, and assist the release of MII-phase oocytes from the cumulus cells by gently pipetting.

4. Collect the denuded oocytes and wash them several times in fresh embryo-culture medium (without Hepes or hyaluronidase), and incubate for ∼2 h at 37 °C in a humidified incubator gassed with 5% CO_2, in preparation for enucleation and nuclear transfer.

Oocyte enucleation:

5. Transfer 20–30 oocytes into (5–10 µl) microdrops of cytochalasin B-supplemented M2 or FHM under oil, in the microinjection chamber.[b]

Protocol 8 continued

6 Place the microinjection chamber containing oocytes onto to the stage of the inverted microscope. Select an oocyte by micromanipulation and, using Nomarski optics, locate the small area of protuberance on the surface of the oocyte that contains the metaphase plate (MII-phase chromosomes). Orient the egg with the holding pipette, and slit the zona pellucida overlying this protuberance using a finely drawn glass microinjection needle (see *Protocol 5*) (as shown in *Figure 3A–C*). After cutting the zona pellucida continue incubating the oocyte in cytochalasin B-supplemented M2 or FHM until all oocytes have been processed in this way.[c]

7 Commence enucleation of an oocyte, secured so that the holding pipette is oriented opposite to the slit in the zona pellucida, by inserting the enucleation pipette through the slit and into the perivitelline space (*Figure 3D*).

8 Remove the MII-phase chromosomes from the oocyte as follows. Place the enucleation pipette in close contact with the ooplasm: the area of translucent cytoplasm containing the MII-phase chromosomes can be observed to move and deform when the enucleation pipette is pushed against it. Aspirate this area into the enucleation pipette, and observe the chromosomes entering the enucleation pipette (*Figure 3E*).[d]

9 Repeat the procedure for the whole batch of oocytes. Pool and wash successfully enucleated oocytes through drops of fresh M2 or FHM.

10 To determine successful enucleation, stain manipulated oocytes by incubation in Hoechst 33342-supplemented medium for 5 min at 37 °C in the dark, wash through drops of fresh medium to remove residual dye, and confirm the absence of the metaphase plate or contaminating chromatin by fluorescence microscopy[e].

[a] Nomarski optics are required to visualize clearly the oolemma and metaphase chromatin.

[b] A short period of incubation (~5 min) in cytochalasin B-supplemented medium is required prior to enucleation, to disrupt elements of the cytoskeleton (specifically the microfilaments) making the oocytes more plastic and less susceptible to damage by micromanipulation.

[c] The areas of protuberance may become difficult to observe following cytochalasin B-treatment of oocytes.

[d] If the chromatin cannot be clearly visualized within the enucleation pipette, the ooplasm should not be used as a recipient cytoplasm.

[e] Staining of the ooplasm of enucleated oocytes is not recommended as a routine practice, because of the toxicity of Hoechst 33342 and UV irradiation.

(Subsequently a small, round fragment of the zona pellucida remains in the enucleation pipette.) Note that after enucleation and before direct injection of karyoplasts by Piezo-pulses, oocytes must be incubated in the absence of cytochalasin B, so that the oolemma can reform.

Until proficiency is acquired in the technique of oocyte enucleation, it is advisable to monitor the progress of enucleation by staining the enucleated oocyte

(A)
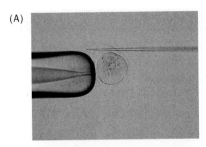

(B)
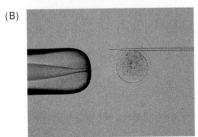

(C)
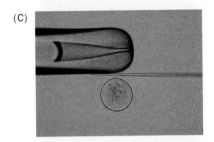

(D)
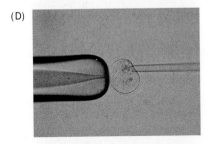

(E)
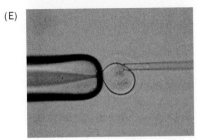

Figure 3 Cutting the zona pellucida of an oocyte in preparation for enucleation. Stages (A) to (C) of the procedure are explained in *Protocol 8*.

Protocol 9
Inactivated Sendai virus-induced fusion

Methods for the production, purification, and inactivation of Sendai virus are given elsewhere (33, 34, 35). The viral preparation for fusion may be stored long term at $-70\,°C$.

Reagents and equipment

- Single-cell suspension of mES cells in M2 or (Protocol 1), cell-cycle synchronized as appropriate (Protocol 7)
- Enucleated MII-phase oocytes (Protocol 8)
- M2 or FHM (Protocol 1)
- Nocodazole-supplemented medium: either M2 or FHM supplemented with 1–3 µg/ml nocodazole (Protocol 7)
- Chemically or UV-inactivated Sendai virus preparation: a cryopreserved aliquot is thawed and diluted in calcium-supplemented BSS to a titre of 2500 U/ml of haemagglutinating activity, and stored for up to a week at 4 °C prior to use
- BSS: prepare a solution of 137 mM NaCl, 5.4 mM KCl, 0.44 mM KH_2PO_4, 0.34 mM $Na_2HPO_4.12H_2O$, and 13 mM Tris base in double-distilled water
- Ca^{2+}-supplement for BSS: a stock solution of 200 mM $CaCl_2.2H_2O$ in double-distilled water is diluted to a working solution of 2 mM in BSS
- Light mineral oil or silicon oil (Protocol 2)
- M16 or KSOM (Protocol 2)
- Cytochalasin B-supplemented medium: 5 µg/ml cytochalasin B (Protocol 8) in M16 or KSOM
- Embryo-transfer capillary (Protocol 3)
- Micromanipulator assembly (Protocol 5)

Method

1. To facilitate micromanipulation, we recommend placing a column of three separate microdrops (~10 µl) on the micromanipulation chamber, the top microdrop contains karyoplasts in M2 or FHM either supplemented with nocodazole for the M-phase donor cells, or without nocodazole for G_1/G_0-phase donor cells; the middle microdrop, a suspension of inactivated Sendai virus; and the bottom microdrop, enucleated recipient oocytes (cytoplasts) in M2 or FHM. Cover the droplets with mineral oil.

2. From the top microdrop take up a single karyoplast into the microinjection pipette, transfer the pipette into the second drop, and draw up (by ~20 µm into the pipette) a small volume of the viral suspension.

3. Transfer the injection pipette into the third microdrop and inject its contents (karyoplast and virus suspension) into the perivitelline space of an enucleated oocyte, through the hole in the zona pellucida made originally for enucleation (Figure 4A, B).

4. Using the injection pipette gently push the karyoplast into direct contact with the oolemma of the cytoplast, to promote cell fusion (Figure 4C).

Protocol 9 continued

5. Repeat the micromanipulations with the remaining cytoplasts.
6. Using a dissecting microscope and an embryo-transfer capillary, collect and transfer all the manipulated embryos (i.e. karyoplast–cytoplast complexes) into fresh M16 or KSOM, wash several times in 1 ml drops of the same medium, and incubate in microdrops.
7. Periodically monitor the complexes for fusion of karyoplasts with cytoplasts over 1 h of incubation, and select the fused complexes for activation (*Protocol 12*)[a]

[a] Optimally, fusion should occur within 1 h of viral infection. Reduce the titre of the virus if cell lysis occurs, or increase it if the fusion rate is low. Aim for a fusion rate of more than 60%.

(A)
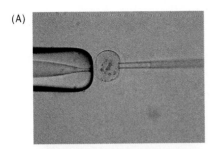

(B)
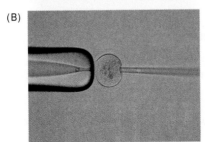

(C)
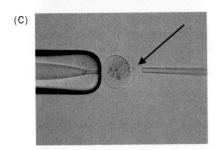

Figure 4 Sendai virus-induced fusion of karyoplast and cytoplast. Stages (A) to (C) of the procedure are explained in *Protocol 9*. Arrow indicates position of karyoplast.

Protocol 10
Electrofusion

Reagents and equipment

- Mouse ES-cell suspension (Protocol 1)
- Enucleated, MII-phase oocytes (Protocol 8)
- Electrofusion medium, calcium-free: either 0.3 M mannitol solution in double-distilled water, or Zimmerman's fusion medium prepared without calcium acetate monohydrate (Protocol 2)[a]
- Cytochalasin B-supplemented medium: 5 μg/ml cytochalasin B in M16 or KSOM (see Protocol 8)
- Embryo-transfer capillary (Protocol 3)
- Micromanipulator assembly (Protocol 5)
- Electrofusion apparatus (see Protocol 2)

Method

1 Microinject single karyoplasts into the perivitelline space of cytoplasts.

2 Equilibrate the karyoplast–cytoplast complexes in electrofusion medium, i.e. until they sink.

3 Place a flat drop of electrofusion medium over the central portion of the platinum electrodes in the electrofusion chamber, and transfer into it the reconstituted oocytes.

4 Orient the complexes relative to the platinum electrodes so that the karyoplasts are parallel to the wires, using an embryo-transfer capillary under mouth control.

5 Deliver two DC pulses of 50 V/mm for 50 μs with a 2 s interval. Repeat the pulses after 20 min.

6 Wash complexes by passing through drops of M16 or KSOM, and incubate in microdrops of the same. Monitor fusion occurring between karyoplasts and cytoplasts over 1 h.

7 Prior to activation, incubate the fused complexes in cytochalasin B-supplemented M16 or KSOM for 1 h, to prevent the dispersal of the chromosomes.

[a] For the electrofusion of mES cells with oocytes, calcium must be absent from the medium to avoid the premature activation of oocytes.

Protocol 11
Nuclear transfer by direct injection

Reagents and equipment

- Enucleated oocytes (Protocol 8)
- Single-cell suspension of mES cells in either M2 or FHM (Protocol 1)
- PVP-medium: M2 or FHM in which BSA is replaced with 10% polyvinyl pyrrolidone (PVP; Cat. No. PVP-360, Sigma)[a]

Protocol 11 continued

- PVA-medium: M2 or FHM in which BSA is replaced with 10% polyvinyl alcohol (PVA; P8136, Sigma)[a]
- Embryo- and cell-transfer capillaries (*Protocol 3*)
- Micromanipulator assembly (*Protocol 5*) adapted with a Piezo-driven microinjection unit e.g. the Piezo Impact Micromanipulator (Cat. No. PMM-150FU, Prime Tech Ltd)
- Light mineral oil (*Protocol 2*)

Method

1. Dispense into the micromanipulation chamber an upper row of two microdrops (5–10 µl) of PVP medium, and a lower row of two microdrops of PVA medium; and cover with mineral oil. The upper two microdrops of PVP medium are for manipulating karyoplasts (the first is reserved for clearing the microinjection pipette if it becomes obstructed with viscous material during micromanipulation, and the second is for containing the karyoplasts). The lower two microdrops of PVA medium are for direct injection (the third is for washing the karyoplasts, and the fourth for the process of direct injection).

2. Place the enucleated oocytes (cytoplasts) into the fourth microdrop of PVA-medium, in preparation for direct injection.

3. Using a cell-transfer capillary, place 100–200 mES cells into the second microdrop of PVP-medium. Repeatedly draw the cells up and down the microinjection pipette in order to destroy their plasma membranes.[b]

4. Rinse the microinjection pipette in the third microdrop (to avoid PVP contamination), and place the karyoplasts into the fourth microdrop of PVA-medium together with the enucleated oocytes, ready for direct injection. Secure an oocyte to the holding pipette and insert the microinjection pipette into the perivitelline space, through the same hole as was made for enucleation.

5. Push the microinjection pipette into the ooplasm as deeply as possible (*Figure 5A*). At this point the oocyte becomes distorted in shape.

6. Deliver a Piezo-pulse, according to the equipment manufacturer's instructions, to enter the oolemma (*Figure 5B*). Once the microinjection pipette breaks through the oolemma and into the ooplasm, and the round shape of the oocyte is restored, inject the karyoplast into the oocyte (*Figure 5C*).

7. Pool and transfer successfully injected oocytes into microdrops of M16 or KSOM; and, prior to activation, culture in cytochalasin B-supplemented M16 or KSOM for 1 h to allow recovery of the oolemma.

[a] Alternatively a complete formulation for Hepes-buffered KSOM, from which BSA may be omitted and PVP or PVA added, is given is *Protocol 8*, Chapter 2.

[b] The diameter of the microinjection pipette should be 5–10 µm, i.e. just smaller than the diameter of the mES cell.

(A)

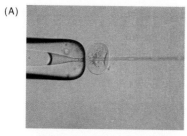

(B)

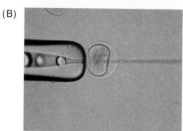

(C)

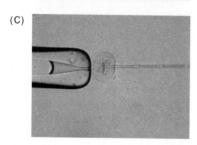

Figure 5 Direct injection of karyoplast into cytoplast by Piezo-driven micromanipulation. Details of the procedure are given in *Protocol 11*. (A) A donor cell is injected into the cytoplast, in the deepest area. (B) Piezo pulses are applied. (C) The karyoplast has been successfully injected.

with a fluorescent, DNA-binding dye such as Hoechst 33342, and observing the oocyte by fluorescence microscopy. It is also practical to perform enucleation simultaneously with fluorescence detection, provided that the exposure time of the oocyte to fluorescence illumination and observation is no longer than 10 s. Longer exposures result in damage specifically to the mitochondrial genome, rendering the oocytes unsuitable to act as cytoplasts.

5.4 Activation of nuclear-transfer embryos

Reconstituted, nuclear-transfer embryos require artificial activation to induce their development. Of the several methods available, chemical activation (with strontium chloride) and electrical activation are most commonly used for mouse nuclear-transfer embryos. Following successful fusion and activation, as denoted by pronucleus formation, nuclear-transfer embryos reaching the morula or blastocyst stages are transferred to the oviducts of day 1 p.c. pseudo-pregnant recipients. Live-born offspring may be delivered by Caesarean section and cross fostering on day 19 p.c., or by natural birth at term. Alternatively, recipients may be sacrificed earlier to examine fetal development.

Protocol 12
Strontium chloride-induced activation

Reagents and equipment

- Postfusion, nuclear-transfer embryos (*Protocols 8–11*)
- M2 or FHM (*Protocol 1*)
- Ca^{2+}- and Mg^{2+}-free embryo-culture medium (M16 or KSOM), which can be obtained as custom-formulation (from Specialty Media)[a]
- Strontium chloride-supplemented embryo-culture medium: stock solution of 100 mM $SrCl_2.6H_2O$ in double-distilled water, diluted immediately prior to use to 10 mM in Ca^{2+}- and Mg^{2+}-free M16 or KSOM

Method

1 Place the fused, nuclear-transfer embryos into Ca^{2+}- and Mg^{2+}-free medium for 1 min, transfer into microdrops of strontium chloride-supplemented medium under oil, and culture for 5 h.[b]

2 Wash nuclear-transfer embryos through drops of M2 or FHM several times to remove residual strontium chloride, and assess activation as follows: for fusions using G_1-phase mES cells, from the formation of one or two pronuclei without a second polar body; and for M-phase mES cells, one pronucleus and a second polar body. These successfully activated embryos are considered to be diploid.

3 Culture the activated embryos in microdrops of M16 or KSOM for several days, until the morula or blastocyst stages are reached, and transfer to pseudopregnant recipients (see Section 2).

[a] Alternatively use the formulation for Hepes-buffered KSOM provided in *Protocol 8*, Chapter 2, omitting Hepes, $CaCl_2.2H_2O$, and $MgSO_4.7H_2O$.
[b] This medium should be supplemented with cytochalasin B for nuclear transfers in which G_1-phase mES cells were used as karyoplasts, to suppress second polar body extrusion after activation.

Protocol 13
Electrical activation

Reagents and equipment

- Postfusion, nuclear-transfer embryos (*Protocols 8–11*)
- Electrofusion medium: either mannitol solution or Zimmerman's fusion medium (*Protocol 2*)
- M2 and M16, or FHM and KSOM (*Protocols 1 and 2*)
- Sterile, glass cavity blocks
- Electrofusion apparatus (*Protocol 2*)

Protocol 13 continued

Method

1. Pass the fused embryos through several drops of electrofusion medium in glass cavity blocks to equilibrate, and to remove residual embryo-culture medium.[a]

2. Transfer 10 embryos at a time into electrofusion medium in the electrofusion chamber, and deliver two DC pulses of 150 V/mm for 50 µs each with a 2 s interval.

3. Wash embryos several times through drops of M2 or FHM, and culture in M16 or KSOM until all embryos are ready for the delivery of the next series of DC pulses.

4. Deliver two DC pulses of 50 V/mm for 50 µs, with a 20 min interval; wash the embryos and repeat the last delivery of two DC pulses.[b]

5. Wash the embryos through drops of M2 or FHM, and culture in microdrops of M16 or KSOM for 5 h.

6. Monitor the cultured embryos for pronucleus formation, denoting successful activation (see *Protocol 12*).

7. Continue to culture the activated embryos in M16 or KSOM for several days, until either the compacted morula or blastocyst stage is reached; and proceed to embryo transfer (see Section 2).

[a] Initially, oocytes float on the surface of the fusion medium; and so wait momentarily for them to sink.

[b] In total by this stage, six DC pulses will have been delivered to the oocytes.

6 Analysis of distribution of mES-cell derivatives in chimeras

To confirm the production of mES cell-derived mice or fetuses, it is recommended to use isozyme, genetic or transgene markers in addition to conventional coat-colour markers. The locus for glucose phosphate isomerase (GPI), which is polymorphic in mouse strains, remains a commonly used and readily assessed marker to determine mES-cell contribution within tissues. GPI isozyme detection relies on electrophoretic separation of samples (36), a relatively quick and simple procedure that is sensitive to a level of 10%. GPI analysis is carried out on cellulose acetate plates using the Titan system supplied by Helena Laboratories.

In addition, β-galactosidase (*lacZ*) and *GFP* are widely used reporter genes. Derivative cells expressing *lacZ* are detected by postfixation staining; but the method is confined to postimplantation embryos, not later than day 13.5 p.c., due to the presence of endogenous β-galactosidase activity at later stages. *GFP* expression may be visualized in samples or in the live animal with appropriate fluorescence illumination, without affecting viability (see http://www.mshri.on.ca/nagy).

7 Perinatal mortality and developmental abnormalities in mES cell-derived mice

Gestational mortality, neonatal mortality, and developmental abnormalities are features associated with the production of mES cell-derived mice, whether by nuclear transfer or by reconstitution of tetraploid or hyperthermically treated embryos. The rates of neonatal mortality are broadly similar for these diverse methods, at over 50% (19); and as neonatal mortality also has been noted in mES cell-derived mice produced (albeit at very low frequency) by diploid-blastocyst injection (19), high-level mES-cell contribution is likely to be a contributory factor to the phenomenon.

For cloning by mES cell-NT, although the efficiency of producing mES cell-derived mice and the incidence of neonatal mortality are dependent on the cell lines used (4, 9, 10, 19), there is no apparent correlation with genetic background apart from a beneficial effect of a hybrid component. According to Rideout et al. (9), the rate of development from blastocyst to term was higher for mES cell lines derived from F_1 hybrid mice (21%) compared with inbred mice (11%), as was the viability of derived clones. Yabuuchi et al. (27) also deduced that mES cell lines from hybrid mice have greater developmental potential in cloning experiments; and observed that the degree of confluence and passage number of the mES cells prior to nuclear transfer are additional, important parameters (11). But even between mES cell lines of the same genetic background, tremendous variation is observed in the resultant efficiency of cloning by mES cell-NT. The current challenge is therefore to ascertain whether the observed neonatal death and developmental abnormalities arise from experimental manipulations, or from intrinsic defects in the cultured mES cell lines. We have ascertained that for blastocysts produced by mES cell-NT, both the ICM and trophectoderm components have reduced developmental potential (39).

It has been established that prolonged culture leads to epigenetic instability in mES cells; and within a single mES cell line, expression patterns of imprinted genes are found to vary between individual mES cell colonies (23). It has been suggested that such epigenetic modification of imprinted genes (such as *H19*, *Igf2*, *Igf2r*, and *U2af1-rs1*) may underlie developmental defects and neonatal mortality, which hypothesis is reinforced by the wide variation in expression observed in the tissues of mice produced by mES cell-NT or tetraploid-embryo reconstitution (23, 37). Irregular expression of imprinted genes has been observed also in mouse fetuses derived from primordial germ cells by nuclear transfer (37), from which it is deduced that germline cells in general cannot correct, but rather maintain, aberrant gene expression following embryo reconstitution or nuclear transfer.

Finally, as abnormal gene expression has been observed in cloned mice produced using both mES and somatic cells as karyoplasts, it is inferred that some defects in gene expression may arise from the nuclear transfer procedure, or to incompleteness of the nuclear reprogramming upon which normal development depends. Other defects are ascribed to the karyoplast cell type, that is mES cell *versus* somatic cell (23). The nuclear reprogramming factor(s) are as yet unknown

(38), but clarifying the reprogramming mechanism will lead eventually to the production of normal animals by nuclear transfer using mES cells as well as somatic cells. Methods for the production of mES cell-derived mice therefore provide excellent systems in which to study the complex scenario where epigenetic modifications and other factors affect development. Once these inherent problems are resolved, the power of these methods for the efficient production of cloned, or entirely mES cell-derived, mice will be enormous.

Acknowledgements

We are grateful to Akiko Yabuuchi, Kimiyo Nakamura, and Tomokazu Amano for their technical methodologies. The work has been partially supported by grants from the Programme for Promotion of Basic Research Activities for Innovative Biosciences (PROBRAIN).

References

1. Evans, M.J. and Kaufman, M.H. (1981). *Nature*, **292**, 154-6.
2. Martin, G.R. (1981). *Proc. Natl. Acad. Sci. USA*, **78**, 7634-8.
3. Tam, P.P. and Rossant, J. (2003). *Development*, **130**, 6155-63.
4. Nagy, A., Gocza, E., Diaz, E.M., Prideaux, V.R., Ivanyi, E., Markkula, M., and Rossant, J. (1990). *Development*, **110**, 815-21.
5. Amano, K., Nakamura, K., Tani, T., Kato, Y., and Tsunoda, Y. (2000). *Theriogenology*, **53**, 1449-58.
6. Wakayama, T., Rodriguez, I., Perry, A.C.F., Yanagimachi, R., and Monbaerts, P. (1999). *Proc. Natl. Acad. Sci. USA*, **96**, 14984-9.
7. Nagy, A., Vintersten, K., Gertsenstein, M., and Behringer, R., eds. (2003). *Manipulating the Mouse Embryo*, Cold Spring Harbor Laboratory Press, New York.
8. Wang, Z.Q., Kiefer, F., Urbanek, P., and Wagner, E.F. (1997). *Mech. Dev.*, **62**, 137-45.
9. Redeout, III W.M., Wakayama, T., Wutz, A., Eggan, K., Jackson-Grusby, L., Dausman, J., Yanagimachi, R., and Jaenisch, R. (2000). *Nature Genetics*, **24**, 109-10.
10. Nagy, A., Rossant, J., Nagy, R., Abramov-Newerly, W., and Roder, J. (1993). *Proc. Natl. Acad. Sci. USA*, **90**, 8424-8.
11. Gao, S., McGarry, M., Ferrier, T., Pallante, B., Gasparrini, B., Fletcher, J., Harkness, L., De Sousa, P., McWhir, J., and Wilmut, I. (2003). *Biol. Reprod.*, **68**, 595-603.
12. Eggan, K. and Jaenisch, R. (2003). *Methods Enzymol.*, **365**, 25-39.
13. Kubiak, J.Z. and Tarkowski, A.K. (1985). *Exp. Cell. Res.*, **157**, 561-5.
14. Wood, S.A., Pascoe, W.S., Schmidt, C., Kemler, R., Evans, M.J., and Allen, N.D. (1993). *Proc. Natl. Acad. Sci. USA*, **90**, 4582-5.
15. Tarkowski, A.K. (1961). *Nature*, **190**, 857-60.
16. Gardner, R.L. (1968). *Nature*, **220**, 596-7.
17. Martin, G.R. and Evans, M.J. (1975). *Proc. Natl. Acad. Sci. USA*, **72**, 1441-5.
18. Alliston, C.Q., Howarth, B. Jr, and Ulberg, L.C. (1965). *Reprod. Fertil.*, **9**, 33-341.
19. Amano, T., Kato, Y., and Tsunoda, Y. (2001). *Zygote*, **9**, 153-7.
20. Amano, T., Kato, Y., and Tsunoda, Y. (2001). *Reproduction*, **121**, 729-33.
21. Wakayama, T., Rodriguez, I., Perry, A.C., Yanagimachi, R., and Mombaerts, P. (1999). *Proc. Natl. Acad. Sci. USA*, **96**, 14984-9.
22. Rideout, W.M. 3rd, Hochedlinger, K., Kyba, M., Daley, G.Q., and Jaenisch, R. (2002). *Cell*, **109**, 17-27.

23. Humpherys, D., Eggan, K., Akutsu, H., Hochedlinger, K., Rideout, W.M. 3rd, Biniszkiewicz, D., and Yanagimachi (2002). *Science*, **95**, 95-7.
24. McGrath, J. and Solter, D. (1983). *Science*, **220**, 1300-2.
25. Tsunoda, Y. and Kato, Y. (2002). Nuclear transfer technologies. In *Transgenic Animal Technology: a Laboratory Handbook*, 2nd edn (ed. C.A. Pinkert), pp.195-231, Academic Press, San Diego.
26. Kimura, Y. and Yanagimachi, R. (1995). *Biol. Reprod.*, **52**, 709-20.
27. Yabuuchi, A.J. (2004). *Reprod. Dev.*, **50**, 263-8.
28. Amano, T., Tani, T., Kato, Y., and Tsunoda, Y. (2001). *J. Exp. Zool.*, **289**, 139-45.
29. Kato, Y. and Tsunoda, Y. (1992). *J. Reprod. Fert.*, **95**, 39-43.
30. Tsunoda, Y. and Kato, Y. (2002). *Differentiation*, **69**, 158-61.
31. Zhou, Q., Jouneau, A., Brochard, V., Adenot, P., and Renard, J.P. (2002). *Biol. Reprod.*, **65**, 412-19.
32. Tsunoda, Y. and Kato, Y. (1993). *J. Reprod. Fertil.*, **98**, 537-40.
33. Neff, J.M. and Enders, J.F. (1968). *Proc. Soc. Exp. Biol. Med.*, **127**, 260-7.
34. Giles, R.E. and Ruddle, F.H. (1973). *In Vitro*, **9**, 103-7.
35. Okada, Y., Suzuki, T., and Hosaka, Y. (1957). *Med. J. Osaka Univ.*, **7**, 709.
36. Eicher, E.M. and Washburn, L.L. (1978). *Proc. Natl. Acad. Sci. USA*, **75**, 946-50.
37. Tsunoda, Y. and Kato, Y. (2002). Donor cell type and cloning efficiency in mammals. In *Principles of Cloning* (eds J. Cibelli, R. Lanza, K. Campbell, and M.D. West), pp. 267-77, Elsevier, Amsterdam.
38. Tani, T., Kato, Y., and Tsunoda, Y. (2003). *Biol. Reprod.*, **69**, 1890-4.
39. Amano, T., Kato, Y., and Tsunoda, Y. (2002). *Cell Tissue Res.*, **307**, 367-70.

Chapter 4
Generating animal models of human mitochondrial genetic disease using mouse ES cells

Grant R. MacGregor,[1,2,3] Wei Wei Fan,[1,4] Katrina G. Waymire,[1,4] and Douglas C. Wallace[1,4,5,6]

[1]Center for Molecular and Mitochondrial Medicine and Genetics; [2]Department of Developmental and Cell Biology; [3]Developmental Biology Center; [4]Department of Biological Chemistry; [5]Department of Ecology and Evolutionary Biology; [6]Department of Pediatrics, Hewitt Hall, University of California, Irvine, Irvine, CA 92697–3940.

1 Background

The use of mouse ES (mES) cells to generate transgenic animals has revolutionized the study of gene action *in vivo*, and has provided animal models for studying the pathophysiology of human genetic disease and therapies thereof. Since the discovery that inherited mitochondrial DNA (mtDNA) mutations can cause age-related degenerative diseases (1), it has been realized that inherited mtDNA mutations are also associated with human adaptation to different environments, a broad spectrum of degenerative diseases, and with cancer. Somatic mtDNA mutations can accumulate during the lifetime of the individual, and have been implicated in the onset and progression of age-related diseases, and of ageing and cancer. Indeed, the importance of somatic mtDNA mutations in ageing has been affirmed by the expression in mice of a mutant form of mtDNA polymerase γ subunit A, which contains a D257A mutation that inactivates the 3′–5′ endonuclease. These animals have an increased mtDNA mutation rate and age prematurely (2). These findings highlight the pressing need for animal models of mitochondrial disease to better understand the role of the mitochondrion in the pathophysiology of disease, and in human genetics.

To date, the vast majority of transgenic mouse models of mitochondrial disease have been generated by modifying nuclear-encoded genes whose products contribute to mitochondrial function (3–8). These strains of mutant mice exhibit certain phenotypes characteristic of mitochondrial disease. However, inherently, they cannot recapitulate the unique genetic features of mtDNA mutations, such as maternal inheritance, bioenergetic threshold expression (see Section 1.1.1), and heteroplasmy, which are central to understanding mitochondrial diseases (9, 10). Heteroplasmy is defined as a mixture of two mtDNA genotypes, for example mutant and wild-type (wt) mtDNA, within a cell; and homoplasmy, as the presence of one genotype only.

Several different experimental approaches have been used in attempts to generate *trans*-mitochondrial mice, that is animals whose cells contain mitochondria with altered genomes. Microinjection of exogenous cytoplasm containing mitochondria with different mtDNA sequences into oocytes produced *trans*-mitochondrial embryos, but the mutation was lost by segregation during pre-implantation development (11). Heteroplasmic mice were also generated by fusion of enucleated cells (termed 'cytoplasts') with one-cell stage embryos, permitting introduction of mtDNAs with either naturally occurring polymorphisms (12, 13) or deletions (14). A third experimental approach to produce *trans*-mitochondrial mice involved microinjection of mES cells harbouring mitochondria with mutations in their DNA into blastocyst-stage embryos. Initial efforts to manipulate whole-animal systems involved the mtDNAs from cultured mouse cells resistant to the mitochondrial ribosome inhibitor, chloramphenicol (CAP). In a number of independent mouse cell lines, CAP resistance (CAP^R) results from a T to C transition at nucleotide pair (np) 2433 (nucleotide (nt) #2433 T > C) near the 3' end of the 16S rRNA gene, among others (15, 16). Several decades ago, chimeric mice were produced using mouse embryonal carcinoma (mEC) cells into which CAP^R mitochondria had been introduced by fusion of the mEC cells with cytoplasts from CAP^R mouse cells; these CAP^R mEC cells were then introduced into blastocysts, and the chimeric embryos transferred to the reproductive tract of pseudopregnant females (17). However, this experimental approach was limited by the extremely low efficiency with which mEC cells contribute to the germline, and there was a lack of data as to whether the CAP^R mtDNAs were incorporated into the tissues of the chimeric animals.

More recently, these limitations have been overcome using mES cells that can contribute to the germline with high efficiency in combination with methods for identifying molecular polymorphisms to trace the origin and fate of the CAP^R mtDNAs. In the contemporary experimental approach, female mES cells are fused to cytoplasts carrying a CAP^R mtDNA, thereby generating cytoplasmic-hybrid mES cells, or 'mES cybrids.' These mES cybrids are tested to confirm that they harbour the mutant mtDNA, and those cybrids with mutant mtDNAs used to generate chimeras (18, 19). By combining this approach with the selective depletion of the mES cells' endogenous mitochondria and mtDNAs prior to fusion, through treatment with rhodamine 6G (R6G) (20, 21), we succeeded in causing the CAP^R mitochondria to be transmitted by chimeric females, through their oocytes, into subsequent generations (22). This has permitted us to make a more detailed analysis of the transmission of both heteroplasmic and homoplasmic mutations, and to assess their *in vivo* consequences.

Our strategy should enable introduction of a wide variety of mouse mtDNA genotypes into the mouse female germline via somatic cell genetics. These include naturally occurring variants from different strains or species of mice and rodents, as well as mutants recovered from cells resistant to inhibitors of mitochondrial oxidative phosphorylation (OXPHOS), such as rotenone, antimycin A, and mycidin (23–28). In addition, the capture of somatic mtDNA mutations that accumulate during normal ageing, by clonal expansion of brain mtDNAs in synaptosome cybrids (29), may permit the introduction of naturally occurring, deleterious somatic mtDNA mutations into mice. Such *trans*-mitochondrial mice would allow the exploration of mtDNA changes in complex genetic processes, such as ageing.

In this chapter we present a brief primer to mitochondrial genetics. We then describe our protocols for the introduction of mtDNA mutations into female mES cells *via* cytoplast fusion to generate mES cybrids, in order to achieve maternal germline transmission.

1.1 Mitochondrial genetics and biochemistry

Mitochondria generate much of the cellular energy by OXPHOS. This process uses five multi-subunit protein complexes that are assembled from a mixture of gene products derived from both (cytoplasmic) mtDNA and nuclear (chromosomal) DNA (nDNA). Hence, mitochondrial energy production involves an interplay of Mendelian and mitochondrial genetics.

1.1.1 The mammalian mitochondrial genome

Mitochondria are derived from a symbiosis between bacteria and protoeukaryotic cells that occurred approximately two to three billion years ago. The ancestral bacterial genome contained all the genes required by a free-living organism that generated its energy by oxidizing fats and carbohydrates from its environment with oxygen, to generate water and ATP. However, early in the consolidation of this symbiotic relationship, many of the bacterial genes were transferred to the nucleus where, in the mammalian cell, they now reside, are replicated, and transcribed. The remainder of the ancestral bacterial genes are compartmentalized within the mitochondria. Messenger RNA from these nuclear-encoded genes is translated on cytosolic ribosomes and the resulting proteins are selectively imported into the mitochondrion, in some cases directed by an amino-terminal mitochondrial-targeting peptide. Today, the mammalian mitochondrial proteome is comprised of approximately 1500 gene products, 37 of which are encoded by the mtDNA (*Figure 1*), while the remainder are encoded by the nDNA. The mitochondrial genome is circular and encodes: seven (ND1, 2, 3, 4L, 4, 5, and 6) of the 46 polypeptides of OXPHOS complex I (NADH dehydrogenase, or 'NDH'); none of the four subunits of complex II (succinate dehydrogenase, SDH); one (cytochrome b, or 'cytb') of the 11 subunits of complex III (bc_1 complex); three (COI, II, III) of the 13 subunits of complex IV (cytochrome c oxidase, or 'COX'); and two (ATP6 and 8) of the 16 polypeptides of complex V (ATP synthase). All

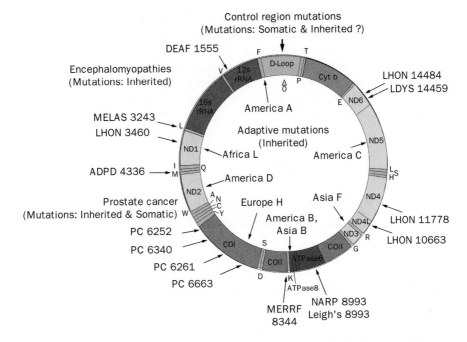

Figure 1 Map of the human mitochondrial genome. D-loop = control region. Letters around the outside perimeter indicate cognate amino acids of the tRNA genes. Other gene symbols are defined in the text. Arrows followed by names of continents, and associated letters on the inside of the circle, indicate the position of polymorphisms that define region-specific mtDNA lineages (i.e. haplogroups). Arrows associated with abbreviations followed by numbers, around the outside of the circle, indicate representative pathogenic mutations, the number being the nucleotide position of the mutation. DEAF = deafness; MELAS = mitochondrial encephalomyopathy, lactic acidosis, and stroke-like episodes; LHON = Leber's hereditary optic neuropathy; ADPD = Alzheimer's and Parkinson's disease; MERRF = myoclonic epilepsy and ragged red fibre disease; NARP = neurogenic muscle weakness, ataxia, and retinitis pigmentosum; LDYS = LHON plus dystonia; PC = prostate cancer.
(http://www.mitomap.org).

remaining mitochondrial gene products are encoded by the nDNA, including the DNA and RNA polymerases, ribosomal proteins, and all enzymes (30).

The mtDNA genome also encodes a 12S and a 16S rRNA, as well as 22 tRNAs whose genes punctuate the larger genes (*Figure 1*). In addition, the mtDNA contains a 'control region' (CR) that harbours the promoters for transcription of the C-rich light- (L-) and G-rich heavy- (H-) strands, as well as the origin of H-strand replication. The origin of L-strand replication is located two-thirds of the way in a clockwise direction from the CR.

Mitochondrial DNA is inherited exclusively through the mother (31). In part this is because the mammalian oocyte contains up to 200 000 copies of the mitochondrial genome while the sperm contains relatively few (on the order of tens to hundreds). Moreover, the sperm mitochondria, which are located in the mid-piece of the flagellum, are ubiquitinated on the prohibitin isoforms of the

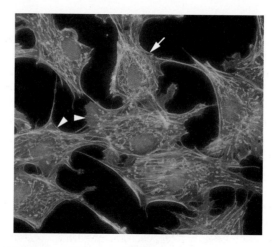

Figure 2 (see Plate 1) Morphology of mitochondria in endothelial cells in culture. Endothelial cells were fixed and stained with DAPI for the presence of DNA (blue), BODIPY FL phallacidin for filamentous actin (green), and Mitotracker Red (CMXRos) for mitochondria (red). In some cells the mitochondria appear as single organelles that can be elongated (arrowheads), while in other cells the mitochondria have a networked appearance (arrow).

mitochondrial membrane, and thus targeted for destruction by the oocyte ubiquitin-proteosome pathway (32–36).

Each somatic cell contains hundreds of mitochondria and thousands of mtDNAs. Mitochondria are frequently found as interconnected organelles within the cytosol of the cell (*Figure 2*). Hence, when a mtDNA mutation arises (whether spontaneously or experimentally), it creates an intracellular mixture of mutant and wt mtDNAs, that is heteroplasmy. During cell division the mutant and wt mtDNAs can be randomly distributed between the daughter cells, resulting in intracellular genetic drift toward either pure mutant or wt genotypes (i.e. homoplasmy), a process referred to as replicative segregation. As the percentage of mutant mtDNAs increases, the energy output available to the cell declines until the cell has insufficient energy to function properly (i.e. the bioenergetic threshold), and ultimately dies (30).

1.1.2 Somatic-cell mitochondrial genetics

The first proof that mtDNA encodes gene products that are important for the mammalian phenotype came from somatic-cell genetic analysis of CAP^R mouse and human cell lines, isolated following treatment with the mutagen, ethidium bromide (EtBr), and selection in CAP. The hereditary factor that carried the CAP^R phenotype was subsequently shown to be cytoplasmic, as resistance could be transferred to a sensitive (CAP^S) cell by fusion with a cytoplast derived from the CAP^R cell. The resulting *trans*-mitochondrial cybrids could be cultured indefinitely with CAP, demonstrating that they had acquired an altered, heritable trait that was independent of the nucleus (37–39). Subsequent studies linked the transfer of CAP^R with that of polymorphisms in mtDNA restriction endonuclease sites (40) and of mtDNA-encoded proteins (41); and that the CAP^R phenotype was the result of mutations in the mtDNA-encoded 16S rRNA (15, 16, 42).

The availability of the CAP^R marker enabled elucidation of the cellular principles of mitochondrial genetics. Using the CAP^R and CAP^S phenotypes as surrogate markers for the mtDNA genotypes of different cells, the fate of the corresponding mtDNAs could be monitored following the mixing that occurred during cell–cell and cytoplast–cell fusions to create heteroplasmy in hybrid cells and cybrids, respectively. These studies revealed that when mtDNAs originating from the same cell line, but with distinct and introduced genetic markers, were mixed by cell fusion, the two mtDNA genotypes could be propagated in culture as a heteroplasmic mixture for a considerable period of time. Eventually, however, the two forms segregated to homoplasmy. If the mtDNAs used were derived from different cell types of the same species, then incompatible combinations could arise and one of the mtDNA genotypes would selectively be excluded (43–45). Hence, it was inferred that the cells could discriminate between different mtDNAs within a mixed cytoplasm. As the genetic difference between the two mtDNA types increased, the selective exclusion of one of the mtDNAs became increasingly pronounced (39, 46–49).

Studies using different types of human cells with CAP^R and CAP^S phenotypes linked to distinctive mtDNA ND3 polymorphisms permitted analysis of the interaction between different mtDNAs. This revealed that mtDNAs that were present within a cell could complement each other in *trans*, suggesting that these organelles can fuse to allow mixing of the mtDNA-encoded products (50, 51). This mixing phenomenon is potentially problematic for studies involving the introduction of heteroplasmy into somatic cells (by embryo reconstitution). To obviate this, two methods have been developed to eliminate mtDNAs from the recipient cells prior to cell fusion. Cells can be exposed to an acute treatment with the dye, rhodamine 6G (R6G), which is toxic to mitochondria (20, 21). Alternatively, the mtDNAs of the recipient cell can be permanently removed by relatively long-term culture in EtBr (in the presence of increased concentrations of glucose, plus pyruvate and uridine), resulting in mtDNA deficient cells, commonly referred to as ρ^0 ('rho zero') cells (52).

The use of ρ^0 cells as recipients in cell-fusion experiments has permitted the production of xeno-mitochondrial cybrids, that is cultured cells in which the nucleus and mtDNAs are derived from different species. Using this approach, mtDNAs of chimpanzee and gorilla have been combined with the human nucleus. This resulted in cells with a partial complex I deficiency, presumably because of interspecies incompatibility of the mtDNA- and nDNA-encoded complex I polypeptides (47, 48). Similarly, xeno-mitochondrial cybrids with rat mtDNA and mouse nuclei have been prepared, which resulted in significant reduction in activities of complexes I–IV and in respiration. This illustrates the close functional coevolution of the nuclear and mitochondrial genomes. By contrast *Mus spretus* mtDNAs were functionally compatible with *Mus musculus* nuclei (48, 49).

1.2 Deleterious mtDNA mutations in hereditary disease

Since the late 1980s, many independent pathogenic mtDNA mutations have been identified and linked to a plethora of degenerative diseases. Mutations involving both mtDNA rearrangements and base substitution are common, and all are associated with diseases having a delayed onset and progressive course. Due to the organization of the mtDNA genome most rearrangement mutations remove at least one functional tRNA, which results in defects in synthesis of mitochondria-encoded proteins. Base-substitution mutations can either alter a tRNA or rRNA molecule, and thereby result in a protein synthesis defect, or alter a polypeptide and cause a specific OXPHOS complex defect. Representative pathogenic mtDNA mutations are shown in *Figure 1*. The tissues most commonly affected by mtDNA mutations involve those with a high-energy demand, including the central nervous system, leading to various forms of blindness, deafness, dementias, movement disorders, etc. Other affected tissues may include the heart, skeletal muscle, renal, and endocrine tissues, with dysfunction of the latter commonly presenting as Type II diabetes. Hence, mtDNA diseases typically affect the same tissues, and give the same pathology as seen in the common age-related diseases and in the ageing process *per se* (30). This concept has been amply confirmed by generating mice that are heteroplasmic for mtDNA mutations.

1.3 Generation, identification, and recovery of mtDNA mutations from mice

To study the effects of mtDNA mixing in mammals with the mouse as the experimental paradigm, methods had to be developed for introducing heterologous mtDNAs into the whole animal. This, in turn, necessitated the identification of variant mouse mtDNAs, as well as the development of procedures to recover the mutant mtDNAs and introduce them into the female germline. Functional mtDNA mutations can be generated by several methods, which are summarized as follows.

1.3.1 Capture of naturally occurring mtDNA variants from ageing animals, and from mice around the world

In addition to variants available in cell lines and mutant mice, the wild mouse population harbours an extensive reservoir of mtDNA sequence variation. Many of the inbred mouse strains were derived from the same female lineage, and thus have very similar mtDNAs (53). In contrast, the genetic diversity of wild mice is great. *Mus musculus* encompasses five subspecies: *domesticus* found in Europe, Africa, Australia, and the Americas; *musculus* in Asia; *bactrianus* in India; *castaneus* in Southeast Asia; and *molossinus* in Japan (54). The corresponding strain differences in mouse mtDNAs have recently been exploited to generate homoplasmic and xeno-mitochondrial mice (55).

That naturally occurring mtDNA variants can be functionally relevant in different mouse strains has been demonstrated by the introduction of New Zealand

Black (NZB/BinJ) mouse mtDNA into BALB/c embryos, or C57BL/6 mtDNA into NZB embryos. This was achieved by fusion of cytoplasmic vesicles from donor oocytes with recipient, fertilized oocytes (12). In the resulting heteroplasmic mice, the NZB mtDNA was invariably selected for in the liver and kidney, while the BALB/c mtDNA was selected for in the blood and spleen; and the mtDNAs did not segregate in postmitotic tissues (13). The molecular basis for this selection is not yet known. Studies of single hepatocytes indicated that the NZB mtDNA had a 14% selective advantage over the BALB/c mtDNA, which apparently was not associated with a difference in mitochondrial respiration. Interestingly, *in vitro* culture of heteroplasmic hepatocytes results in the reversal of the segregation pattern to favour BALB/c-derived mtDNA (56). This same directionality in mtDNA segregation was observed when heteroplasmic, female BALB/c mice were crossed with male *Mus musculus castaneus*. The segregation rate of the NZB and BALB/c mtDNAs in liver, kidney, and spleen was found to vary among the F_2 generation progeny, which facilitated mapping of nDNA loci that modulated the tissue-specific segregation patterns. Accordingly, quantitative trait loci (QTL) were mapped to chromosome 5 for liver segregation, chromosome 2 for kidney, and chromosome 6 for spleen. The chromosome 5 locus was associated with marker D5Mit25 with a LOD score of 31.5, clearly demonstrating that the tissue-specific segregation patterns can be associated with specific chromosomal loci (57).

Further evidence that the NZB mtDNA sequence variation was of phenotypic relevance was obtained by backcrossing the NZB mtDNA onto a CBA/H nuclear background, and *vice versa*. This resulted in mice with differences in cognitive capacity and neuroanatomy, indicating that mtDNA polymorphisms can affect cortical development and function (58).

1.3.2 Generation of cultured cells resistant to inhibitors of mitochondrial respiration

A variety of mouse cell lines have been isolated that have mutations in specific mtDNA genes that encode polypeptides. Lines resistant to the complex I inhibitor, rotenone, were derived with frame-shift mutations in both the ND5 (59) and ND6 (60) genes. Similarly, lines resistant to a number of inhibitors of cytb, including Antimycin A (24, 26), HQNO (25, 27), myxothiazol (23), and stimatellin (26), have been reported. Lines carrying cytb mutations have been isolated using negative selection procedures (61), while spontaneous mutations have been identified in mtDNA COX genes (62).

Thus, diverse procedures have been used to isolate mitochondrial mutations in cultured cells. Those with cytoplasmic – as opposed to nuclear – mutations are then identified using the cybrid transfer technique. Cells that harbour the mtDNA 16S rRNA gene CAP^R mutation have a mitochondrial protein synthesis defect similar to those seen in patients with mtDNA, rRNA, and tRNA mutations. The CAP^R mutation reduces mitochondrial protein synthesis, leading to a partial inhibition of complexes I, III, and IV (18).

1.3.3 Generation of random mtDNA mutations in cells or animals using various mutator gene strategies

An alternative method for selecting mtDNA mutations in cultured cells involves rescue of somatic mutations from either ageing mice or mice harbouring mtDNA mutator genes. Age-related somatic mtDNA mutations have been retrieved from both human and mouse adult brain cells by fusion of mitochondria-containing synaptosomes with ρ^0 cells. Fresh brain tissue is isolated and homogenized, and the synaptosomes isolated using Percoll™ gradients (*Protocol 4*). The synaptosomes are then fused with ρ^0 cells generated from cell lines of the same species, and the cybrids selected for growth in the absence of pyruvate and/or uridine, which selects for cells that have acquired functional mitochondria. Such synaptosome cybrids can then be screened for defects in OXPHOS, and for mutations in the mtDNA (29).

To increase the probability of isolating desirable mutations, cytoplasts or synaptosomes may be generated from cells or mice, respectively, that display increased rates of mtDNA mutation. For cultured cells this may arise from expression of a mitochondrial polymerase γ subunit A transgene that encodes an inactivated 3′–5′ exonuclease, for example the D198A missense mutation in human polymerase γ (63). Human cell lines expressing such a polymerase γ mutation display very high mtDNA mutation rates, resulting in the accumulation of 1:1700 mtDNA point mutations during 3 months of continuous culture. Further culture results in the inactivation of the mutant polymerase (64). Such induced mtDNA mutations may be recovered by fusion of the cytoplasts of the mutator cells with ρ^0 cells. Similarly, somatic mtDNA mutations can be rescued from the brains of transgenic mice in which the endogenous polymerase γ has been replaced by a polymerase lacking the proofreading capacity (2).

The somatic mtDNA mutation rate can also be elevated by increasing production of mitochondrial reactive oxygen species (ROS). This has been achieved in our laboratories either through inactivation of the *ANT1* and *2* genes in the brains of mice (4, 65, 66), or through inactivation of the mitochondrial manganese superoxide dismutase (MnSOD) (3, 7, 67).

1.3.4 Genetic engineering of desirable mtDNA mutations for introduction into cells

Ideally, experimental strategies would enable the introduction of specific genetic alterations into the mtDNAs of mice. Various attempts have been made to accomplish mtDNA transformation, but no practical procedure for mtDNA-mediated gene transfer currently exists (68). However, one promising approach is to generate a peptide nucleic acid (PNA), in which a peptide backbone substitutes for a sugar phosphate backbone, conjugated with an amino-terminal mitochondrial-targeting peptide. The PNA can then be annealed to an oligonucleotide homologous to the mtDNA sequence of interest. When cells are exposed to PNA-oligonucleotide, the complex is taken up and rapidly targeted to the mitochondrion, in whose matrix the oligonucleotide is deposited (69). By engineering a base substitution within the oligonucleotide to constitute the desired mutation, the oligonucleotide may serve as a primer for mtDNA replication and thereby generate the desired mutation. Additional studies are required to demonstrate the practicality of this method.

In summary, the above experimental approaches have been successful to varying degrees in generating, identifying and recovering mtDNA mutations, and introducing them into cultured cells. Next, methods were required for introduction of these mutations into mES cells with subsequent production of lines of *trans*-mitochondrial mice.

2 Introduction of mtDNA mutations into the female mouse germline

To date, several experimental strategies have been attempted to introduce genetically distinct mtDNAs into the germline of female mice, but only two basic procedures have been successful in generating '*trans*-mitochondrial' mice: (i) fusion of (enucleated cell) cytoplasts bearing mutant mtDNAs with undifferentiated, female mES cells, followed by microinjection of blastocysts with the mES cybrids and embryo transfer; and (ii) fusion of cytoplasts from mutant cells directly with mouse single-cell embryos, followed by embryo transfer. The former method has facilitated the generation of mouse strains bearing deleterious base-substitution mutations (70), while the latter has been used to produce mouse strains harbouring mtDNA deletions (14). For (ii), detailed methods are beyond the scope of this chapter, and the reader is referred to the original manuscripts (11, 14, 29). Here, we describe protocols used in our laboratories for generating *trans*-mitochondrial mice via mES cells, and for their characterization. These comprise methods for the following:

(a) Identification of mutations in mtDNA (*Protocol 1*).

(b) Recovering mitochondria containing these mutations into cells that lack mtDNA, as an intermediate step allowing maintenance and expansion of the mutation in a convenient (e.g. fibroblast) cell line (*Protocols 2–5*).

(c) Preparation of female mES cells to receive mtDNA with genetic polymorphisms, by mitochondrial depletion (*Protocol 6*).

(d) Transfer of the mtDNAs into female mES cells (*Protocols 5 and 6*); selection (*Protocol 7*) and characterization (*Protocol 8*) of the resulting mES cybrids; and the subsequent production of lines of *trans*-mitochondrial mice (*Protocol 9 and 10*).

(e) Detection of alterations in phenotype, physiology, and biochemistry that might be encountered in the *trans*-mitochondrial mice (Section 3).

2.1 Screening for mutations in mtDNA

A variety of methods have been devised for screening for mtDNA mutations. These include detection of mtDNA rearrangements by Southern blot or PCR (71); of mtDNA point mutations by alteration of a restriction enzyme site (1); of heteroduplexes by denaturing-gel electrophoresis (72–74) or denaturing-high pressure liquid chromatograph (D-HPLC) (75-77); and of base substitutions, or insertion deletion mutations, by primer extension (29). Here, we focus on methods for the detection of the mtDNA sequence variants associated with CAP^R, and of the naturally occurring differences between the strain NZB mitochondrial genotype and the mouse 'common haplotype' (53).

Protocol 1
Characterization of mutations in the mouse mitochondrial genome

Reagents and equipment

- Total genomic DNA, extracted from the cells or tissue of interest (biopsies, cultured cells, synaptosomes, or cybrids) by proteinase K digestion followed by phenol/chloroform extraction and ethanol precipitation
- PCR thermocycler
- ABI 3100 automated DNA sequence analyser

Methods

Primers, amplification conditions, purification of products:
Amplify by PCR those regions of the mtDNA to be sequenced using appropriate primers, purify resulting fragments using Centricon 100 separators (from Amicon), and subject to cycle sequencing using AmpliTaq FS DNA polymerase with fluorescent dye-terminator chemistry and an ABI DNA Sequencer, according to the manufacturer's instructions (Prism Ready Reaction Cycle Sequencing Kit, Applied Biosystems).

PCR- and restriction enzyme polymorphism-based analysis of NZB and CAP^r mtDNA:
The mtDNA from strain NZB mice and the wt or 'common haplotype' (53) can be distinguished using a naturally occurring *Bam* HI polymorphism that is absent from NZB mtDNA.

Primers (5'-3') CGGCCCATTCGCGTTATTC (forward primer, F_w) and AGGTTGAGTAGAGTGAGGGA (reverse primer, R_v) will amplify a 1127 base pair (bp) fragment using 30 cycles of PCR (30 s denaturation (94 °C), annealing (55 °C), and extension (72 °C)) with *Taq* polymerase (Roche). Digest the PCR product with *Bam* HI, and electrophorese through a 1.7% agarose gel. The NZB mtDNA lacks the *Bam* HI recognition site at np 4277. Consequently, *Bam* HI digestion produces 165 and 962 bp fragments from NZB mtDNAs, but 165, 251, and 711 bp fragments from the 'common haplotype' mtDNA.

To detect the CAP^R mtDNA, we utilize a *Tai* I polymorphism generated by the CAP^R mutation, T2433C. The mtDNA region of the CAP^R mutation is amplified by PCR (21) using primers (5'-3') TTAACGGCCGCGGTATCCTG (F_w), representing heavy strand nt 1999-2018, and TTGTAAGGCTCTATTTC (R_v), representing nt 2599-2580 (18); and digest the 601 bp PCR product with *Tai* I (MBI). Separate the fragments through a 1.8% agarose gel, with the CAP^R mtDNAs giving a 434 bp fragment and wt mtDNAs a 502 bp fragment.

2.2 Preparation of ρ^0 cells for use as recipient cells for mtDNA transfer

The impact of a specific mitochondrial haplotype on the function of a cell or organ can depend upon the haplotype of the corresponding nuclear genome. For example the effect of an identical mtDNA mutation on a cell's function might be quite different when analysed in primary embryonic fibroblasts isolated from two separate strains of mice. This problem can be circumvented by analysing the effects of diverse mtDNA polymorphisms in cells with a consistent nuclear genetic background. To achieve this, in 1974 we developed a technique for transferring mitochondria and their genomes from one nucleated cell to another, by enucleation of a donor cell and fusion of the resulting mitochondria-containing cytoplast with a recipient cell to form a cybrid (37, 38).

Subsequently, our cybrid technique was refined to incorporate the depletion of the endogenous mitochondria and mtDNA in the recipient cell. This has the effect of biasing the postfusion complement of mitochondria towards the donor genotype, and thereby reduces the initial degree of heteroplasmy. Mitochondrial depletion to produce mtDNA-deficient cells can be accomplished either by acute poisoning of the mitochondria with R6G (20, 21, 78) or through long-term culture in EtBr (52). Rho-zero cells can be isolated from a variety of species, and can be used to recover mtDNA variants from the same species, as well as to analyse the consequence of combining mtDNA from one species with the nuclear genome of another (i.e. in xeno-mitochondrial cybrids).

In the method we use to isolate ρ^0 cells (52, 78), cells are plated at low density and exposed to EtBr, which inhibits mtDNA replication but not nDNA replication. Consequently, the cells lose the ability to perform OXPHOS, although they can still perform glycolysis. As loss of the electron-transport chain causes the accumulation of cytosolic NADH and inhibits dihydroorate dehydrogenase activity for pyrimidine biosynthesis, ρ^0 cells require exogenous pyruvate to reoxidize NADH to NAD$^+$ (through the generation of lactate by lactate dehydrogenase), and exogenous uridine for conversion to thymidine (GUP medium, see *Protocol 2*). We have used this method to generate the LMEB4 mouse ρ^0 cell line from the parental cell line, LM(TK$^-$) (78).

Protocol 2 is intended for the derivation of ρ^0 cells from established fibroblast cell lines to act as intermediate recipients of mtDNAs of interest, and subsequently as donors of mitochondria and mtDNAs for mES cells. We do not recommend the generation of ρ^0 mES cells because we believe that long-term culture in EtBr would degrade their pluripotency. Rather, for mES cells we destroy the resident mitochondria and mtDNA immediately prior to fusion with mtDNA donor cytoplasts, by treatment of the mES cells with the mitochondrial poison, R6G (*Protocol 6*) (21, 78, 79).

In order to successfully transfer mitochondria and mtDNAs into the mitochondrially depleted cells it essential to be able to select *against* the unenucleated mitochondrial donor cells, and *for* the host cell, postcytoplast fusion. (This is to prevent overgrowth of cultures with contaminating fibroblasts.) It is

Protocol 2
Preparation of ρ^0 cells

Reagents and equipment

- Parental mouse fibroblast cell line, e.g. LM(TK$^-$) (80) and its coderivative of L929 cells, LA9 cells (81)
- EtBr: ethidium bromide (Cat. No. E-7637, Sigma). Caution: ethidium bromide is a weak mutagen and carcinogen, and so appropriate precautions should be taken when handling and disposing of this chemical
- DMEM: Dulbecco's Modified Eagle's Medium containing 4.5 mg/ml glucose (Cat. No. 10-017-CV; Mediatech)
- DMEM + GUP medium: DMEM supplemented with 50 μg/ml uridine (tissue-culture grade; Cat. No. U-3003, Sigma) and 1 mM pyruvate, (tissue-culture grade; Cat. No. P-8574, Sigma)[a]
- Heat-inactivated fetal calf serum (hi-FCS), e.g. defined-grade serum from Hyclone
- Trypsin/EDTA solution: 0.25% trypsin and 1 mM EDTA in HBSS (Cat. No. 25200-056, Invitrogen)
- Dulbecco's phosphate buffered saline, Ca^{2+}- and Mg^{2+}-free (PBS; Cat. No. 14190-144, Invitrogen)
- 250 mm and 100 mm tissue-culture dishes, e.g. from Nunc
- Pyrex glass cloning rings (Bioptechs)

Method

1. Plate 5×10^6 mouse fibroblast cells into a 100 mm dish containing 20 ml DMEM + GUP supplemented with 10% hi-FCS, and with 250 ng/ml EtBr. Culture the cells in this medium for 28 d.

2. Harvest the culture by trypsinization, ensuring that a single-cell suspension is achieved. Replate at a density of 1×10^4 cells per 100 mm dish in DMEM + GUP supplemented with 10% hi-FCS, and 100 ng/ml EtBr. Maintain the cells (with periodic changes of this medium) for up to an additional 10 weeks, until discrete (clonal) colonies are observed.[b,c]

3. Isolate the colonies using cloning rings, harvest by trypsinization, and replate in DMEM + GUP supplemented with 10 % hi-FCS, and 100 ng/ml EtBr. Use of conditioned medium may improve cloning efficiency.

4. After colonies have been expanded into clonal sublines and cryopreserved, replate aliquots of each subline in duplicate and assess their ability to grow in medium with or without pyruvate (1 mM) and uridine (50 μg/ml): those clones lacking mtDNA should be unable to grow in medium lacking these metabolites.[d]

5. Verify the absence of mtDNA in these ρ^0 clones by molecular analysis for the presence of mtDNA (e.g. by quantitative PCR or Southern blot).

[a] Glucose is used by the cells to generate ATP via glycolysis, pyruvate for reoxidation of NADH to NAD^+, and uridine to provide pyrimidines.

Protocol 2 continued

[b] The population doubling time for ρ^0 cells should be substantially longer than for the parental cell line (e.g. 10 h for LM(TK$^-$) cells, compared with 72 h for the LMEB4 ρ^0 derivative cell line (29)).

[c] The ρ^0 cells will acidify the medium significantly quicker than the parental cells. As ρ^0 cells rely upon glycolysis for production of ATP, they produce a large amount of lactic acid.

[d] The requirement of ρ^0 cells for pyruvate and uridine for growth is used to select *against* ρ^0 cells when generating cybrid cell lines.

Protocol 3

Preparation and isolation of cytoplasts

Reagents and equipment

- Donor, mouse cell line containing mitochondrial genomic polymorphism(s)
- Ficoll® 400 (Cat. No. F-8016, Sigma)
- Percoll™ solution (Cat. No.17–0891, Amersham Biosciences)
- PSA solution: 500 × stock solution of antibiotics penicillin, streptomycin, and amphotericin B (Cat. No. R-004–10, Cascade Biologics)
- Serum-free, PSA-supplemented medium: culture medium (e.g. DMEM) containing 1 × PSA (i.e. 100 U/ml penicillin, 100 U/ml streptomycin, and 0.25 µg/ml amphotericin B)
- Complete culture medium for donor cells, e.g. DMEM supplemented with 10% hi-FCS
- Cytochalasin B solution: prepare a 2 mg/ml stock solution of cytochalasin B (Cat. No. C-6762, Sigma) in DMSO (Cat. No. D-2650, Sigma)
- Milli-Q® or glass-distilled water
- Mannitol buffer: 0.3 M mannitol (Cat. No. M-9546, Sigma) in distilled water, pH 7.3, sterilized by filtration through a 0.22 µm membrane
- Hoechst 33258 (Cat. No. 09460, Polysciences) and Rhodamine 123 (Cat. No. R-8004, Sigma)
- UV light-treated Parafilm™, or autoclaved aluminium foil
- Ultracentrifuge, including suitable swing-out rotor (e.g. SW41Ti) and tubes for Ficoll® gradient-mediated enucleation
- Super speed centrifuge (e.g. Sorvall RC5B) with fixed angle SS-34 rotor, and 30 ml polysulfonate centrifuge tubes (e.g. from Nalgene), for Percoll™ gradient-mediated enucleation
- Fluorescence microscope with appropriate excitation filters

Methods

Ficoll® gradient-mediated enucleation:

1. Prepare a 50% (w/v) stock solution of Ficoll® 400 by adding 125 g Ficoll® to 125 ml water, and stirring or rocking overnight with low heat. Sterilize the solution by filtration, which due to its high viscosity may require up to 24 h.

Protocol 3 continued

2. Using sterile technique, prepare 20 ml of 25% (w/v) Ficoll® solution by combining 10 ml serum-free, PSA-supplemented medium with 10 ml of 50% (w/v) Ficoll® stock solution. Using this solution and medium prepare additional dilutions as indicated in Table 1.[a]

3. Add cytochalasin B to a final concentration of 20 µg/ml in all Ficoll® solutions, including the remaining 25% (w/v) Ficoll®.

4. Prepare the step gradients on the afternoon before the day of enucleation. Cover the tubes with sterilized Parafilm™, or foil, and store overnight in a cell-culture incubator at 37 °C. This will permit minor diffusion to occur at the interface between step gradients, which will reduce trauma to the cells. Also allow the swing-out ultracentrifuge rotor to equilibrate overnight at 37 °C.

5. For each gradient, resuspend 4×10^7 freshly harvested donor cells in 3 ml of 12.5% (w/v) Ficoll® containing 10 µg/ml cytochalasin B, and overlay the top layer of the step gradient. Add medium to just below the top of the tube.

6. Centrifuge tubes for 60 min at 77 000 g_{av} and 31 °C.

7. Gently remove each tube, aspirate the upper band that contains the cytoplasts (see Figure 3), and resuspend in 20 volumes of complete culture medium.

8. Pellet the cytoplasts by centrifugation for 10 min at 650 g, and wash and resuspend in 10 ml mannitol buffer.

Percoll™ gradient-mediated enucleation:

1. Harvest 2×10^7 viable cytoplast-donor cells by removal from the substratum and centrifuge at 300 g for 3 min. Thoroughly resuspend the pellet in 20 ml of a 1:1 mixture of complete medium and Percoll™ that has been pre-equilibrated overnight in a cell-culture incubator.[b]

2. Transfer the mixture to a 30 ml polysulfonate centrifuge tube, and add 200 µl of cytochalasin B solution to give a final concentration of 20 µg/ml.

3. Centrifuge for 70 min at 44 000 g, with the centrifuge set to maintain a rotor temperature of 37 °C.[c]

4. Following centrifugation, collect the band located one-third of the height from the bottom of the tube, along with the hazy band immediately above that is rich in cytoplasts. If the enucleation has been successful, this mixture of fractions should contain few, if any, intact cells.

5. Dilute the cell suspension 10 fold with complete medium, collect the cells by centrifugation at 650 g for 10 min, and resuspend in 10 ml medium. The mixture of washed cytoplasts, karyoplasts, and whole cells is then ready for electrofusion (78).

Analysis of efficiency of enucleation by fluorescence microscopy:

Fractions from the gradients can be stained using Hoechst 33258 and Rhodamine 123 fluorescent dyes, which label nuclei and mitochondria, respectively. This allows

Protocol 3 continued

assessment of the efficiency of enucleation, and verification of the expected distribution of cytoplasts and karyoplasts in the gradients.

1. Label fractions by incubating cells in culture medium containing 0.5 µg/ml final concentration each of Hoechst 33258 and Rhodamine 123, for 15 min at 37 °C.

2. Wash the cells twice with fresh medium, without dyes.

3. Place an aliquot of the cells onto a clean microscope slide, apply a coverslip, and immediately view cells at 400 × or 630 × magnification using a microscope equipped with fluorescence filter sets. To visualize Hoescht 33258-stained nuclei, use a DAPI (blue) filter set; and to visualize Rhodamine 123-stained mitochondria, a rhodamine (red) filter set.[d]

[a] When pipetting Ficoll® suspensions ensure that pipettes are completely evacuated and that individual solutions are mixed to homogeneity.

[b] This preincubation helps to reduce alkalinity caused by the Percoll™, which otherwise may lower cell viability.

[c] A fixed-angle rotor is adequate, and we routinely use a Sorvall RC5B super speed centrifuge with SS-34 rotor at 19 000 rpm, set at 25 °C.

[d] Fraction 1 (see Figure 3) should be rich in intact rhodamine-staining cytoplasts, and should contain very few nuclei or intact cells. This fraction produces cybrids readily following fusion with ρ^0 cells. The band(s) of intermediate buoyant density comprises a heterogeneous mixture of larger cytoplasts and many intact cells. This fraction also can be used to generate cybrids. The lowest fraction, at the interface between the 17 % and 25 % (w/v) Ficoll® solutions, is comprised predominantly of karyoplasts (i.e. nuclei). Approximately one-third of these karyoplasts contain no mitochondria while the remaining two-thirds retain some.

consequently necessary for the cytoplasmic donor cells to have appropriate nuclear selectable markers. One readily available selective system is that incorporating thymidine kinase (TK) and hypoxanthine phosphoribosyltransferase (HPRT) (described further in Section 2.8). Thus, the mitochondria and mtDNAs of interest can be transferred (Protocol 5) from an initial source (e.g. brain synaptosomes, Protocol 4) into a mouse L cell whose nuclear genome is either HPRT-deficient (HPRT$^-$) or TK-deficent (TK$^-$), and which lacks mtDNA (ρ^0). The cybrids can be positively selected in 5-bromodeoxyuridine (BrdU) if the cell line is TK$^-$, or in 6-thioguanine (6TG) if HPRT$^-$, and cultured in medium that lacks uridine and/or pyruvate to inhibit ρ^0 cells that have not acquired functional mitochondria from a cytoplast fusion (Protocol 7). These (TK$^-$ or HPRT$^-$) L-cell cybrids can then be used as mitochondrial donors for the mES cells. They are enucleated (Protocol 3), and their cytoplasts fused to the R6G-treated mES cells (Protocol 5). The desired mES cybrids are then selected in HAT medium (see Section 2.8) in which the TK$^+$ and HPRT$^+$ mES cells will proliferate but any contaminating, unenucleated L cells will die (being either TK$^-$ or HPRT$^-$); and in medium lacking uridine to select for cells with functional mitochondria (Protocol 7).

2.3 Preparation of cytoplasts from cell lines harbouring a mtDNA polymorphism

Two methods have been developed to enucleate either suspension or adherent mammalian cells. When there is no requirement to separate cytoplasts from karyoplasts, the simpler method of isopycnic Percoll™ gradient centrifugation may be used. However when these fractions must be separated, the use of Ficoll® step-gradient centrifugation is appropriate (82).

2.4 Generation of brain synaptosomes from mice harbouring mtDNA polymorphisms

Synaptosomes are derived from neuronal synaptic boutons, which pinch off during homogenization of brain tissue to form membrane-bound cytoplasmic particles that include mitochondria. When synaptosomes are fused to a ρ^0 cell, the brain mtDNAs continue to be replicated and amplified by the normal mechanisms. Synaptosomes are isolated using a rapid Percoll™ step-gradient centrifugation technique (83) (*Protocol 3*).

Table 1 Preparation of Ficoll® suspension step gradients

Final % (w/v) Ficoll®	Volume of 25 % (w/v) Ficoll® (ml)	Volume of serum-free, PSA-supplemented medium (ml)
17	3.4	1.6
16	3.2	1.8
15	3.0	2.0
12.5	5.0	5.0

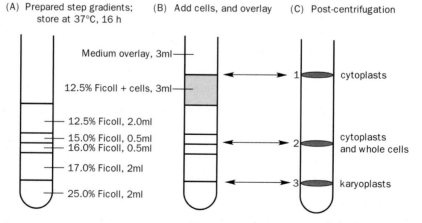

Figure 3 Enucleation of cells using Ficoll® step-gradient centrifugation. The cartoon depicts the appearance of the centrifuge tubes (A) before loading, (B) after loading and immediately before centrifugation, and (C) after centrifugation. The middle band in (C) (labelled '2') can vary in appearance depending upon the particular cell line used, and sometimes appears blurred or split into two bands.

Protocol 4

Production of brain synaptosomes

Reagents and equipment

- Adult mice containing the mtDNA variants of interest
- Percoll™ solution (Cat. No.17–0891, Amersham Biosciences)[a]
- Sterile dissection tools and razor blades
- 70% ethanol in water
- SED solution: 0.32 M sucrose, 1 mM EDTA, and 0.25 mM dithiothreitol (DTT), adjusted to pH 7.3 with dilute HCl, and filter sterilized
- Culture or Petri dishes
- Glass Dounce Tissue Homogenizer, 40 ml capacity (Cat. No. 1984–10040, Bellco Glass Inc.) with large clearance pestle for initial processing, and small clearance pestle for cell disruption
- Sterile centrifuge tubes
- Mannitol buffer (*Protocol 3*)
- Refrigerated bench-top centrifuge
- Ultracentrifuge

Method

1. Sacrifice the mice using a humane method, in accordance with local and national regulatory agencies.[b]
2. Working quickly and carefully, thoroughly swab the body of each mouse with 70% ethanol and transfer the carcass to a tissue-culture hood.
3. Decapitate the mouse using a strong pair of scissors, and transfer the head into a culture dish.
4. Carefully cut the skull open using a sharp pair of small, pointed scissors: begin at the rear of the skull on the dorsal side and work forward, cutting down the mid-sagittal plane. Using a pair of forceps, grasp one flap of the skull at the midline and reflect the skull. Repeat for the contra-lateral side.
5. Using a wide spatula and sterile technique, remove the brain and immediately place into SED solution, prechilled on ice. Allow the brain to cool for 10 min.
6. Place the brain in a sterile culture dish on ice, and mince the tissue with scissors.
7. Transfer the minced tissue into a sterile homogenizer containing nine volumes of ice-cold SED solution.
8. Homogenize the tissue with ten strokes using the loose pestle, followed by ten strokes using the tight pestle.
9. Centrifuge the 10% homogenate for 10 min at 1000 g and 4 °C, to remove nuclei and unbroken cells.
10. Carefully transfer the supernatant into a fresh tube, dilute 1:1 with SED solution, and maintain on ice.

> **Protocol 4** continued
>
> 11 Prepare six tubes to contain Percoll™ step gradients, as follows: dispense 2 ml of 23% Percoll™ into a sterile centrifuge tube, then carefully layer further suspensions of Percoll™ in the following order; 2 ml of 15%, 2 ml of 10%, and 2 ml of 3%, each prepared in sterile SED solution.
>
> 12 Carefully layer 2 ml tissue-homogenate supernatant onto each of six gradients, and centrifuge for 5 min at 32 500 g and 4 °C, excluding acceleration and braking time.
>
> 13 Using sterile technique, carefully aspirate and combine the fractions at the 23%/15% Percoll™ interface.
>
> 14 Dilute pooled fractions 10 fold with sterile mannitol buffer and centrifuge for 15 000 g for 15 min at 4 °C. Carefully remove the mannitol wash. The loose pellet of synaptosomes is used for electrofusion with ρ^0 cells, immediately.
>
> [a] This is supplied as a sterile, 250 ml suspension of 15–30 nm silica particles coated with non-dialysable polyvinyl pyrrolidone. Unopened, Percoll™ has a shelf life of 2 years at r.t.; and, undiluted, can be re-sterilized by autoclaving for 30 min at 121 °C.
>
> [b] When a fusion is performed using synaptosomes isolated from brains immediately postmortem, the yield of cybrids is $1/10^4$ ρ^0 cells (29). Similar efficiencies of cybrid formation are obtained using brains that have been removed immediately postmortem and subsequently stored on ice for up to 4 h. However, if mice are stored at 4 °C for 4 h postmortem before removal of brains, the yield of cybrids is reduced 100 fold (29).

2.5 Recovery of mtDNA mutations by electrofusion of cytoplasts or synaptosomes with ρ^0 or mES cells

To induce the fusion of cyotoplasts or synaptosomes with recipient cells (either ρ^0 cells or mES cells that have been treated with R6G), the technique of electrofusion using standard electroporation apparatus has proven effective and reliable. During electroporation, cells in suspension are exposed to a defined electric field for a short period of time (milliseconds). At appropriate field strength, the direct current (DC) pulse overcomes the field potential of the cell's lipid bilayer. This causes transient breakdown of the lipid bilayer with formation of pores that reseal within a few seconds. If cells and cytoplasts are in physical contact when the electric field is applied, they may fuse. Contact between the cells and cytoplasts can be enhanced, by applying a constant alternating current (AC) before delivery of the (DC) electric field pulse. The AC causes dielectrophoresis, which leads to the cells and cytoplasts becoming aligned in a manner similar to 'pearls on a chain'. Consequently, electrofusion is often performed in two stages; bringing cells into contact with AC, followed by a DC pulse that stimulates their fusion. Some investigators use an additional exposure to AC following the DC pulse as this may stabilize and compress the hybrids, thereby increasing the efficiency of the process.

A variety of commercially available apparatus are available for electrofusion. We have successfully used the pulse generator, ECM 2001, from BTX

ANIMAL MODELS OF HUMAN MITOCHONDRIAL DISEASE

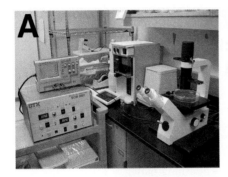
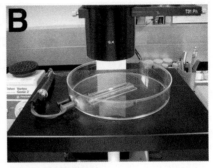
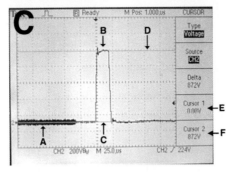

Figure 4 Apparatus used for electrofusion. (A) The BTX ECM 2001 pulse generator sits on a cart beside a bench in a cell-culture facility. The BTX Enhancer 400™ oscilloscope is located on top of the pulse generator. On the right side of the photo is an inverted microscope, which is used to observe cells during fusion. (B) A close-up view of the electrofusion microslide, located inside a 150 mm plastic culture dish. Note, the dish lid has been removed for photography; it normally covers the dish not only to maintain sterility but also to provide protection to the operator from possible electric shock during electrofusion. (C) A typical read-out from the BTX Enhancer 400™ oscilloscope after an electrofusion. The conditions consisted of an AC alignment phase (arrow A) followed by a single DC pulse at 800 V for 25 μsec (arrow B). The actual voltage across the electrodes can be recorded by aligning two cursors (arrows C and D) with the baseline and peak, respectively, of the voltage pulse. Note that the difference between these values (arrows E and F) is 872 V, that is approximately 10% greater than the setting on the pulse generator.

Techonologies Inc in conjunction with an Enhancer 400™ oscilloscope (*Figure 4A*). The oscilloscope serves an extremely important function as it records the actual voltage transferred across the electrodes of the sample chamber. This value usually varies significantly from the value set on the pulse generator (*Figure 4C*).

Conditions for efficient electroporation depend upon many variables, including time and field strength of AC and DC voltages, composition of the buffer used, the cell type and density, and the apparatus used. The conditions listed below have been successfully used to generate both fibroblast and mES (*trans*-mitochondrial) cybrids. However, the reader should consider these only as a starting point from which to begin an evaluation of appropriate conditions for their particular cell lines, using their equipment and buffers.

Protocol 5

Electrofusion of synaptosomes or cytoplasts with ρ^0 cells or R6G-treated mES cells

Reagents and equipment

- Suspension of cytoplasts (*Protocol 3*) or synaptosomes (*Protocol 4*) in mannitol buffer (*Protocol 3*)
- Suspension of ρ^0 cells (*Protocol 2*) (52, 78) or R6G-treated mES cells (*Protocol 6*) (21, 78, 70) in mannitol buffer
- Electrofusion apparatus, e.g. BTX ECM 2001 Pulse generator (*Figure 4A*), with BTX Enhancer 400™ Oscilloscope, and BTX Microslide Model No. 453 with 3.2 mm electrode gap (*Figure 4B*) (from BTX Technologoies Ltd)
- Washing buffer: 0.3 M mannitol (Cat. No. M-9546, Sigma) in Milli-Q® water, pH 7.0, and filter sterilized
- Haemocytometer

Methods

Electrofusion:

1. Combine cytoplasts or synaptosomes with the suspension of ρ^0 cells (or R6G-treated mES cells): for synaptosomes, combine material emanating from one pellet (*Protocol 4*, step 14) with 2×10^6 ρ^0 cells; while for cytoplasts, combine roughly 10 fold more cytoplasts than ρ^0 cells (e.g. 3×10^7 cytoplasts and 2×10^7 ρ^0 cells).
2. If using a reusable electrofusion microslide chamber, sterilize e.g. by flooding the microslide with 70% ethanol and allowing to air-dry in a laminar-flow cabinet.[a]
3. Connect the microslide to the electrodes inside a 150 mm diameter plastic Petri dish that also has been sterilized by immersion in 70% ethanol (*Figure 4B*).[b]
4. Using sterile technique, bring the volume of the combined ρ^0 cells and cytoplasts to a quantity appropriate for the sample chamber being used, e.g. 0.6 ml for BTX Microslide Model No. 453.[c]
5. Subject the cells to electrofusion. Using the apparatus described above and for fusion of mouse or human ρ^0 cells with either cytoplasts or synaptosomes, the conditions we use are an initial alignment for 20 s with AC at 35 V (field strength 0.11 kV/cm), followed immediately by two square-wave DC pulses at 800 V (2.5 kV/cm). We do not apply a postfusion AC field.[d]
6. Using the oscilloscope, record the voltage and time of the AC and DC pulses (see *Figure 4C*).
7. Remove the fusion mixture from the microslide by gently pipetting, and transfer into a dish containing prewarmed and pregassed medium.[e,f]
8. Gently disperse the postfusion mixture by repetitive pipetting through a wide bore pipette. Do not swirl the dish, as the centrifugal force can cause the cells to cluster on the circumference.

ANIMAL MODELS OF HUMAN MITOCHONDRIAL DISEASE

Protocol 5 continued

9 If necessary apply selective medium to the cells (*Protocol 7*), culture, and examine the dishes for growth of (clonal) colonies on a daily basis. Depending upon the cell type, colonies should become visible between 8 and 14 d postfusion.[g]

[a] Do not use UV light to sterilize the plastic dish housing the electrofusion microslide, as this will cause the plastic to degenerate over time.

[b] Use silicone adhesive to keep the electrode connections in place.

[c] Differences in the volume used can have significant impact on conditions for electrofusion.

[d] The reader is strongly encouraged to perform pilot experiments to determine optimum conditions for cell fusion. This is best achieved by viewing the cell and cytoplast mixture using an inverted microscope whilst applying the AC and DC pulses (*Figure 4B*). Determine the minimum time and amplitude of AC voltage required for alignment, and observe the fusion and postfusion processes. We have found that efficient formation of cybrids requires electrofusion conditions resulting in 10–50% cell death.

[e] Some protocols call for the cells and cytoplasts to be left undisturbed in the electrofusion chamber for a period of time postfusion (e.g. 3 min), the rationale being to permit stabilization of cybrids. However, depending upon the buffer used for electrofusion, cell viability may be improved by transferring the mixture into fresh medium promptly after fusion. The impact of such variables on the efficiency of electrofusion is dependent on the specific conditions, equipment, and cell types used, and these should be evaluated empirically by the investigator. But as a generalization for mES cybrids, we recommend plating after a 2 min recovery period.

[f] When generating mES cybrids, the dish should contain an appropriate feeder layer (see Section 2.6).

[g] Use of conditioned medium can improve cell survival in the initial stages.

2.6 Culture of mES cells

Due to the maternal inheritance of the mitochondrial genome, experiments described here necessitate the use of female mES cell lines to allow germline transmission of mtDNA mutations. Methods for *de novo* mES cell line derivation and maintenance, and a discussion on female lines, are provided in Chapter 2. Our protocols were optimized using the female mES cell line, CC9.3.1, which was isolated from a 129SvEv-*Gpi1^c* embryo (A. Bradley, personal communication), and selected after screening several candidate lines for the presence or absence of a Y chromosome by PCR amplification of *Sry* (84). Crucially, pilot experiments demonstrated that this line would contribute to the germline efficiently and produce normal, fertile females. Line CC9.3.1 has a karyotype of 39X, with a normal chromosome G-banding pattern (unpublished observations).

We routinely culture line CC9.3.1 and its sublines on feeder layers prepared from the STO-derivative fibroblast line, SNL76/7 (85). Feeder cells are mitotically inactivated by mitomycin C treatment (Chapter 2, *Protocol 3*), and plated onto gelatinized culture dishes (Chapter 6, *Protocol 1*). The incubation of feeder layers in mES-cell medium for at least 1 d prior to coculture with mES cells promotes conditioning with LIF. To enhance cell survival, we refeed mES-cell cultures 30–60 min prior to harvesting by trypsinization.

2.7 Depletion of mitochondria in mES cells by treatment with R6G

When generating mES cybrids we recommend using mES cells in which the mitochondria have been acutely inactivated by exposure to R6G, since long-term growth of mES cells in EtBr was found in preliminary experiments to induce differentiation and thereby compromise germline transmission. R6G inhibits OXPHOS, causing an irreversible collapse of the mitochondrial membrane potential that renders cells respiration deficient.

Treatment of mES cells with sufficient R6G to inhibit colony-forming capacity by a factor of at least 10^{-5} is necessary to permit the isolation of cybrids within a frequency range of $10^{-3} - 10^{-4}$. However, some of the mES cybrids recovered will still be heteroplasmic for donor and recipient mitochondria. This can be an advantage if the mtDNA that is being introduced into the mES cell, and then the mouse, harbours highly deleterious mutations since the residual, normal mES cell mtDNAs can partially complement the donor mtDNA's mitochondrial defect, permitting the mES cybrids to contribute successfully to the female germ-line. This strategy offers additional opportunities for experimental analysis. For example formation of mES cybrids with 'neutral' (i.e. relatively innocuous) polymorphisms permits investigation of patterns of segregation of mitochondrial genomes in different tissues during embryonic development, and in adult homeostasis. Furthermore, cybrids formed using cytoplasts whose mitochondrial genomes contain mutations that are anticipated to have adverse affects on physiological function can still contribute to the formation of a viable chimeric animal, by virtue of the heteroplasmy; and analysis of the distribution of mES cell-derived tissues in the chimera can provide insights into the fitness of those heteroplasmic cells to contribute to a specific organ system.

2.8 Culture and selection of *trans*-mitochondrial mES cybrids

An essential requirement for studies involving the production of cybrids is a culture system that enables their selection and maintenance. Such systems have been developed based on differences in metabolic capacity compared with parental cells, commonly the capacity to synthesize purines and pyrimidines (81, 78).

Aminopterin blocks all reactions involving dihydrofolate reductase (DHFR), which is involved in the *de novo* synthesis of inosine monophosphate (IMP) and hence all purines. DHFR activity is also required for methylation of dUMP to form dTMP. In the presence of aminopterin, therefore, cells are dependent upon extraneous sources of purines and thymidine for metabolism. Hypoxanthine serves as an extraneous source of purines, as it can be converted by HPRT to form IMP, which can subsequently be converted to AMP and GMP. And thymidine can be converted by TK to TMP. Normal (i.e. $HPRT^+$ and TK^+) cells can be cultured in medium containing hypoxanthine, aminopterin, and thymidine

Protocol 6
Depletion of mitochondria in mES cells by R6G treatment

The specific conditions used will vary depending upon the cell line, and on batches of R6G and FCS. However, in general, we use conditions that give low (e.g. 0.1%) survival of mES cells per well. As a starting point, it is suggested to try a range of R6G concentrations between 0.15 and 1.5 µg/ml.

Reagents and equipment

- Suspension of female ES cells in culture medium
- Feeder layers in six-well dishes
- DMEM + GUP medium (*Protocol 2*)
- Mouse ES-cell medium: high-glucose (4.5 g/L) formulation DMEM (Cat. No. 12100-046, Invitrogen) supplemented with 15% mES batch-tested FCS (Cat. No. SH30070, HyClone), 1 × glutamine (Cat. No. 25030-081, Invitrogen), 0.1 mM β-mercaptoethanol (Cat. No. M-3148, Sigma), 1 × minimal-essential amino acids (Cat. No. 11140-050, Invitrogen), and 100 U/ml penicillin and streptomycin (Cat. No. 15140-122, Invitrogen)[a]
- Mouse ES-cell medium, supplemented with 50 µg/ml uridine plus or minus 1 mM pyruvate
- R6G solution: stock solution of 1 mg/ml rhodamine-6G (Cat. No. R-4127, Sigma) in 3% ethanol
- PBS (*Protocol 2*), prewarmed to 37 °C
- Inverted phase-contrast microscope.
- Methylene blue solution (Chapter 2, *Protocol 4*)
- Mannitol buffer (*Protocol 5*)

Methods

Optimization of R6G treatment for mES cells:

1. Plate mES cells onto feeder layers in six-well dishes at a density of 1×10^5 mES cells per well, in uridine-supplemented mES-cell medium (to promote pyrimidine metabolism prior to its inhibition with R6G), and culture for 24 h.

2. Add R6G at final concentrations of 0.5, 1.0, 1.5, 2.5, and 5.0 µg/ml in duplicate wells, and culture for 72 h.

3. Aspirate the R6G-containing medium, gently wash the cells twice with prewarmed PBS, and add fresh mES-cell medium without R6G. Culture for 10–12 d.

4. Examine the dishes for the presence and morphology of mES-cell colonies using phase-contrast microscopy, before fixing and staining with methylene blue (Chapter 2 *Protocol 4*).

5. Count the number of undifferentiated mES-cell colonies in each well. Choose the concentration of R6G that results in a cloning efficiency of less than 10^{-5}.

Treatment of mES cells with R6G:
This treatment is undertaken prior to the electrofusion of mES cell with donor cytoplasts, to produce mES cybrids.

Protocol 6 continued

6. Plate female mES cells at a density of $2 \times 10^5/cm^2$ on feeder layers in DMEM+GUP medium. Culture for 24 h.

7. Add R6G to each dish to give the optimized concentration (e.g. 1 µg/ml), and incubate cultures for up to 72 h.[b,c]

8. Aspirate medium, wash cells gently with warmed PBS twice, and replace with fresh, prewarmed medium without R6G but containing 1 mM pyruvate and 50 µg/ml uridine. Culture for 30–60 min.

9. Harvest the cells using trypsin/EDTA solution (*Chapter 2*), wash twice with warmed PBS, and resuspend finally in mannitol buffer for electrofusion.

[a] We routinely culture our SNL 76/7 and derivative feeder cells in the same medium, except supplemented with 10% rather than 15% FCS.

[b] The ultimate test of the efficacy of R6G treatment is the proportion of colonies from an mES cybrid fusion that have the mES cell's nucleus but contain the donor cell's mtDNAs, either in the homoplasmic or heteroplasmic state. If the R6G treatment is inadequate, then most of the colonies will comprise mES cells that have escaped R6G inhibition without donor cytoplast fusion, and have returned to the original mES mtDNA genotype. Obviously such cells are useless for generating mtDNA mutant mice, and if too frequent will swamp the desired cybrids, making it impractical to isolate any useful cybrid clones with ease.

[c] The initial morphology of mES cells treated with R6G and maintained in DMEM+GUP remains relatively normal for 2–3 d post R6G treatment. However, these cells should not be able to proliferate due to the complete inhibition of their mitochondrial function. Hence, they ultimately granulate and die.

(HAT medium). However, cells that are deficient in HPRT or TK cannot, as both the *de novo* and salvage pathways are blocked.

Cultured cells can also be selected for a deficiency in HPRT or TK. HPRT-deficient (HPRT$^-$) cells are resistant to the toxic purine analogue, 6-thioguanine (6TG), while TK-deficient (TK$^-$) cells are resistant to 5-bromodeoxyuridine (BrdU). Hence, selection systems exist for both the presence and absence of HPRT or TK function, and accordingly can be used to select for recipient nuclei or against donor nuclei following cybrid formation, thereby eliminating contaminating donor cells from the cybrid population.

Strategies also exist to select for and against mitochondrial function. For example, a T>C transition at np 2433 within the mtDNA 16S rRNA gene renders cells partially respiratory deficient (18) and confers resistance to CAP (15, 42). Similarly, ρ^0 cells, whose mitochondria lack DNA, and are therefore incapable of OXPHOS, are dependent on glycolysis for energy production, on pyruvate specifically to maintain the redox balance, and on uridine to generate pyrimidines. These cells will not survive without glucose, uridine, and pyruvate (GUP); but they can be rescued by fusion with cytoplasts containing functional mitochondria and mtDNAs from a cytoplast donor cell. The ρ^0 cells proliferate relatively slowly, even in GUP medium (see footnotes, *Protocol 2*).

Protocol 7
Selection and maintenance of mES cybrids

The composition of culture medium is altered to select for growth of cybrids, and against growth of parental cell lines. For example when recovering mitochondria from synaptosomes via fusion with LMEB4 (TK$^-$) ρ^0 cells, the cybrids are plated in medium that is supplemented with neither pyruvate nor uridine to block the replication of the (LMEB4) ρ^0 cells, thus permitting the preferential proliferation of the LMEB4 cybrids that contain mitochondria.

Subsequently, when transferring the mutant mitochondria and mtDNAs from the LMEB4 (TK$^-$) cybrids to the R6G treated mES cells, fused cells are plated in medium containing HAT (to eliminate any non-enucleated LMEB4 (TK$^-$) cells). Here, the medium is supplemented also with pyruvate and uridine for 24 h postfusion to sustain the R6G treated mES cells until the cytoplast mitochondria can become established in the cybrid mES cells. Then the medium is changed to DMEM plus HAT without pyruvate and/or uridine to select against any TK$^-$ or HPRT$^-$ donor cells that escaped enucleation, and to encourage the growth of the mES cells that have acquired functional mitochondria.

Reagents and equipment

- Mouse ES cybrids following electrofusion (*Protocol 5*)
- Feeder layers in multi-wells or dishes
- Mouse ES-cell medium (*Protocol 6*)
- HAT medium: mES-cell medium supplemented with 100 µM hypoxanthine, 0.4 µM aminopterin, and 16 µM thymidine (Cat. No. H-0262, Sigma)[a]
- TK$^-$ selection-medium: mES-cell medium supplemented with 50 µg/ml 2,5-bromo-deoxyuridine (BrdU; Cat. No. 858811, Sigma)
- TG-selection medium: mES-cell medium supplemented with 10 µg/ml 6-thioguanine (6TG; Cat. No. A-4660, Sigma)
- CAP-selection medium: mES-cell medium supplemented with 50 µg/ml chloramphenicol (Cat. No. C-7795, Sigma)
- GUP medium (*Protocol 2*).

Method

1. Plate fused mES cybrids onto fresh feeder layers. If selecting against HPRT$^-$ or TK$^-$ donor cells, culture the cybrids in HAT medium. Include supplementation with pyruvate and uridine for the first 24 h postfusion only.

2. Individual mES-cell colonies should become visible by day 5 after fusion, and can be picked for further expansion and analysis between days 7 and 9.[b]

[a] In older publications, HAT medium was supplemented with 10 µM glycine as conversion of serine to glycine involves DHFR, which is inhibited by aminopterin. However, as FCS is extremely rich in glycine, this supplementation is unnecessary.

[b] As mES cells divide rapidly it is important to monitor growth on a daily basis, and to pick colonies for expansion before they become too large and differentiated. The morphology of female mES cybrid colonies of different sizes is shown in *Figure 5*. Note the differentiation that occurs at the periphery of colonies with over-growth (*Figure 5C*).

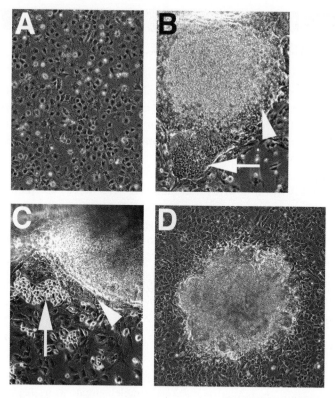

Figure 5 Morphology of mES cybrids. (A) Morphology of SNL76/7 feeder cells 1 week after plating. (B) A large and clonal mES cybrid colony. Note that morphology remains typical of mES cells, including the multilayered, high density of cells in the cybrid colony, and the outgrowth of cells at the periphery (arrowhead and arrow, respectively). (C) Spontaneous differentiation of mES cybrids, at the periphery of an overgrown colony, into primitive endoderm cells (arrow) with concomitant deposition of Reichert's membrane-like ECM (arrowhead). (D) An optimally sized mES cybrid colony for subcloning. Note the lack of differentiation around the periphery of this colony. Magnifications (A–C) 100×; (D) 50×.

2.9 PCR-based allele genotyping of the nuclear genome of cybrids

When generating mouse cybrid lines, it is useful to be able to discriminate between the nuclear genomes (nDNA) of the two cell types used to construct the cybrids. This is especially important when drugs or other compounds cannot be used in the culture medium to select for cybrids postfusion, for example when the recipient cells have been treated with R6G and concomitantly have a finite probability of escaping the mitochondrial inactivation and forming non-cybrid colonies. In the case of mES cybrids, it is also useful to be able to monitor for the presence of contaminating feeder-cell nDNA. This can be done using simple PCR assays involving two independent microsatellite markers that together permit identification of nDNA of strain 129 mES cells, either SNL76/7 or STO feeder cells, and the LA9 fibroblast cell line.

Protocol 8
Genotyping the nDNA of cybrids

Reagents and equipment

- Genomic DNA extracted from clonal, cybrid lines[a]
- PCR thermocycler
- Suitable thermostable DNA polymerase
- PCR buffer (1X): 16 mM $(NH_4)_2SO_4$, 66 mM Tris-HCl pH 8.8, 3 mM $MgCl_2$, 0.05% Tween-20
- Agarose- or polyacrylamide-gel electrophoresis apparatus
- Mouse microsatellite primer pairs D3Mit256 and D8Mit289 (Table 2), purchasable from Invitrogen

Method

1. Perform a PCR using: 5–20 ng genomic DNA, 0.25 µM each primer, and 0.8 mM dNTPs.
2. PCR cycle conditions are: 94 °C, 2 min; followed by 35 cycles of 94 °C for 30 s, 57 °C for 45 s, and 72 °C for 1 min; with a final extension period at 72 °C for 3 min.
3. Separate reaction products using electrophoresis through a 3.5% agarose gel (1.75% each NuSieve and SeaKem agaroses (from FMC, or equivalent) or a 10% non-denaturing polyacrylamide gel. The sizes of amplification products are listed in the Table 2.

[a] We routinely use a negative control of no DNA, and positive controls of pure DNA from

Table 2 Expected sizes (in base pairs) of PCR products obtained using specified microsatellite markers and mouse DNA

Primer pair	Sequences (5' to 3', F_w/R_v)	Strain 129	Strain C57BL/6	Feeder cells (SNL76/7 or STO)	LA9 fibroblasts
D3Mit256	TACATTGCTTTTTGCTTTGA GTCGAATGTTATCAGAATTTGCA	195	–	274	274
D8Mit289	AAAAAGAAAAGAAGGCTTAGTAATGTG CTTGCTATTCATTGCAAAATTCC	142	–	142	103
D10Mit10	CCAGTCTCAAAACAACAACAAAA TTGCACCTAGATTGCCTGA	128	180	–	–

2.10 Use of X-linked GFP transgenic mice to select female blastocysts for microinjection with *trans*-mitochondrial mES cells

Injection of male (XY) mES cells into a population of mouse blastocysts of presumptively equally mixed sex usually produces a large bias towards male offspring, as colonization of the indifferent gonad in an XX embryo by XY mES cells is sufficient to convert the sex of that embryo from female to male. However, the converse does not generally occur, that is XX mES cells injected into an XY blastocyst rarely induces sex conversion. As they lack a Y chromosome, derivatives of XX mES cells cannot produce spermatozoa. Moreover, as mitochondria are inherited maternally, male chimeras fail to transmit mitochondrial genomes to their offspring. Thus, without prior selection of blastocysts only half of the chimeric mice generated by injection of XX mES cells would be expected to be female, and therefore to be of use in establishing lines of mice with *trans*-mitochondrial genomes. Thus, the ability to select XX blastocysts for microinjection doubles the efficiency of production of female chimeras. This can be accomplished efficiently by harvesting blastocysts from female mice that have mated with males transgenic for an X-linked green fluorescence protein (*GFP*) reporter gene that is active in somatic tissues (86).

2.11 PCR-based allele genotyping of mice

Injection of 129-derived mES cells into C57BL/6 embryos produces chimeras that are identifiable by agouti coat colouration on a black background. This method can be used to assess the relative contribution of mES cells to a chimera from around 7 days postnatal.

We have generated chimeric mice using mES cells that have a mutation in the mitochondrial genome, which confers CAP^R to cells in culture. Many of these

Protocol 9
Selection of female blastocysts for microinjection

Reagents and equipment

- Male transgenic mice of line Tg(GFPX)4Nagy/J, which carry an X-linked GFP-expressing transgene (Stock Number 003116, Jackson Laboratory)
- Strain C57BL/6 female mice, 8–12 weeks old, for blastocyst production
- Suitable microscope equipped with fluorescence (FITC or GFP filter sets) and phase/Nomarski/Hoffman optics, with which to screen blastocysts
- Flushing medium (e.g. serum-free DMEM with 20 mM Hepes, pH 7.4)

- Embryo culture medium, without phenol red to enable visualization of GFP activity (e.g. BMOC-3; Cat. No. 11126–034, Invitrogen)
- Sterile, 50 mm glass-bottom dishes (Cat. No. P50G-1.5-14-F, MatTek Corporation)
- Gelatinized culture dish minus feeder cells
- PCR thermocycler, plus reagents for conducting PCR
- Optional–GFP-goggles, which may be homemade or commercially purchased (Cat. No. GFsP-5, BLS Ltd)

Methods

Maintenance of Tg(GFPX)4Nagy/J line of mice:
This line is conveniently maintained by mating hemizygous GFP females ($X^{GFP}X^m$) with non-transgenic males (X^PY). Female ($X^{GFP}X^P$) offspring are used to propagate the

Protocol 9 continued

transgenic line, while male ($X^{GFP}Y$) offspring are used as stud males for production of host $X^{GFP}X^m$ blastocysts. All offspring are screened for the presence of the X-linked GFP transgene by PCR (see below). Alternatively, GFP-positive animals can be identified prior to development of fur by illuminating the pups with a blue-light source and viewing them with appropriate emission filters for enhanced GFP. The line is maintained on strain C57BL/6, which facilitates eventual analysis of chimerism from the extent of 129-derived agouti coat colour on a black coat-colour background.

PCR-based genotyping of X-linked GFP transgene:
The following PCR assay gives two products, one specific for the GFP transgene and the other for a region of the interleukin-2 gene. The latter amplification product serves as an internal control for the PCR reaction. The protocol was obtained from the Jackson Laboratory website (http://jaxmice.jax.org/pub-cgi/protocols).

IMR0042	CTAGGCCACAGAATTGAAAGATCT
IMR0043	GTAGGTGGAAATTCTAGCATCATCC
IMR0872	AAGTTCATCTGCACCACCG
IMR1416	TCCTTGAAGAAGATGGTGCG

Primers IMR0042 and IMR0043 amplify a 342 bp product from the IL-2 gene.
Primers IMR0872 and IMR1416 amplify a 173 bp product from the GFP transgene.
PCR conditions: 2 mM $MgCl_2$, 0.2 mM dNTPs, 1.0 μm each primer, 0.5 U Taq polymerase, 5–20 ng DNA template.
PCR cycling information: following initial denaturation, conduct 35 cycles of 94 °C for 30 s, 60 °C for 30 s; and 72 °C for 30 s.

Isolation of $X^{GFP}X^m$ blastocysts for microinjection:

1. Mate $X^{GFP}Y$ males with XX females to generate $X^{GFP}X$ and XY blastocysts.

2. Recover blastocysts from pregnant females at 3.5 d p.c. (87, 88), suspend in medium lacking phenol red, place in a glass-bottom dish, and visualize GFP activity by fluorescence microscopy (*Figure 6*).

3. The GFP fluorescence in $X^{GFP}X$ embryos is strong within the epiblast, but reduced in the trophectoderm as a result of inactivation of the paternal X chromosome (89). Use only the green-fluorescent (i.e. female) blastocyts for injection with mES cells.

Microinjection of $X^{GFP}X^m$, expanded blastocysts:

5. Prepare mES cells for microinjection by preplating onto a gelatinized dish minus feeder cells 1 h prior to injection, to select for healthy cells (87, 88). Discard supernatant cells, harvest loosely adherent cells by gentle washing with medium, and select from this fraction mES cells that are small and have a smooth appearance for microinjection, using 10–15 cells per blastocyst.

6. Transfer successfully injected blastocysts into the uterus of day 2.5 pseudopregnant recipients (87, 88).

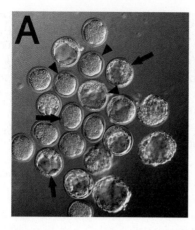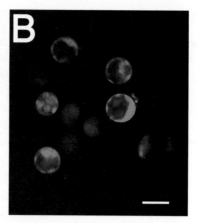

Figure 6 (see Plate 2) Identification of female Tg(GFPX)4Nagy/J preimplantation embryos. Hemizygous TgN(GFPX)4Nagy male mice were mated with C57BL/6 females, and preimplantation embryos were recovered at E2.5 and E3.5. With this cross only female embryos will inherit the X-linked GFP transgene from the male parent. (A) DIC image of a mixture of morulae and blastocysts. Several GFP-positive ($X^{pGFP}X^m$) female embryos are indicated by an arrow, while GFP-negative (X^mY) embryos are denoted by an arrowhead. (B) Corresponding image of the same embryos examined using GFP-optics. (Scale bar = 100 μM).

chimeras were severely runted, and died during embryogenesis or as neonates (70). This necessitated development of a straightforward method to identify chimeras that was independent of coat colouration. We accomplish this by PCR assay for the *D10Mit10* microsatellite marker that discriminates nDNA of strains C57BL/6 and 129 mice (*Protocol 10*).

3 Phenotypic analysis of *trans*-mitochondrial, chimeric mice

In our experience, mice harbouring mitochondrial defects due to either nDNA or mtDNA mutations exhibit many of the same phenotypic, physiological and biochemical abnormalities as seen in patients with mitochondrial disease. Hence, we analyse these animals comprehensively for metabolic, ophthalmological, neurological, and exercise physiology defects.

(a) *Physiological status:* The viability and behavioural characteristics of the mice are reviewed daily during cage surveys. Serum and urine are collected and analysed for organic acids, amino acids, and carnitine and acyl-carnitine derivatives using standard laboratory procedures (90–92).

(b) *Ophthalmological and neurological evaluation:* Mice are evaluated for potential phenotypic effects on the visual, auditory, motor, and cognitive systems. Ophthalmological analysis of mutant animals includes slit-lamp analysis (for cataracts), ophthalmoscopy, electroretinograms, and retinal histology (93, 94). Auditory evoked brainstem response (ABR) is recorded at 8, 16, and 32 kHz over a range of intensities, using vertex, ear, and dorsal

Protocol 10
PCR-based allele genotyping of mice

Reagents and equipment (see also Protocol 8)

- PCR thermocycler
- Suitable thermostable DNA polymerase and buffer (Protocol 8)
- Agarose or polyacrylamide gel electrophoresis apparatus
- D10Mit10 primer pair (5' to 3', F_w/R_v) CCAGTCTCAAAACAACAACAAAA TTGCACCTAGATTGCCTGA

Method

1. Extract total genomic DNA from a mouse tail biopsy. Use a negative control of no DNA, plus positive controls consisting of a mixture of strains 129 and C57BL/6 genomic DNAs in different ratios.[a]

2. Perform PCR using reaction components and cycling conditions given in *Protocol 8*, Step 1; and analyse separate reaction products as in *Protocol 8* Step 2. The size of the amplification product of the strain 129 allele is 128 bp, and of the strain C57BL/6 allele is 180 bp (*Table 2*).

[a] We have also had success in isolating total nucleic acids from tail biopsies using Direct PCR Lysis Reagent (ViaGene Biotech). This offers significant savings on labour and time as phenol-chloroform extractions and ethanol precipitation are not required.

electrodes (94). Ataxias and motor disturbances are detected by the rotor-rod ataxia test (95), activity levels are monitored using the open-field activity test (96), and learning and memory are monitored using the Morris water task (97).

(c) *Exercise physiology*: Mitochondrial physiology and cardiac function are evaluated by a non-invasive exercise stress test. Mice are exercised on an enclosed, variable speed and angle treadmill system (Columbus Instruments) using a constant-flow atmosphere of known composition. The VO_2 and VCO_2 of the outflow are monitored using paramagnetic sensors; and when the CO_2 exhaled begins to rise rapidly, this indicates that the mouse has undergone the transition from aerobic to anaerobic exercise, the 'anaerobic threshold' Differences in anaerobic threshold provide an effective indicator of relative defects in mitochondrial energy production, as the transition occurs when the animal's muscle must switch from primary use of the Type I oxidative to the Type II glycolytic muscle fibres (4, 98).

(d) *General and cellular pathology*: The organs of the normal and mutant mice are characterized both morphologically and histochemically. Formalin-fixed sections are stained with haematoxilin and eosin for light microscopy.

Isopentane-frozen sections are histochemically stained for SDH, COX, NADH-tetrazolium, acidic and basic ATPase, Gomori Modified Trichrome, and Oil-Red O. Glutaraldehyde-fixed samples are analysed by electron microscopy (99). Finally, apoptotic cells are detected in tissues by TUNEL (terminal-deoxynucleotidyltransferase (TdT)-mediated dUTP-biotin nick end labelling) staining (Roche).

(e) *Mitochondrial functional assays:* Mitochondria are isolated from tissues by differential centrifugation. The specific activities of the OXPHOS enzymes, respiration rates, $\Delta\psi$, H_2O_2 production, and mtPTP are measured to assess mitochondrial biochemical function (78).

Assays for all of the ETC complexes are routinely performed using kinetically optimized enzyme protocols (100–102, 78). Mitochondrial respiration is analysed using a Clark oxygen electrode and Yellow Springs Monitoring System. NADH-linked (site I) and FAD-linked (site II) substrates are added as substrates to respiring mitochondria, with or without ADP. Uncoupled respiration is measured after addition of DNP (2,4-dinitrophenol) or FCCP (78). The $\Delta\psi$ is measured by the uptake of tetraphenyl phosphonium cation (TPP+) using a TPP+ sensitive electrode (103–105). Mitochondrial ROS generation is assessed by measuring the mitochondrial release of H_2O_2 through the coupled oxidation of p-hydroxyphenylacetate (PHPA) (106). The mtPTP is assessed by monitoring mitochondrial swelling or the loss of mitochondrial membrane potential (ΔpH). The mtPTP transition can be initiated in response to Ca^{2+} challenge, uncoupling, oxidative stress, and other inducers, and can be modulated by a variety of effectors (66).

4 Concluding comments

Virtually all of the research on the genetics and the pathophysiology of human age-related disease has focused on Mendelian genes, and generally tissue-specific structural genes. However, by definition the age-related diseases (including the metabolic syndromes, neurodegenerative diseases, most cancers, and ageing itself) have a delayed onset and progressive course. That is, the individual starts with a normal genotype and phenotype, which gradually over time shift to abnormality. Therefore, all age-related diseases must be influenced by a genetic clock, which accumulates genetic changes over many years until there are insufficient non-defective genes to sustain a normal phenotype. This implies the quantitative accumulation of hundreds of mutations within the same critical cellular function. However, each Mendelian chromosomal gene is present in the cell in only two copies, giving only three genetic and biochemical states (+/+, +/−, and −/−). Hence, the accumulation of nDNA mutations cannot provide a credible ageing clock.

By contrast, the mtDNA is present in thousands of copies per cell, and each mtDNA encodes 37 genes that are essential for cellular energy production. Moreover, since the mitochondrion generates most of the cellular reactive oxygen species (ROS) as a toxic by-product of energy production, and these ROS damage the mtDNAs over the life of the individual, the slow erosion of mitochondrial energy production as a consequence of the accumulation of somatic mtDNA mutations can provide the needed genetic clock (30). This proposal is supported by the fact that many of the common clinical problems observed in the elderly and in the age-related degenerative diseases have been linked to various mtDNA mutations in family studies, including Type II diabetes (107–110) and the broader metabolic syndrome (111), blindness (1), myopathy and epilepsy (100, 112), deafness (113, 114), dementias including Alzheimer's and Parkinson's disease (115, 116), movement disorders including Parkinson's disease and dystonia (115, 117, 118), cardomyopathy (119, 120), and cancer (121). Furthermore, mitochondrial energy production has been shown to decline with age in association with the accumulation of somatic mtDNA mutations in the brain (122), heart (120), muscle (123, 124), skin fibroblasts (73), Alzheimer's diseased brains (125, 126), and Huntington's disease brains (127). Thus, there is now convincing evidence that the accumulation of somatic mtDNA mutations provides the ageing clock, and that mitochondrial defects underlie a wide spectrum of age-related degenerative diseases.

Initial studies where nDNA genes affecting mitochondrial function were targeted have provided direct experimental support for the concept that the accumulation of mtDNA mutations provides the ageing clock. The introduction in mice of a mutant mtDNA polymerase γ that lacked the normal proofreading capacity resulted in a decreased lifespan in association with the appearance of multiple, classical ageing phenotypes, and the greatly increased accumulation of somatic mtDNA mutations (2). Conversely, introduction of a mitochondrially targeted catalase gene into the mitochondria of mice increased the mean lifespan by about 20%, and the maximum life span by 10%, in association with reduced sensitivity to mitochondrial oxidative stress and a reduction in the age-related accumulation of mtDNA mutations (128). Thus, these mouse studies substantiate the conclusion that the ageing clock is, at least in part, the product of the accumulations of somatic mtDNA mutations.

However, to validate and further define the role of mtDNA mutations in the predisposition to age-related diseases, we need mouse models harbouring defined mtDNA mutations at different percentages of heteroplasmy. Mouse models of mtDNA diseases are essential: (a) to investigate the role of mtDNA-encoded energy defects in the age-related degenerative diseases; (b) for the development of new metabolic and genetic therapeutics for mtDNA diseases; (c) for investigation of the importance of uniparental inheritance of mtDNA; and (d) the analysis of the clinical implications of subverting uniparental inheritance during assisted reproduction and therapeutic cloning. Confirmation that mtDNA mutations are both necessary and sufficient to cause age-related disease was achieved by our production of the CAP^R mouse (18, 70). However, to determine the full range of phenotypes that can result from mtDNA deficiency we will need to generate mice harbouring a wide variety of mtDNA protein, tRNA, rRNA, and regulatory mutations, and then to age these mice to analyse the effects of the interplay between inherited and somatic mtDNA mutations. Once mouse lines are developed which harbour deleterious mtDNA mutations, these can be used as models to develop new metabolic (6, 129) and genetic therapeutics (69) for mtDNA disease.

The ability to mix mtDNAs from two different maternal mtDNA lineages, as we have achieved by combining the synaptosome mtDNAs from NZB mice with the germline mtDNAs from 129 mice (70), provides an unprecedented opportunity to investigate why nature has gone to such lengths to assure exclusively maternal inheritance of the mtDNA (31). One possibility (as the mtDNA encodes key genes for the electron transport and proton pumping for multiple respiratory complexes, and mutations in these genes can change the rates of electron flux and efficiency of proton pumping) is that recombination between very different mtDNA lineages could result in a high frequency of dysfunctional OXPHOS systems. By ensuring uniparental inheritance of the mtDNA, the only way that it can change is by sequential mutations such that each new mutation would be tested by selection on a coherently functional background. If the new mutations compromise that background, it will be quickly eliminated by natural selection as a genetic disease. If the mutation is advantageous, then it will flourish and be added to the integrated system.

If, as we expect, we find that the mixing of different mtDNAs within a female germline does result in new combinations which are deleterious, this will have major implications for certain current practices in assisted reproduction, and for therapeutic cloning. In an effort to increase the fecundity of the oocytes of older women during assisted reproduction, a practice has been introduced in which the cytoplasm with its mitochondria from a younger oocyte is injected into the cytoplasm of the older oocyte (130, 131). As of 2001, this 'ooplasm transfer' has been used in the conception and birth of nearly 30 infants worldwide. Analysis of the mtDNAs from two such 1-year-old children revealed that both were heteroplasmic for the donor and recipient mtDNAs (132). Thus, the human experiment is being conducted without any prior investigation of the implications of such procedures in a model mammalian system.

A similar concern arises with therapeutic cloning in which a somatic cell from an older donor is fused to an enucleated oocyte to generate human ES cells, for eventual cell or tissue transplantation. Unfortunately this procedure also mixes mtDNAs from different lineages, which could be problematic for several reasons. First, it has already been shown in the mouse that variant mtDNA-encoded polypeptides are displayed on the cell surface by the *H-2M3* (MHC) class I gene product, and can instigate an immune response in allotypic animals (133). Hence, the therapeutically cloned and transplanted cells could be subject to immune rejection by the patient to be treated. Second, the mtDNAs from the somatic cell have already aged, and thus are likely to harbour deleterious somatic mtDNA mutations that could compromise the cloned cells' longevity. Third, the two mtDNAs could recombine, which could result in the generation of novel, aberrant peptides as well as compromised OXPHOS function. Since one of the effects of compromised OXPHOS function is cancer, this could render the cloned cells pathogenic.

All of these concerns highlight the pressing need for the development of new mouse models to study mtDNA genetics. It is likely that the strategies we have presented for transferring mtDNAs between mice via mES cells will be useful for this purpose, as well as for a wide range of additional and novel biomedical applications.

Acknowledgements

The authors thank Ms Marie Lott for her assistance in preparation of the bibliography. This work has been supported by NIH grants HD45913 to GRM, and NS21438, NS41850, AG13154, and HL64017 to DCW, plus an Ellison Foundation Senior Investigator Award to DCW.

References

1. Wallace, D.C., Singh, G., Lott, M.T., Hodge, J.A., Schurr, T.G., Lezza, A.M., Elsas, L.J., and Nikoskelainen, E.K. (1988). *Science*, **242**, 1427–30.
2. Trifunovic, A., Wredenberg, A., Falkenberg, M., Spelbrink, J.N., Rovio, A.T., Bruder, C.E., Bohlooly, Y.M., Gidlof, S., Oldfors, A., Wibom, R., Tornell, J., Jacobs, H.T., and Larsson, N.G. (2004). *Nature*, **429**, 417–23.
3. Li, Y., Huang, T.T., Carlson, E.J., Melov, S., Ursell, P.C., Olson, J.L., Noble, L.J., Yoshimura, M.P., Berger, C., Chan, P.H., Wallace, D.C., and Epstein, C.J. (1995). *Nat. Genet.*, **11**, 376–81.
4. Graham, B.H., Waymire, K.G., Cottrell, B., Trounce, I.A., MacGregor, G.R., and Wallace, D.C. (1997). *Nat. Genet.*, **16**, 226–34.
5. Larsson, N.G., Wang, J., Wilhelmsson, H., Oldfors, A., Rustin, P., Lewandoski, M., Barsh, G.S., and Clayton, D.A. (1998). *Nat. Genet.*, **18**, 231–6.
6. Melov, S., Schneider, J.A., Day, B.J., Hinerfeld, D., Coskun, P., Mirra, S.S., Crapo, J.D., and Wallace, D.C. (1998). *Nat. Genet.*, **18**, 159–63.
7. Melov, S., Coskun, P., Patel, M., Tunistra, R., Cottrell, B., Jun, A.S., Zastawny, T.H., Dizdaroglu, M., Goodman, S.I., Huang, T., Miziorko, H., Epstein, C.J., and Wallace, D.C. (1999). *Proc. Natl. Acad. Sci. USA*, **96**, 846–51.
8. Esposito, L.A., Kokoszka, J.E., Waymire, K.G., Cottrell, B., MacGregor, G.R., and Wallace, D.C. (2000). *Free Radical Biol. Med.*, **28**, 754–66.
9. Wallace, D.C., Brown, M.D., and Lott, M.T. (1996). Mitochondrial genetics. In *Emery and Rimoin's Principles and Practice of Medical Genetics*, Vol. 1 (eds D.L. Rimoin, J.M. Connor, R.E. Pyeritz, A.E.H. Emery), pp. 277–332, Churchill Livingstone, London.
10. Wallace, D.C. (1999). *Science*, **283**, 1482–8.
11. Pinkert, C.A., Irwin, M.H., Johnson, L.W., and Moffatt, R.J. (1997). *Transgenic Res.*, **6**, 379–83.
12. Jenuth, J.P., Peterson, A.C., Fu, K., and Shoubridge, E.A. (1996). *Nat. Genet.*, **14**, 146–51.
13. Jenuth, J.P., Peterson, A.C., and Shoubridge, E.A. (1997). *Nat. Genet.*, **16**, 93–5.
14. Inoue, K., Nakada, K., Ogura, A., Isobe, K., Goto, Y., Nonaka, I., and Hayashi, J.-I. (2000). *Nat. Genet.*, **26**, 176–81.
15. Blanc, H., Adams, C.W., and Wallace, D.C. (1981). *Nucleic Acids Res.*, **9**, 5785–95.
16. Kearsey, S.E. and Craig, I.W. (1981). *Nature*, **290**, 607–8.
17. Watanabe, T., Dewey, M.J., and Mintz, B. (1978). *Proc. Natl. Acad. Sci. USA*, **75**, 5113–7.
18. Levy, S.E., Waymire, K.G., Kim, Y.L., MacGregor, G.R., and Wallace, D.C. (1999). *Transgenic Res.*, **8**, 137–45.
19. Marchington, D.R., Barlow, D., and Poulton, J. (1999). *Nat. Med.*, **5**, 957–60.

20. Ziegler, M.L. and Davidson, R.L. (1981). *Somatic Cell Genet.*, **7**, 73-88.
21. Trounce, I. and Wallace, D.C. (1996). *Somat. Cell Mol. Genet.*, **22**, 81-5.
22. Sligh, J.E., Levy, S.E., Allard, P.M., Waymire, K.G., MacGregor, G.R., Heckenlively, J.R., and Wallace, D.C. (2000). *Am. J. Hum. Genet.*, **67**, A19.
23. Howell, N., Bantel, A., and Huang, P. (1983). *Somatic Cell Genet.*, **9**, 721-43.
24. Howell, N., Huang, P., Kelliher, K., and Ryan, M.L. (1983). *Somatic Cell Genet.*, **9**, 143-63.
25. Howell, N., Appel, J., Cook, J.P., Howell, B., and Hauswirth, W.W. (1987). *J. Biol. Chem.*, **262**, 2411-14.
26. Howell, N. and Gilbert, K. (1988). *J. Mol. Biol.*, **203**, 607-18.
27. Howell, N. (1990). *Biochem.*, **29**, 8970-7.
28. Hofhaus, G. and Attardi, G. (1995). *Mol. Cell. Biol.*, **15**, 964-74. [Published erratum appears in *Mol. Cell. Biol.*, (1995) **15**, 3461.
29. Trounce, I., Schmiedel, J., Yen, H.C., Hosseini, S., Brown, M.D., Olson, J.J., and Wallace, D.C. (2000). *Nucl. Acids Res.*, **28**, 2164-70.
30. Wallace, D.C. and Lott, M.T. (2002). Mitochondrial genes in degenerative diseases, cancer and aging. In *Emery and Rimoin's Principles and Practice of Medical Genetics*, 4th edn (eds D.L. Rimoin, J.M. Connor, R.E. Pyeritz, B.R. Korf), pp. 299-409, Churchill Livingstone, London.
31. Giles, R.E., Blanc, H., Cann, H.M., and Wallace, D.C. (1980). *Proc. Natl. Acad. Sci, USA*, **77**, 6715-9.
32. Sutovsky, P., Moreno, R.D., Ramalho-Santos, J., Dominko, T., Simerly, C., and Schatten, G. (1999). *Nature*, **402**, 371-2.
33. Shitara, H., Kaneda, H., Sato, A., Iwasaki, K., Hayashi, J., Taya, C., and Yonekawa, H. (2001). *FEBS Lett.*, **500**, 7-11.
34. Sutovsky, P., McCauley, T.C., Sutovsky, M., and Day, B.N. (2003). *Biol. Reprod.*, **68**, 1793-800.
35. Thompson, W.E., Ramalho-Santos, J., and Sutovsky, P. (2003). *Biol. Reprod.*, **69**, 254-60.
36. Sutovsky, P., Van Leyen, K., McCauley, T., Day, B.N., and Sutovsky, M. (2004). *Reproductive Biomedicine Online*, **8**, 24-33.
37. Bunn, C.L., Wallace, D.C., and Eisenstadt, J.M. (1974). *Proc. Natl. Acad, Sci, USA*, **71**, 1681-5.
38. Wallace, D.C., Bunn, C.L., and Eisenstadt, J.M. (1975). *J. Cell Biol.*, **67**, 174-88.
39. Wallace, D.C. and Eisenstadt, J.M. (1979). *Somatic Cell Genet.*, **5**, 373-96.
40. Wallace, D.C. (1981). *Mol. Cell. Biol.*, **1**, 697-710
41. Oliver, N.A., Greenberg, B.D., and Wallace, D.C. (1983). *J. Biol. Chem.*, **258**, 5834-9.
42. Blanc, H., Wright, C.T., Bibb, M.J., Wallace, D.C., and Clayton, D.A. (1981). *Proc. Natl. Acad. Sci. USA*, **78**, 3789-93.
43. Wallace, D.C., Pollack, Y., Bunn, C.L., and Eisenstadt, J.M. (1976). *In Vitro*, **12**, 758-76.
44. Bunn, C.L., Wallace, D.C., and Eisenstadt, J.M. (1977). *Somatic Cell Genet.*, **3**, 71-92.
45. Wallace, D.C., Bunn, C.L., and Eisenstadt, J.M. (1977). *Somatic Cell Genet.*, **3**, 93-119.
46. Giles, R.E., Stroynowski, I., and Wallace, D.C. (1980). *Somatic Cell Genet.*, **6**, 543-54.
47. Kenyon, L. and Moraes, C.T. (1997). *Proc. Natl. Acad. Sci. USA*, **94**, 9131-5.
48. Dey, R., Barrientos, A., and Moraes, C.T. (2000). *J. Biol. Chem.*, **275**, 31520-27.
49. McKenzie, M. and Trounce, I. (2000). *J. Biol. Chem.*, **275**, 31514-9.
50. Oliver, N.A. and Wallace, D.C. (1982). *Mol. Cell. Biol.*, **2**, 30-41.
51. Wallace, D.C. (1982) Cytoplasmic inheritance of chloramphenicol resistance in mammalian cells. In *Techniques in Somatic Cell Genetics* (ed. J.W. Shay), pp. 159-87, Plenum Press, New York.
52. King, M.P. and Attardi, G. (1989). *Science*, **246**, 500-3.
53. Ferris, S.D., Sage, R.D., and Wilson, A.C. (1982). *Nature*, **295**, 163-5.

54. Guenet, J.L. and Bonhomme, F. (2003). *Trends Genet.*, **19**, 24–31.
55. McKenzie, M., Trounce, I.A., Cassar, C.A., and Pinkert, C.A. (2004). *Proc. Natl. Acad. Sci. USA*, **101**, 1685–90.
56. Battersby, B.J. and Shoubridge, E.A. (2001). *Hum. Mol. Genet.*, **10**, 2469–79.
57. Battersby, B.J., Loredo-Osti, J.C., and Shoubridge, E.A. (2003). *Nat. Genet.*, **33**, 183–6.
58. Roubertoux, P.L., Sluyter, F., Carlier, M., Marcet, B., Maarouf-Veray, F., Cherif, C., Marican, C., Arrechi, P., Godin, F., Jamon, M., Verrier, B., and Cohen-Salmon, C. (2003). *Nat. Genet.*, **35**, 65–9.
59. Bai, Y., Shakeley, R.M., and Attardi, G. (2000). *Mol. Cell. Biol.*, **20**, 805–15.
60. Bai, Y. and Attardi, G. (1998). *EMBO J.*, **17**, 4848–58.
61. Acin-Perez, R., Bayona-Bafaluy, M.P., Fernandez-Silva, P., Moreno-Loshuertos, R., Perez-Martos, A., Bruno, C., Moraes, C.T., and Enriquez, J.A. (2004). *Mol. Cell*, **13**, 805–15.
62. Acin-Perez, R., Bayona-Bafaluy, M.P., Bueno, M., Machicado, C., Fernandez-Silva, P., Perez-Martos, A., Montoya, J., Lopez-Perez, M.J., Sancho, J., and Enriquez, J.A. (2003). *Hum. Mol. Genet.*, **12**, 329–39.
63. Longley, M.J., Ropp, P.A., Lim, S.E., and Copeland, W.C. (1998). *Biochemistry*, **37**, 10529–39.
64. Spelbrink, J.N., Toivonen, J.M., Hakkaart, G.A., Kurkela, J.M., Cooper, H.M., Lehtinen, S.K., Lecrenier, N., Back, J.W., Speijer, D., Foury, F., and Jacobs, H.T. (2000). *J. Biol. Chem.*, **275**, 24818–28.
65. Esposito, L.A., Melov, S., Panov, A., Cottrell, B.A., and Wallace, D.C. (1999). *Proc. Natl. Acad. Sci. USA*, **96**, 4820–5.
66. Kokoszka, J.E., Waymire, K.G., Levy, S.E., Sligh, J.E., Cai, J., Jones, D.P., MacGregor, G.R., and Wallace, D.C. (2004). *Nature*, **427**, 461–5.
67. Kokoszka, J.E., Coskun, P., Esposito, L., and Wallace, D.C. (2001). *Proc. Natl. Acad. Sci. USA*, **98**, 2278–83.
68. Chinnery, P.F. (2004). *Reproductive Biomedicine Online*, **8**, 16–23.
69. Flierl, A., Jackson, C., Cottrell, B., Murdock, D., Seibel, P., and Wallace, D.C. (2003). *Mol. Therapy*, **7**, 550–7.
70. Sligh, J.E., Levy, S.E., Waymire, K.G., Allard, P., Dillehay, D.L., Nusinowitz, S., Heckenlively, J.R., MacGregor, G.R., and Wallace, D.C. (2000). *Proc. Natl. Acad. Sci. USA*, **97**, 14461–6.
71. Shoffner, J.M., Lott, M.T., Voljavec, A.S., Soueidan, S.A., Costigan, D.A., and Wallace, D.C. (1989). *Proc. Natl. Acad. Sci. USA*, **86**, 7952–6.
72. Michikawa, Y., Hofhaus, G., Lerman, L.S., and Attardi, G. (1997). *Nucl. Acids Res.*, **25**, 2455–63.
73. Michikawa, Y., Mazzucchelli, F., Bresolin, N., Scarlato, G., and Attardi, G. (1999). *Science*, **286**, 774–9.
74. Wong, L.J., Liang, M.H., Kwon, H., Park, J., Bai, R.K., and Tan, D.J. (2002). *Clin. Chem.*, **48**, 1901–2.
75. O'Donovan, M.C., Oefner, P.J., Roberts, S.C., Austin, J., Hoogendoorn, B., Guy, C., Speight, G., Upadhyaya, M., Sommer, S.S., and McGuffin, P. (1998). *Genomics*, **52**, 44–9.
76. Oefner, P.J. and Underhill, P.A. (1998). DNA mutation detection using denaturing high performance liquid chromatography (DHPLC). In *Current Protocols in Human Genetics*, Vol. 19 (eds N.C. Dracopoli, J.L. Haines, B.R. Korf, D.T. Moire, C.C. Morton, C.C. Seidman, J.G. Sidman, D.R. Smith), pp. 7.10.1–2, John Wiley and Son, New York.
77. Jones, A.C., Austin, J., Hansen, N., Hoogendoorn, B., Oefner, P.J., Cheadle, J.P., and O'Donovan, M.C. (1999). *Clin Chem.*, **45**, 1133–40.
78. Trounce, I.A., Kim, Y.L., Jun, A.S., and Wallace, D.C. (1996). *Methods Enzymol.*, **264**, 484–509.
79. Levy, S.E., Chen, Y., Graham, B.H., and Wallace, D.C. (2000). *Gene*, **254**, 57–66.
80. Kit, S., Dubbs, D.R., Piekarski, L.J., and Hsu, T.S. (1963). *Exp. Cell Res.*, **31**, 297–312.

81. Littlefield, J.W. (1964). *Nature*, **203**, 1142-4.
82. Wigler, M.H. and Weinstein, I.B. (1975). *Biochem. Biophys. Res. Commun.*, **63**, 669-74.
83. Dunkley, P.R., Heath, J.W., Harrison, S.M., Jarvie, P.E., Glenfield, P.J., and Rostas, J.A. (1988). *Brain Research*, **441**, 59-71.
84. Koopman, P., Gubbay, J., Vivian, N., Goodfellow, P., and Lovell-Badge, R. (1991). *Nature*, **351**, 117-21.
85. McMahon, A.P. and Bradley, A. (1990). *Cell*, **62**, 1073-85.
86. Wang, J., Mager, J., Chen, Y., Schneider, E., Cross, J.C., Nagy, A., and Magnuson, T. (2001). *Nat. Genet.*, **28**, 371-5.
87. Hogan, B., Beddington, R., Constantini, F., and Lacy, E. eds (1994). *Manipulating the Mouse Embryo, a Laboratory Manual*. Cold Spring Harbor Laboratory Press, Cold Spring Harbor.
88. Nagy, A., Gertsenstein, M., Vinterson, K., and Behringer, R., eds (2003). *Manipulating the Mouse Embryo, a Laboratory Manual*. Cold Spring Harbor Laboratory Press, Cold Spring Harbor.
89. Hadjantonakis, A.K., Cox, L.L., Tam, P.P., and Nagy, A. (2001). *Genesis*, **29**, 133-40.
90. Hommes, F.A., ed. (1991). *Techniques in Diagnostic Human Biochemical Genetics*. A Laboratory Manual. Wiley-Liss, New York.
91. Roberts, J.R., Narasimhan, C., Hruz, P.W., Mitchell, G.A., and Miziorko, H.M. (1994). *J. Biol. Chem.*, **269**, 17841-6.
92. Mitchell, G.A., Ozand, P.T., Robert, M.F., Ashmarina, L., Roberts, J., Gibson, K.M., Wanders, R.J., Wang, S., Chevalier, I., Plochl, E., and Miziorko, H. (1998). *Am. J. Hum. Genet.*, **62**, 295-300.
93. Heckenlively, J.R., Winston, J.V., and Roderick, T.H. (1989). *Doc. Ophthalmol.*, **71**, 229-39.
94. Heckenlively, J.R., Chang, B., Erway, L.C., Peng, C., Hawes, N.L., Hageman, G.C., and Roderick, T.H. (1995). *Proc. Natl. Acad. Sci. USA*, **92**, 11100-4.
95. Miller, L.G., Greenblatt, D.J., Barnhill, J.G., and Shader, R.I. (1988). *J. Pharmacol. Exp. Ther.*, **246**, 170-6.
96. Miller, L.G., Galpern, W.R., Byrnes, J.J., Greenblatt, D.J., and Shader, R.I. (1992). *J. Pharm. Exper. Therap.*, **261**, 285-9.
97. Owen, E.H., Logue, S.F., Rasmussen, D.L., and Wehner, J.M. (1997). *Neuroscience*, **80**, 1087-99.
98. DeRose, J.J., Jr, Banas, J.S., Jr, and Winters, S.L. (1994). *Prog. Cardiovasc. Dis.*, **36**, 475-84.
99. Dubowitz, V., Sewry, C.A., and Fitzsimons, R.B., eds (1985). *Muscle Biopsy: A Practical Approach*. Baillière Tindall, London.
100. Wallace, D.C., Zheng, X., Lott, M.T., Shoffner, J.M., Hodge, J.A., Kelley, R.I., Epstein, C.M., and Hopkins, L.C. (1988). *Cell*, **55**, 601-10.
101. Zheng, X., Shoffner, J.M., Voljavec, A.S., and Wallace, D.C. (1990). *Biochim. Biophys. Acta*, **1019**, 1-10.
102. Trounce, I., Neill, S., and Wallace, D.C. (1994). *Proc. Natl. Acad. Sci. USA*, **91**, 8334-8.
103. Kamo, N., Muratsugu, M., Hongoh, R., and Kobatake, Y. (1979). *J. Membr. Biol.*, **49**, 105-21.
104. Rottenberg, H. (1984). *J. Membr. Biol.*, **81**, 127-38.
105. LaNoue, K.F., Jeffries, F.M., and Radda, G.K. (1986). *Biochemistry*, **25**, 7667-75.
106. Kwong, L.K. and Sohal, R.S. (1998). *Arch. Biochem. Biophys.*, **350**, 118-26.
107. Ballinger, S.W., Shoffner, J.M., Hedaya, E.V., Trounce, I., Polak, M.A., Koontz, D.A., and Wallace, D.C. (1992). *Nat. Genet.*, **1**, 11-5.
108. Van den Ouweland, J.M., Lemkes, H.H.P., Ruitenbeek, W., Sandkjujl, L.A., deVijlder, M.F., Struyvenberg, P.A.A., van de Kamp, J.J.P., and Maassen, J.A. (1992). *Nat. Genet.*, **1**, 368-71.

109. Ballinger, S.W., Shoffner, J.M., Gebhart, S., Koontz, D.A., and Wallace, D.C. (1994). *Nat. Genet.*, **7**, 458–9.
110. van den Ouweland, J.M., Maechler, P., Wollheim, C.B., Attardi, G., and Maassen, J.A. (1999). *Diabetologia*, **42**, 485–92.
111. Wilson, F.H., Hariri, A., Farhi, A., Zhao, H., Petersen, K.F., Toka, H.R., Nelson-Williams, C., Raja, K.M., Kashgarian, M., Shulman, G.I., Scheinman, S.J., and Lifton, R.P. (2004). *Science*, **306**, 1190–4.
112. Shoffner, J.M., Lott, M.T., Lezza, A.M., Seibel, P., Ballinger, S.W., and Wallace, D.C. (1990). *Cell*, **61**, 931–7.
113. Hutchin, T., Haworth, I., Higashi, K., Fischel-Ghodsian, N., Stoneking, M., Saha, N., Arnos, C., and Cortopassi, G. (1993). *Nucl. Acids Res.*, **21**, 4174–9.
114. Prezant, T.R., Agapian, J.V., Bohlman, M.C., Bu, X., Oztas, S., Qiu, W.Q., Arnos, K.S., Cortopassi, G.A., Jaber, L., Rotter, J.I., Shohat, M., and Fischel-Ghodsian, N. (1993). *Nat. Genet.*, **4**, 289–94.
115. Shoffner, J.M., Brown, M.D., Torroni, A., Lott, M.T., Cabell, M.R., Mirra, S.S., Beal, M.F., Yang, C., Gearing, M., Salvo, R., Watts, R.L., Juncos, J.L., Hansen, L.A., Crain, B.J., Fayad, M., Reckord, C.L., and Wallace, D.C. (1993). *Genomics*, **17**, 171–84.
116. Hutchin, T. and Cortopassi, G. (1995). *Proc. Natl. Acad. Sci. USA*, **92**, 6892–5.
117. Jun, A.S., Brown, M.D., and Wallace, D.C. (1994). *Proc. Natl. Acad. Sci. USA*, **91**, 6206–10.
118. Simon, D.K., Mayeux, R., Marder, K., Kowall, N.W., Beal, M.F., and Johns, D.R. (1999). *Neurology*, **54**, 703–9.
119. Corral-Debrinski, M., Stepien, G., Shoffner, J.M., Lott, M.T., Kanter, K., and Wallace, D.C. (1991). *JAMA*, **266**, 1812–6.
120. Corral-Debrinski, M., Shoffner, J.M., Lott, M.T., and Wallace, D.C. (1992). *Mutation Research*, **275**, 169–80.
121. Petros, J.A., Baumann, A.K., Ruiz-Pesini, E., Amin, M.B., Sun, C.Q., Hall, J., Lim, S., Issa, M.M., Flanders, W.D., Hosseini, S.H., Marshall, F.F., and Wallace, D.C. (2005). *Proc. Natl. Acad. Sci. USA*, **102**, 719–24.
122. Corral-Debrinski, M., Horton, T., Lott, M.T., Shoffner, J.M., Beal, M.F., and Wallace, D.C. (1992). *Nat. Genet.*, **2**, 324–9.
123. Murdock, D.G., Christacos, N.C., and Wallace, D.C. (2000). *Nucl. Acids Res.*, **28**, 4350–5.
124. Wang, Y., Michikawa, Y., Mallidis, C., Bai, Y., Woodhouse, L., Yarasheski, K.E., Miller, C.A., Askanas, V., Engel, W.K., Bhasin, S., and Attardi, G. (2001). *Proc. Natl. Acad. Sci. USA*, **98**, 4022–7.
125. Corral-Debrinski, M., Horton, T., Lott, M.T., Shoffner, J.M., McKee, A.C., Beal, M.F., Graham, B.H., and Wallace, D.C. (1994). *Genomics*, **23**, 471–6.
126. Coskun, P.E., Beal, M.F., and Wallace, D.C. (2004). *Proc. Natl. Acad. Sci. USA*, **101**, 10726–31.
127. Horton, T.M., Graham, B.H., Corral-Debrinski, M., Shoffner, J.M., Kaufman, A.E., Beal, B.F., and Wallace, D.C. (1995). *Neurology*, **45**, 1879–83.
128. Schriner, S.E., Linford, N.J., Martin, G.M., Treuting, P., Ogburn, C.E., Emond, M., Coskun, P.E., Ladiges, W., Wolf, N., Van Remmen, H., Wallace, D.C., and Rabinovitch, P.S. (2005). *Science*, **308**, 1909–11.
129. Melov, S., Doctrow, S.R., Schneider, J.A., Haberson, J., Patel, M., Coskun, P.E., Huffman, K., Wallace, D.C., and Malfroy, B. (2001). *J. Neurosci.*, **21**, 8348–53.
130. Cohen, J., Scott, R., Schimmel, T., Levron, J., and Willadsen, S. (1997). *Lancet*, **350**, 186–7.
131. Cohen, J., Scott, R., Alikani, M., Schimmel, T., Munne, S., Levron, J., Wu, L., Brenner, C., Warner, C., and Willadsen, S. (1998). *Hum. Reprod.*, **4**, 269–80.
132. Barritt, J.A., Brenner, C.A., Malter, H.E., and Cohen, J. (2001). *Hum. Reprod.*, **16**, 513–6.
133. Lindahl, K.F., Hermel, E., Loveland, B.E., and Wang CR (1991). *Annu. Rev. Immunol.*, **9**, 351–72.

Chapter 5
Controlling the differentiation of mouse ES cells *in vitro*

Michael V. Wiles[1] and Gabriele Proetzel[2]

[1]The Jackson Laboratory, 600 Main Street, Bar Harbor, Maine 04609–1500, USA; [2]Scil Proteins, Heinrich-Damerow-Str. 1, 06120 Halle, Germany.

1 Introduction

In this chapter we outline the concepts and practical considerations in understanding and developing defined culture systems for the controlled *in vitro* differentiation of mouse embryonic stem (mES) cells.

When looking at the history of mES-cell *in vitro* differentiation, one is struck by the level of alchemy that was introduced, and at times still is used, to obtain desired cell types. However, a fundamental difference between alchemy and science is that science is based on understanding; in designing ES-cell differentiation experiments it is essential that we have a total grasp of the question, and use reagents that are defined and reliable. In examining the differentiation of ES cells into specific, differentiated cell types we must define the system by using reagents that allow interpretation of the results and, equally important, ensure reproducibility. Ideas can thus be proposed and tested, data interpreted, and the information used by others to extend the work. It is only by progressing from alchemy to science that the full potential of ES cells can be exploited and developed coherently.

The emergence of *in vitro* cell culture techniques occurred in the early 20th century (1, 2) and rapidly lead to the establishment of the first human tumour cell lines, for example HeLa (3). These achievements reflected considerable advances in understanding the biophysical and biochemical parameters for cell survival and maintenance. In working with cells *in vitro* it is evident that the key to success is the immediate microenvironment; that is, cell culture systems must provide nutrients, survival factors, protection from negative effects of the atmosphere, and dilution of waste products. However, eukaryotic cells and their culture still represent an intricate biological system of a subtle complexity far beyond our current understanding. Therefore we must even now recognize that any analytical cell culture-based approach is still a balance between what we understand and what remains to be discovered.

The history of *in vitro* cell differentiation began during the 1970s, when embryonal carcinoma (EC) cell lines were first isolated. EC cells show an amazing ability to differentiate *in vitro*, and can perhaps contribute to the formation of multiple cell types in chimeras. However, in general EC cells exhibit limited, or possibly predetermined, patterns of differentiation *in vitro*. In the early 1980s mouse ES (mES) cells were isolated (see Chapter 1). These cells are regarded as pluripotent, showing a far greater capacity to differentiate *in vitro* and *in vivo* than EC cells. Linked with this greater differentiation capacity of mES cells comes more stringent cell culture requirements, to sustain their proliferation as undifferentiated and pluripotent cells (4). The maintenance and differentiation of ES cells from diverse species currently presents some of the greatest challenges to researchers studying stem cells and their controlled differentiation.

Stem cells, by definition, renew and/or differentiate. The first mES cell lines were isolated and were maintained in an undifferentiated state by coculture with feeders cells and/or in the presence of cell-conditioned media (5). With the discovery that leukaemia inhibitory factor (LIF) prevents mES-cell differentiation in culture (at least for mES cells derived from the 129 mouse substrains), mES cell culture became simpler (6). Under conventional culture conditions using medium supplemented with batch-selected fetal calf serum (FCS) and LIF or similar defined factors (e.g. CNTF, Il-11 or OSM (7–10), and see Chapter 2), mES cells are maintained poised between self-renewal and differentiation, with self-renewal being favoured. In the absence of LIF and in the presence of FCS, differentiation dominates over self-renewal. It is the ability specifically to control the nature and outcome of this differentiation process, producing any desired cell type upon command, that is the Holy Grail for many stem-cell researchers.

In vitro differentiation of mES cells is performed most frequently via their aggregation, forming clumps of cells known as 'embryoid bodies' (EBs). These structures can contain experimental sources of intermediate progenitor cells for the three germ layers. In EBs, mES cells undergo transitions to ectoderm, mesoderm, and to visceral and parietal endoderm, mimicking to varying degrees events at gastrulation (from day 4.5 to 7.5 p.c. in the mouse). However, the factors and events occurring within these complex structures are poorly defined and not well understood. Understanding is further complicated as these structures are highly dynamic in nature, with cellular differentiation, survival, and differential expansion of differentiated derivatives all being sensitive to numerous parameters, most of which are not known. Our approach to unravelling these complexities has been to refine *in vitro* culture systems, allowing controlled manipulation of germ-layer differentiation from mES cells (11–15).

Independently during the mid 1980s, Tom Doetschman and Anna Wobus studied *in vitro* differentiation of mES cells. Both showed that although EB-mediated induction of mES-cell differentiation appeared to be simple, the efficiency of differentiation and the nature of cell types obtained were difficult to regulate or predict (16, 17). The lack of repeatable, controlled *in vitro* differentiation severely limited the use of mES cells as a tool in understanding embryonic development; it also limited their exploitation in pharmaceutical testing and the burgeoning field of cell therapy.

More recently, we demonstrated that mES cells could be routinely and efficiently differentiated into mesoderm and various haematopoietic lineages in media containing 10–15% batch-selected FCS (11, 12). Additionally, in the case of mesoderm and subsequent haematopoiesis, the sequence of events mimics aspects of embryonic development *in vivo* (8). The critical, simple medium additive that made this approach reproducible was monothioglycerol (MTG), or β-mercaptoethanol also was shown to be effective; that is, an *active* reducing agent was shown to be crucial to the survival of differentiating cells. These findings indicate that free-radical scavenging is an essential environmental need for reproducible mES-cell *in vitro* differentiation. These data also revealed that a principal 'driver' of mES-cell differentiation *in vitro* is serum (i.e. *batch-selected* FCS), present in the medium at 10–15%. It was consequently shown that serum effects are generally overriding, so that the addition of many growth factors appears to have little or no influence on the initial development of either mesoderm or subsequent haematopoietic progenitors. However once progenitors develop, exogenous factors do influence their survival and also the subsequent expansion of differentiated cell types, such as erythrocytes, macrophages, and mast cells, all of which can be specifically enhanced (12, 18).

The finding that FCS effects are so dominant in mES cells during *in vitro* differentiation was the stimulus to devise a cell culture system where all components of the medium are defined and described completely; that is, a chemically defined medium (CDM) (13–15) in which FCS and (later) bovine serum albumin (BSA) were replaced by chemically defined reagents.

Differentiation experiments using CDM allowed, for the first time, separation of serum effects from those of specific, exogenously added growth factors. Using this approach we conclusively demonstrated that mesoderm is induced by the TGF-β superfamily members, bone morphogenetic protein-2, -4, -7 (BMP-2, BMP-4, BMP-7), or activin A (*Figure 1*). Furthermore, we found that these factors initiated a process that resembles primitive-streak formation, followed rapidly by haematopoietic cell formation and expansion (13–15). Using CDM, important precedents were established: (i) mES cells, whilst cultured as EBs in basal CDM, are responsive to induction of differentiation by treatment with exogenous growth factors; and (ii) in this system, mES cells can be *directed* towards neuroectodermal or mesodermal pathways (13–15).

A constructive lesson on the masking effect of serum (or plasma-derived serum) can be drawn from the studies of EB-mediated haematopoiesis. Using FCS to supplement media, differentiation can be driven towards mesoderm and, depending upon the serum batch, erythroid–myeloid pathways. Evidence from non-mammalian systems (e.g. *Xenopus* and Zebrafish) had indicated that BMP is crucial in mesoderm induction; however, in FCS-based mES-cell differentiation systems, treatment with this factor appeared to have no observable effect. By using serum-free, defined medium it was possible to demonstrate unequivocally

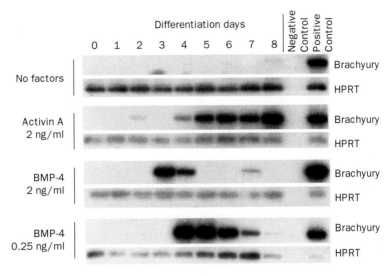

Figure 1 Timecourse analysis of expression of a mesoderm-specific gene in mES cells developing as EBs in CDM plus BMP-2, BMP-4, or activin A (13–15, 29). Line CCE mES cells were seeded for EB formation in basal CDM, or in CDM plus 0.25 ng/ml BMP-4, 2 ng/ml BMP-4, or 2 ng/ml activin A. Semiquantitative RT-PCR for *T* (*Brachyury*) expression was used to monitor mesoderm development. In this example the approximate amounts of cDNA were adjusted using a housekeeping gene, *hypoxanthine guanine phosphoribosyl transferase* (*HPRT*). PCR products were separated electrophoretically, Southern blotted, and hybridized using specific probes. 'Day 0' used undifferentiated mES cells. The negative controls were water, and the positive control cDNA from EBs cultured for 4 d in FCS-supplemented medium (27). In EBs cultured with 2 ng/ml activin A, expression of *T* became detectable by 2 d of differentiation, was significantly elevated by 5 d, and continued to be expressed over the period examined. With 2 ng/ml BMP-4, *T* expression was strongly induced at 3–4 d and then diminished; this is reminiscent of the pulse of mesodermal gene expression occurring during primitive-streak formation *in vivo*. Interestingly, at the lower concentration of BMP-4, 0.25 ng/ml, induction of *T* expression occurred later, at 4 d, and took longer to diminish.

a critical role for BMP-4 or BMP-2 in mesoderm, and subsequent erythropoietic, differentiation. Thus, it is the 'inherent' factors in FCS that can induce these differentiation processes as a matter of course under conventional *in vitro* culture conditions (15).

Nakayama *et al.* extended these results, confirming that under serum-free conditions BMP-4 is essential for erythroid–myeloid differentiation; and also found that the FLT1 ligand, VEGF, acts synergistically with BMP-4, enhancing BMP-4-dependent lympho-haematopoietic cell production (19). This group published that BMP-4 causes differentiation of mES cells, cultured as EBs under serum-free conditions, into non-committed mesodermal precursor cells (possibly the presumptive haemangioblast); and that subsequent differentiation into lympho-haematopoietic progenitor cells requires the sequential activity of VEGF. This contradicted other reports, where FCS-supplemented medium was used and where an inhibitory effect of VEGF on erythropoietic differentiation had been reported.

In summary, although developmentally important genes are being identified in increasing numbers, the actual mechanisms of cellular differentiation, the precise effects of growth factor signalling molecules, and the nature of sequential progenitor cells, have yet to be fully elucidated. Defined culture systems offer an experimental framework in which these highly complex processes can begin to be successfully dissected.

2 Experimental rationale

When devising experiments involving mES-cell *in vitro* differentiation, one is confronted by a multiplicity of protocols. It is therefore essential to define the specific question and/or the task being addressed, and whether or not it is necessary to employ a completely defined culture system. Although we regard the defined culture system described below as effective, it is a 'bare-bones', minimalist, and technically challenging approach, placing the cells in a very basic environment. If one wishes to understand growth factor actions and signalling pathways, such an approach is necessary. If, however, the production of a particular cell type is the only required end point, with the mechanism of action not being of direct interest, it may be appropriate to use other simpler, prescribed, partially defined factors and reagents.

The contrast between a partially defined and a fully defined system is exemplified by two examples outlined here. In the first, the object was to obtain reconstituting haematopoietic stem cells (HSC) from mES cells (20). Many groups have attempted this and although primitive erythropoiesis had been often reported, no one had achieved the goal of developing reconstituting HSC from mES cells *in vitro*. Kyba *et al.* (20) demonstrated that *HoxB4*, a homoeotic gene normally expressed in HSC and implicated in their self-renewal, could have a role in the formation of HSC from mES cells. Their approach involved developing embryoid bodies (EBs) over 6 days in medium containing transferrin, monothioglycerol, and 15% batch-selected FCS. The resulting EBs were disrupted to a cell suspension, and transfected with a *HoxB4*-expressing retrovirus. The resulting virally transformed cells were grown on a feeder layer of the 'haematopoietic stromal' cell line, OP9, in the presence of 10% batch-selected FCS, and finally assayed for haematopoietic reconstituting activity in irradiated mice. The significance of this work is that *they succeeded* in obtaining reconstituting HSC, something that countless groups had attempted for many years without success. However, the precise mechanism and controlling sequence of events that gave rise to the reconstituting HSC could not be defined using this approach as it stands. On the other hand, as reviewed by Proetzel and Wiles (21), it is possible to remove *all* undefined components from mES-cell *in vitro* differentiation media, including BSA, and dictate the development of mesoderm and primitive erythropoiesis by addition of BMP-4. This approach clearly demonstrates the role of BMP-4 in mesoderm formation, but the cells are really in a 'simplistic' environment, and hence limited. To summarize, the first approach *achieved* a specific goal, while the second *identified* a mechanism of action. For different goals one needs different approaches.

An operational compromise between these two extremes is to supplement defined media with complex, premixed cocktails. These, often proprietary, cell-culture supplements have been designed for use in specialized cell assays. Judicious use of such cocktails can lead to excellent, moderately interpretable, and reproducible results. For example Austin Smith's group used defined media replacing FCS with cell culture supplements N2 and B27 for mES-cell differentiation studies (22). Their work clearly showed that LIF is insufficient to maintain mES cell self-renewal in the absence of serum, and that undefined serum components involved in the maintenance of mES cell self-renewal in culture could be substituted by BMP-4. Again, this demonstrates the confounding effects serum can have on the analysis of mES cell culture and differentiation. The media supplement N2 was originally developed for neural stem-cell expansion, and supplement B27 was developed and optimized for the survival of hippocampal neurons (23). Both these supplements are available commercially (from StemCell Technologies, or Invitrogen). The downside of using such cocktails is that of their numerous constituent ingredients, many are probably not involved in mES-cell differentiation, leaving open the question – which are relevant?

Proprietary supplements, such as Sigma's SRM and Invitrogen's KSR, do have a place in defined differentiation systems – provided that the supplier at least publishes a list of their ingredients, or is responsive to enquiries. (Constituents are sometimes available in patents: for KSR, see Price PJ, Goldsborough MD, and Tilkins ML (1998), 'Embryonic stem cell serum replacement; International Patent Application WO98/30679'). Where, of necessity, one is obliged to use supplements containing undisclosed ingredients, it is strongly recommended that any data obtained be interpreted with care. Even when the formulation is made public, some supplements are found to contain crude protein fractions, for example yeast or meat hydrolysates. These additives should always be avoided, as they are as undefined as serum. From listed ingredients the most intractable and hardest to eliminate are the pseudodefined ingredients, the most ubiquitous being bovine serum albumin (BSA). BSA is not a pure protein (although inexpensive, recombinant, and pure protein may appear in time). Currently, its elimination from media can be expensive and frustrating, and for the majority of work its convenience perhaps outweighs its disadvantages. If BSA has to be avoided, one alternative which we have used successfully for mES-cell *in vitro* differentiation is to substitute it with 0.1% polyvinyl alcohol (PVA) (14, 24).

3 Methods

3.1 General considerations for controlled mES-cell differentiation

The protocols given here should not be regarded as 'written in stone' – although functional as they stand, they are intended as starting points. When changes are made, however, it is vital that the intent of the approaches be grasped and the variations in methodology be meticulously recorded, ensuring repeatable and consistent results.

When undertaking serum-free cell culture it is important to understand that cells are being grown in an environment lacking the 'chemical-buffering' capacities of FCS. Consequently, the cells are far more sensitive to any toxic substances that might inadvertently be introduced into the culture system. To minimize this hazard it is crucial to use reagents of the highest quality, and avoid contact with any potentially contaminated glassware etc. Good record keeping (including the source of reagents, purchase date, storage conditions, and batch numbers) is essential to assist when troubleshooting is necessary. To date, successful *in vitro* differentiation has been achieved in CDM (see below) for mES cell lines derived from 129 substrains, including lines CCE from 129/Sv/Ev (25), D3 from 129/Sv (16), and E14.1 from 129/Ola (26). Attempts at *in vitro* differentiation using the same reagents but with a strain C57BL/6-derived mES cell line were unsuccessful, with the cells dying within a few days in basal CDM (MVW, unpublished). This difference probably reflects strain-dependent (genetic) variation in the ability of mES cells to survive in CDM. It may be relevant that 129 substrain-derived mES cells have a considerably higher plating efficiency than C57BL/6-derived lines in standard media (MVW, own observations), and therefore may be more robust in culture.

For the procedures outlined here, mES cells are routinely cultured in Dulbecco's modified Eagle's medium (DMEM) supplemented with 15% FCS (batch selected for optimal mES-cell survival), 150 μM monothioglycerol, and 1000 U/ml LIF (also see Chapter 2). As the presence of contaminating feeder cells will affect mES-cell differentiation, mES cell lines are either adapted to grow feeder free (27) or subcultured without feeder cells for two passages prior to differentiation experiments in CDM. Directly before use in CDM differentiation tests, mES cells are thoroughly washed with CDM to remove both residual serum and excess LIF. It cannot be overemphasized that a uniformly high standard of mES cell culture must be maintained at all times, and that standard protocols must be established and followed as routine to ensure their success.

The key features of CDM usage can be summarized as follows:

- The basic formulation of CDM provides a limited stimulatory environment, containing just three factors: insulin, transferrin, and LIF at a very low concentration.
- EBs formed in basal CDM differentiate and then begin to die after 7–9 days of culture, suggesting that the medium provides a selective environment for differentiated cells and, thereby, a system for testing cell survival and expansion.
- Cells in EBs may be rescued by culture, either during the CDM culture phase or subsequently with additional growth factors to promote specific survival, for example NGF for neural cells, or even FCS for more general rescue.

In establishing an *in vitro* differentiation system it is essential to consider cell interactions, both paracrine and autocrine; for example, exogenous growth factor BMP-4 induces its own synthesis in mES cells during mesoderm differentiation. Furthermore, cellular differentiation is influenced profoundly by the culture system, with variables including the method of culture (monolayer *vs.* EB; and EB culture in suspension *vs.* hanging drop), the initial plating densities of EBs and outgrowths, as well as the frequency of media changes. For mesoderm development, for example, plating density has direct consequences for the efficiency of differentiation (15). Collectively, these effects are thought to be attributable to medium detoxification, conditioning, and autocrine effects. For example FGF5, Nodal, BMP-2, and BMP-4 are expressed at low, variable levels in mES cells (15), and the concentration of these and other endogenous factors is thus dependent upon cell density and frequency of media changing. Any such differences between culture systems should therefore be considered when comparing and interpreting the resulting data.

3.2 Recommendations for CDM preparation and experimental design

(a) Wherever possible plastic disposable containers, tubes, and pipettes should be used for reagent handling and preparation. This avoids the potential introduction of toxic substances via laboratory glassware etc.

(b) When preparing reagents, use water of a high level of purity. It is recommended that pure water be purchased (e.g. from Invitrogen); the extra cost is outweighed by the risk of poor results from the use of lower-quality water.

(c) We strongly advise preparing CDM at the time of use, avoiding any possible deleterious effects due to storing the complete medium; for example, loss of activity of the reducing agent over time, or degradation of other unstable ingredients.

(d) When making stock solutions of reducing agents, monothioglycerol (MTG) or β-mercaptoethanol, it is *essential* that these dilutions be prepared on the day of use.

(e) Care must be taken when diluting growth factors to maintain their activity; therefore follow carefully the supplier's instructions and use only the recommended diluents.

(f) Although reagents are chemically defined and hence (theoretically) uniform between batches, it is good laboratory practice to record batch numbers for each component.

(g) When making stock solutions, care must be taken not to introduce, inadvertently, undefined components; for example by using FCS in a diluent.

(h) The active concentration of novel factors should be assessed over a range of concentrations (e.g. at 0.1, 0.3, 1, 3 and 10 Units); some factors, for example

Protocol 1
Preparation of mES cells for differentiation in CDM

Mouse ES cells should be in optimal condition (i.e. undifferentiated and proliferating exponentially), and at ~30% confluence at the time of use in differentiation experiments. This will help ensure experimental reproducibility. All cell incubations may be conducted in a humidified incubator, gassed with CO_2 in air, and set at 37 °C.

Reagents and equipment

- A 129 strain-derived mES cell line, e.g. CCE (18), either adapted to feeder-free culture in gelatinized dishes/flasks (12, 14) or sub-cultured twice without feeder cells
- Culture medium, for mES cells: Dulbecco's modified Eagle's medium (DMEM; Cat. No. 11965-092, Invitrogen) supplemented with 15% FCS (batch-selected for optimal mES cell growth and survival), 150 μM monothioglycerol (MTG; Cat. No. M 6145, Sigma), and 1000 U/ml LIF (Cat. No. ESG1107, Chemicon International)
- PBS, pH 7.2 (e.g. Cat. No. 20012043, Invitrogen)
- Tissue culture-grade water (e.g. Cat. No. 15230, Invitrogen)
- Gelatin solution: 0.1% (w/v) cell culture-tested gelatin (e.g. Cat. No. G9391, Sigma) in PBS, dissolved and sterilized by autoclaving.[a]
- Basal CDM, for 100 ml combine: 49 ml Iscove's Modified Dulbecco's Medium with Glutamax-I (IMDM; Cat. No. 31980, Invitrogen); 49 ml Ham's F12 nutrient mixture with Glutamax-I (Cat. No. 31765-035, Invitrogen); 1 ml autoclaved solution of 10 % (w/v) polyvinyl alcohol (PVA; Cat. No. P-8136, Sigma) in distilled water; and 300 μl of 150 mM stock solution of monothioglycerol (MTG), prepared by adding 27 μl MTG (Cat. No. M 6145, Sigma) to 2 ml IMDM/F12[b,c]
- Lipid and growth factors, for 100 ml basal CDM: 1 ml of 100 × stock solution of chemically defined synthetic lipid concentrate (Cat. No. 11905-031, Invitrogen); 20 μl of 10 U/μl diluted stock solution of LIF (a working stock *must* be prepared from that supplied; Cat. No. ESG1107, Chemi-Con International); 0.5 ml stock solution of 30 mg/ml transferrin (Cat. No. 0652202, Roche Applied Science); and 70 μl stock solution of 10 mg/ml insulin, either reconstituted from powder (Cat. No. I 2767, Sigma) or diluted from liquid form (Cat. No. 003-0110SA, Invitrogen)[d]
- Trypsin-EDTA solution: 0.25% trypsin in 1 mM EDTA (e.g. Cat. No. 25200, Invitrogen)
- Trypsin inhibitor solution: prepare a 1 mg/ml solution of trypsin inhibitor (Cat. No. T-6522, Sigma) in CDM
- Enzyme-free dissociation buffer (Cat. No. 13151-014, Invitrogen): this may be substituted for the trypsin-EDTA solution and trypsin inhibitor
- Benchtop centrifuge
- Disposable plastic transfer pipettes
- Centrifuge tubes
- Haemocytometer

Method

1. Subculture feeder-free mES cells on gelatinized dishes 1 day prior to *in vitro* differentiation tests.[a]

Protocol 1 continued

2. On the day of differentiation, wash cell cultures twice with CDM, and replace culture medium with basal CDM. Incubate cells for 30 min at 37 °C.[e]

3. Trypsinize mES cells to a single-cell suspension by washing once with trypsin–EDTA solution, followed by addition of sufficient of the same solution to cover the cells, and incubate for 2–3 min at 37 °C. For example, for a T25 flask of mES cells use 3 ml trypsin–EDTA solution for washing, and 1 ml more for the second, disaggregation step. (Alternatively, enzyme-free dissociation buffer can be used instead of trypsin-EDTA solution, avoiding the use of trypsin inhibitor.)

4. For a T25 flask of mES cells, add 5 ml CDM to the trypsinized cells, draw cells up and down a transfer pipette to obtain a single-cell suspension, and centrifuge the suspension at 1000 rpm for 5 min to pellet the cells.

5. Resuspend mES cells in 2 ml basal CDM containing trypsin inhibitor, where appropriate.

6. Pellet cells by centrifugation, resuspend in 5 ml basal CDM without trypsin inhibitor, count the cells using a haemocytometer, and adjust the suspension to a working cell density appropriate for use in the differentiation protocols.[f]

[a] Where needed, plates are gelatinized by coating surface with 0.1% gelatin and removing the excess immediately. Plates are ready for immediate use.

[b] The final concentrations of constituents are: 1:1 IMDM/F12, 0.1% (w/v) PVA, and 450 μM MTG.

[c] This medium must be filter-sterilized before adding lipid and protein supplement (and growth factors, where appropriate) to complete the basal CDM formulation.

[d] The final concentrations of additives are 2 U/ml LIF, 15 μg/ml transferrin, and 7.0 μg/ml insulin.

[e] This step is essential to eliminate residual FCS, which could induce mesoderm differentiation; and LIF, excess of which will perturb differentiation.

[f] Methods for accurately assessing mES cell density and viability are given in Chapter 6, Protocol 1.

members of the TGF-β family, show striking concentration-dependent kinetics.

(i) If medium needs to be filter-sterilized, this must be done *before* the addition of any proteins (for example growth factors) or lipids.

(j) We employ, and recommend, low oxygen levels (7%) in cell incubations. Although not essential to the success of the protocols outlined here, reduced oxygen tension will produce higher rates of cell survival and more robust differentiation (27).

Protocol 2
Mouse ES-cell differentiation in CDM: suspension culture

Reagents and equipment

- Single-cell suspension of mES cells (*Protocol 1*)
- 35 mm bacteriological-grade Petri dishes (e.g. Cat. No. 627–102, Greiner)
- Basal CDM (*Protocol 1*)
- Transfer pipette

Method

1 For the mES cell line CCE, seed 6000 cells into a 35 mm Petri dish, in 1 ml CDM +/− test factors.[a]

2 Set each of the CDM suspension cultures within a larger, empty Petri dish, and carefully place in the cell culture incubator, together with a few unlidded dishes containing sterile water to reduce evaporation from the CDM cultures.

3 Incubate cultures, depending on the assay, for 1–8 d. Medium can be changed carefully if need be, to avoid acidification, and is recommended for cultures maintained for more than 7 d. To do this, tip the dish slightly allowing the developing cell aggregates to settle, carefully remove the bulk of the medium, and replace with fresh CDM +/− test factors.

4 Record all details of cell density, media changes, and frequency of observations; i.e. if cultures are examined too often (e.g. daily), this may impact on the system.

[a] At this cell density and using the CCE cell line, typically 10–20 EBs form within 5 d of incubation (11). For other mES cell lines, the optimal starting cell number may differ and should be determined empirically.

Protocol 3
Mouse ES-cell differentiation in CDM: hanging-drop culture

(See also Chapters 6 and 9 for hanging-drop culture)

Reagents and equipment

- Single-cell suspension of mES cells (*Protocol 1*)
- Basal CDM (*Protocol 1*)
- 35 mm bacteriological-grade Petri dishes (*Protocol 2*)

Protocol 3 continued

Method

1. Adjust the mES cell suspension to a convenient density, e.g. 250–2500 cells/ml in CDM.[a]
2. Carefully place ~20 separate 20 µl drops of the mES cell suspension (i.e. containing 5–50 cells/drop) onto the base of a 35 mm bacteriological-grade dish.[a]
3. Replace the lid onto the 35 mm dish, and invert rapidly. Keep the dish level. The individual drops should now be hanging vertically downwards.[b]
4. Arrange the dishes as in *Protocol 2*, step 2.
5. Incubate the hanging-drop CDM cultures for 24–48 h, and inspect the cultures periodically through the inverted base. In each drop a single aggregate of cells, the embryoid body (EB), will develop, its size reflecting the initial cell density.
6. After 48 h, the cells will have formed a uniform EB. Reinvert the 35 mm dish to its upright position, and flood the dish with 1 ml CDM +/− test factors. By this method a consistent density of uniform EBs will be produced.
7. Incubate cultures for up to 8 d, depending on the assay, and assess the extent of differentiation (see Section 4).
8. If the medium becomes acidified it should be changed (the frequency depending upon EB density) as in *Protocol 2*, step 3; and this is recommended especially for cultures maintained for more than 7 d.

[a] Steps 1 and 2 control the size and plating density of EBs which develop.
[b] This step may take a little practice but is simpler than it sounds!

Protocol 4

Mouse ES-cell differentiation in CDM via EB formation followed by attachment culture

Here, the treatment of EBs with growth factors in the latter stages of culture can be used to induce or allow the selective expansion and/or survival of specific, desired cell types.

Reagents and equipment

- Mouse ES cells induced to form EBs and to differentiate, according to either *Protocol 2* or *3*
- Sterile, 1.5 ml conical Eppendorf tubes
- Wide-bore pipette
- CDM (*Protocol 1*)
- Gelatin solution (*Protocol 1*)
- Gelatin-coated tissue-culture dishes (*Protocol 1*).
- Culture medium, containing 5–10% FCS (*Protocol 1*).

> **Protocol 4 continued**
>
> **Method**
>
> 1. Transfer EBs into a 1.5 ml Eppendorf tube, and allow to settle.
> 2. Using a wide-bore pipette, transfer the settled EBs into a gelatin-coated tissue-culture dish, limiting the amount of medium transferred. Depending upon EB size, plate at ~1 EB per 20–100 mm^2.
> 3. Add fresh medium to the EB cultures: either culture medium containing 5–10% FCS, or CDM +/− growth factors.
> 4. Incubate cultures; EBs will attach to the substrate, and cells will grow and migrate out from the EB over 24–72 h.[a]
> 5. Visually assess the extent of differentiation after 2–4 d of attachment culture.
> 6. Depending on the starting cell numbers and subsequent EB density, medium should be changed at least weekly.
>
> [a] EBs show less propensity to attach and spread when cultured in CDM.

3.3 Neuroectoderm differentiation

Using basal CDM alone as outlined in *Protocol 2*, EBs will develop neuroectodermal derivatives as indicated by *PAX6* expression, which rises abruptly after ~5 d of culture with little or no mesoderm formation (as assessed by *T* (*Brachyury*) or *GSC* (*goosecoid*) expression). It has been suggested that this is a 'default' differentiation mechanism occurring in the absence of exogenous growth factors (14, 15, 22). If CDM-EBs are plated after 7–9 d, they can be rescued with FCS-containing medium (e.g. mES cell culture medium, minus LIF), and neuronal cells will be found to predominate in the culture 4–10 d later. Alternatively, specific neurotrophic factors may be added, such as NGF in CDM.

3.4 Mesoderm and haematopoietic differentiation

By addition of BMP -2, -4, -6, or -7 (2–10 ng/ml) to CDM as outlined in *Protocol 2*, mesoderm differentiation will be induced. Under these culture conditions, EBs also will develop more rapidly than in CDM alone, with no cell death being apparent. We showed previously that genes indicative of mesoderm are induced by treatment with these BMPs, rapidly followed by haematopoietic differentiation, all occurring within 3–4 d of culture initiation. These factors appear to induce coordinate expression of genes associated with, and denoting, primitive-streak formation and gastrulation, such as *T, GSC, NODAL,* and *FGF-5* (15). By 2–3 d of EB culture, expression of these genes increases >20 fold and, depending upon the initial concentration of factor, will then subside over the following few days.

3.5 Mesoderm and neuroectoderm differentiation

If CDM is supplemented with activin A (at 2–20 ng/ml) following *Protocol 2*, there follows simultaneous induction within 5–6 d of culture (15) of both dorso-anterior-like mesoderm and neuroectoderm.

4 Evaluation of mES-cell differentiation

The methods for the evaluation of differentiation is highly dependent upon the logic behind why the tests are being conducted. For example if the differentiation system is well established and large numbers of cells with a defined phenotype are produced, relatively simple tests can be used, including morphological assessment of EB differentiation. Examples include visualization of red colouration due to erythropoietic activity, or of areas of beating cells indicative of cardiac muscle. If more experimental, or rare, events need to be monitored, more sophisticated and additional, sensitive tests are required. An elegant and highly informative method involves the use of reporter constructs, for example green fluorescent protein-tagged transgenes or other, precise lineage markers. These markers also facilitate specific cell type isolation from a complex mixture of cells. Functional assays may also be established, either alone or in combination with lineage markers; for example haematopoietic-reconstitution competitive-repopulation assays for the development of HSC (28).

The use of whole-mount approaches in combination with reporter genes and/or *in situ* hybridization (ISH) can also be used to provide precise, localized information on gene expression, helping define and understand the changes occurring at the molecular level during mES-cell differentiation. The reader is referred to Chapters 6 and 9 in this volume for detailed methods for monitoring the status of differentiated cells, which include ISH, RT-PCR, and immunocytochemistry. Microarray analysis is also an informative tool for obtaining a global view of a change in a cell culture. A combination of analytical approaches is recommended so that the phenotypes of individual cells, as well as the overall efficiencies of differentiation, can be examined.

A brief, preliminary list of positive marker genes used for RT-PCR includes: for undifferentiated mES cells, *POU5F1* (formerly *Oct4*) and *ZFP42* (previously known as *REX1*); for mesoderm, *T* and *GSC*; for endoderm, *GATA-4* and *AFP*; and for neuroectoderm and neural derivatives, *SOX1*, *PAX6*, and *EGR2*. An example of experimental results obtained using RT-PCR monitoring for mesoderm induction using *T* as a marker gene is shown in *Figure 1* (14).

5 RNA isolation

Total RNA can be isolated from EBs differentiated for various times and following different treatments, and analysed by, for example, RT-PCR, qRT-PCR (Taqman), or expression arrays. A protocol for total RNA isolation is given here that is effective for either single or multiple EBs consisting of 50 to 10 000 cells in

Protocol 5

Total RNA isolation from limiting amounts of mES cells/EBs

Reagents and equipment

- 1–100 EBs, equivalent to 50–10^5 cells (Protocols 2 and 3)
- Trizol (Cat. No. 15596-026, Invitrogen)
- Isopropanol
- RNase-free glycogen, at 4 mg/ml (Cat. No. 0901393, Roche Applied Science)
- RNAse-free water (Cat. No. 9930, Ambion): we do not recommend DEPC-treated water
- 75% ethanol in RNAse-free water
- Chloroform
- Dry ice
- 1.5 ml, conical microcentrifuge tubes
- Centrifuge, preferably cooled to 4 °C

Method

1. Collect EBs and gently pellet in a tube, by microcentrifugation at ∼1000 rpm for 15 s. Carefully remove as much medium as possible.
2. Add 100 µl of Trizol to the sample, mix well, and incubate for 5 min at r.t.[a]
3. Add 20 µl chloroform per 100 µl Trizol suspension, and vortex well.
4. Centrifuge the sample at 12 000 g for 12–15 min, preferably at 4 °C.
5. Transfer the aqueous phase to a fresh microcentrifuge tube, adding a half volume of isopropanol (i.e. 50 µl per 100 µl starting Trizol); mix well, and incubate for 10 min at r.t. If using very limited amounts of RNA (1–20 EBs) it is suggested that RNase-free glycogen be added (4–5 µg).[b]
6. Centrifuge the sample at 12 000 g for 10 min, preferably at 4 °C, to pellet RNA.
7. Remove the supernatant carefully, and wash the RNA pellet once with 75% ethanol.
8. Centrifuge the sample for 5 min, preferably at 4 °C, and remove all ethanol. (A second, short centrifugation may be helpful.) Air-dry the pellet for ∼10 min at r.t.
9. Resuspend the RNA pellet in 50 µl RNAse-free water, and rapidly freeze on dry ice. If more than 10^6 cells were in the starting sample, use 100 µl RNAse-free water/10^6 cells; for less than 10^5 cells (as in a single EB), either resuspend in 20 µl water or use the pellet directly in the next stage.
10. If required, and if there is a sufficient quantity, assess RNA quality by gel electrophoresis (or equivalent). Alternatively, and for cDNA applications, go directly to the DNaseI treatment (Protocol 6).

[a] At this stage the sample can be stored at −20 °C to −70 °C.
[b] Do not add too much glycogen as it may inhibit cDNA synthesis.

Protocol 6
DNAse I treatment of RNA samples[a]

Reagents and equipment

- RNA sample, at concentration 1–200 ng/μl (*Protocol 5*)
- RNAse-free DNAseI (Cat. No. 2222, Ambion)
- DNase I buffer solution: 10 × DNase buffer (Cat. No. 8170G, Ambion)
- RNAsin® ribonuclease inhibitor (Cat. No. 9PIN251, Promega)
- $MgCl_2$ (100 × stock) solution: 1 M $MgCl_2$ in Tris pH 7.5
- Phenol/chloroform mixture
- 1 M NaCl solution
- RNase-free glycogen (*Protocol 5*)
- Absolute ethanol
- RNAse-free water (*Protocol 5*)
- 75% ethanol in nuclease-free water
- Dry ice
- Centrifuge, preferably cooled to 4 °C

Method

1. Thaw RNA sample quickly.
2. Per 100 μl RNA solution add 1 μl of stock 1 M $MgCl_2$ solution (final concentration of 10 mM), 20 U of RNAsin®, and 20–40 U of RNAse-free DNAseI; and incubate samples for 30 min at 37 °C.
3. Add an equal volume of phenol/chloroform mixture to the sample, and centrifuge sample at 12 000 g for 2 min.
4. Transfer the aqueous phase into a fresh tube, and add 0.1 volume of 1 M NaCl (final concentration 0.1 M), 1 μl RNase-free glycogen (4 μg), and 2.5 volumes of absolute ethanol.
5. Freeze the sample for at least 30 min on dry ice (or store at −70 °C).
6. Centrifuge the sample at 12 000 g for 20 min, preferably at 4 °C.
7. Carefully remove the supernatant–take care as the RNA pellet may be nearly invisible. Wash the pellet with 75% ethanol.
8. Centrifuge the sample at 10 000 g for 3 min, preferably at 4 °C; remove the ethanol supernatant carefully, and briefly air-dry the pellet. For <10^5 cells in the starting sample, leave the RNA in pellet form for cDNA synthesis; and for >10^5 cells, resuspend the pellet in approximately 100 μl RNAse-free water per ∼10^6 starting cells.

[a] As an alternative to this protocol, the 'Turbo-DNA-free' kit (Cat. No. 1907, Ambion) works well.

total (29). If RNA is limiting, commercial systems can be used that will provide (apparently linear) reproducible total (cDNA) amplification, for example Clontech's 'SMART' technology kit. Such approaches allow essentially unlimited numbers of experiments to be performed on a very limited amount of starting material, for example a single EB. For larger numbers of EBs (comprising $>10^5$ cells) the 'RNeasy' kit by QIAGEN has been found to be satisfactory.

6 Conclusions

In many ways, mES-cell *in vitro* differentiation mimics aspects of the developing embryo from ~E3.5 to ~E9.0. This period of embryonic development is highly dynamic, with the fate of groups of cells being under the direction of a multitude of environmental cues, ultimately leading to the formation of the complete organism. These systems are the result of evolutionary selection, and hence are highly interwoven networks exhibiting high degrees of developmental homeostasis (30). The use of cell culture approaches to dissect these processes represents a simplification and probably, to a degree, a distortion of the *in vivo* reality. Nevertheless, such approaches can be used as a starting point to unravel the principal control mechanisms involved in early development. By using defined culture systems we are beginning to both understand and direct mES-cell differentiation into desired cell types. Eventually such approaches will revolutionize stem-cell research, leading to direct medical utility (31).

Acknowledgments

The authors would like to thank Dr Barbara Knowles of the Jackson Laboratory for critical reading and advice in the writing of this chapter.

References

1. Harrison, R.G. (1907). *Proc. So.c Exp. Bio. Med.*, **4**, 140–3.
2. Carrel, A. (1912). *J. Exp. Med.*, **15**, 516–28.
3. Gey, G.O., Coffman, W.D., and Kubicek, M.T. (1952). *Cancer Res.*, **12**, 364–5.
4. Nagy, A., Gocza, E., Diaz, E.M., Prideaux, V.R., Ivanyi, E., Markkula, M., and Rossant, J. (1990). *Development*, **110**, 815–21.
5. Smith, A.G. and Hooper, M.L. (1987). *Dev. Biol.*, **121**, 1–9.
6. Smith, A.G., Heath, J.K., Donaldson, D.D., Wong, G.G., Moreau, J., Stahl, M., and Rogers, D. (1988). *Nature*, **336**, 688–90.
7. Behringer, R.R. (1994). *Curr. Top. Dev. Biol.*, **29**, 171–87.
8. Piquet Pellorce, C., Grey, L., Mereau, A., and Heath, J.K. (1994). *Exp. Cell Res.*, **213**, 340–7.
9. Lin, Z., He, Q., and Zhao, Y. (1994). Chondrolaryngoplasty. *Chung Hua Cheng Hsing Shao Shang Wai Ko Tsa Chih*, **10**, 55–7.
10. Koshimizu, U., Taga, T., Watanabe, M., Saito, M., Shirayoshi, Y., Kishimoto, T., and Nakatsuji, N. (1996). *Development*, **122**, 1235–42.
11. Keller, G., Kennedy, M., Papayannopoulou, T., and Wiles, M.V. (1993). *Mol. Cell. Biol.*, **13**, 473–86.
12. Wiles, M.V. and Keller, G. (1991). *Development*, **111**, 259–67.
13. Wiles, M.V. and Johansson, B.M. (1997). *Leukemia*, **3**, 454–6.
14. Wiles, M.V. and Johansson, B.M. (1999). *Exp. Cell Res.*, **247**, 241–8.
15. Johansson, B.M. and Wiles, M.V. (1995). *Mol. Cell Biol.*, **15**, 141–51.
16. Doetschman, T.C., Eistetter, H., Katz, M., Schmidt, W., and Kemler, R. (1985). *J. Embryol. Exp. Morphol.*, **87**, 27–45.

17. Wobus, A.M., Holzhausen, H., Jakel, P., and Schoneich, J. (1984). *Exp. Cell Res.*, **152**, 212-9.
18. Faust, N., Bonifer, C., Wiles, M.V., and Sippel, A.E. (1994). *DNA Cell Biol.*, **13**, 901-7.
19. Nakayama, N., Lee, J., and Chiu, L. (2000). *Blood*, **95**, 2275-83.
20. Kyba, M., Perlingeiro, R., and Daley, G. (2002). *Cell*, **109**, 29-37.
21. Proetzel, G. and Wiles, M. (2002). *Methods Mol. Biol.*, **185**, 17-26.
22. Ying, Q. and Smith, A. (2003). *Methods Enzymol.*, **365**, 327-41.
23. Brewer, G., Torricelli, J., Evege, E., and Price, P. (1993). *J. Neurosci. Res.*, **35**, 567-76.
24. Kane, M.T. and Bavister, B.D. (1988). *J. Exp. Zool.*, **247**, 183-7.
25. Robertson, E., Bradley, A., Kuehn, M., and Evans, M. (1986). *Nature*, **323**, 445-8.
26. Fisher, J.P., Hope, S.A., and Hooper, M.L. (1989). *Exp. Cell Res.*, **182**, 403-14.
27. Wiles, M.V. (1993). *Methods Enzymol.*, **225**, 900-18.
28. Harrison, D. (1980). *Blood*, **55**, 77-81.
29. Ruiz, P., Haasner, D., and Wiles, M.V. (1997). Use of Polymerase Chain Reaction (PCR) on limited amounts of material. In *Immunological Methods* (eds I. Lefkovits and B. Pernis), pp. 287-305, Academic Press, California, USA.
30. Waddington, C.H. (1942). *Nature*, **150**, 563-5.
31. Schuldiner, M., Yanuka, O., Itskovitz-Eldor, J., Melton, D., and Benvenisty, N. (2000). *Proc. Natl. Acad. Sci. USA*, **97**, 11307-12.

Chapter 6

In vitro differentiation of mouse ES cells into muscle cells

Yelena S. Tarasova, Daniel R. Riordon, Kirill V. Tarasov and Kenneth R. Boheler

Laboratory of Cardiovascular Science, National Institute on Aging, NIH, 5600 Nathan Shock Drive, Baltimore, MD, USA.

1 Introduction

Following the derivation of ES cell lines from mouse embryos (1, 2), Wobus *et al.* (3) and Doetschman *et al.* (4) observed that mouse ES (mES) cells when grown in suspension culture spontaneously form three-dimensional cell aggregates, termed embryoid bodies (EBs), which generate multiple differentiated cell types. These aggregates were later shown to give rise to, *in vitro*, cell lineages including those typical of striated and non-striated muscle. This chapter provides current techniques for the production of muscle derivatives from mES cells via EBs, and for their functional analysis.

In vertebrates, striated cardiac and skeletal muscle cells are derived from mesoderm. Commitment of mesodermal cells to the cardiogenic lineage is established during and shortly after gastrulation, at which time cardiogenic cells are organized in bilateral epithelial sheets that are part of the visceral mesoderm lining the developing coelomic cavity (5). Skeletal muscle forms in the vertebrate limb from progenitor cells that originate in somites; these cells delaminate from dermomyotome, and migrate into the limb bud where they proliferate, express myogenic determination factors, and differentiate into skeletal muscle (6). Non-striated smooth-muscle cells (SMC) originate both from neuroectoderm (neural crest), which contains cells committed to the SMC lineage, and from mesodermal sources (endothelium, epicardium) that have undergone limited differentiation (7).

Mouse ES cells provide a number of advantages for the study of muscle cell lineages, compared with other cell types: under carefully controlled conditions they maintain a stable karyotype and can be cultured without apparent limit, whilst retaining pluripotentiality. Cardiomyocytes normally cannot be established as cell lines, and so mES cells offer an *in vitro* system for their production as well as for their analysis in the context of 'development' (8). Moreover, mES cells are amenable to genetic manipulation (e.g. involving homologous recombination or gene over-expression); and in those cases where gene targeting leads to early embryonic lethality in mice, the system represents an *in vitro* alternative for the study of modified genes during cell differentiation into specified muscle lineages.

2 Culture of mES cell lines for differentiation into muscle cells

Maintenance of high-quality, pluripotent ES cell lines is critical to the success of all *in vitro* differentiation protocols. The general principles of mES cell culture have been described (9, 10, and see Chapter 2), but several independent variables consistently influence the potential of cell lines to differentiate into muscle cells. Numerous mES cell lines or their clonal derivatives have proven highly effective in the formation of muscle cells *in vitro* (e.g. D3, R1, BLC6, GTR1, CCE, B117, AB1, AB2.1, HM1, and E14.1), whereas others have not. Reasons for this discrepancy are not always clear, but are most likely due to genetic and epigenetic changes occurring during cell-line derivation or continued passage, or the use of culture techniques that are incompatible with muscle-cell differentiation.

For efficient differentiation into muscle lineages, mES cells should be routinely cultured on feeder layers of mitotically inactivated mouse fibroblasts, either in the presence or absence of leukaemia inhibitory factor (LIF), according to the recommendations of the laboratory of origin. Good-quality, mES cell-qualified, fetal-calf serum (FCS) is prerequisite to long-term culture of mES cells (Chapter 2, *Protocol 4*); but not all of these batches of FCS are conducive to *in vitro* differentiation into muscle cells. *The failure adequately to test serum represents a major reason why efficient muscle-cell differentiation may not be achieved.* Thus, in addition to assaying FCS for the capacity to support mES-cell proliferation, we routinely test serum for the capacity to support muscle differentiation, after five passages of mES cells in selected batches.

Other factors also must be taken into consideration for successful differentiation to be induced. Media components (glucose concentration, amino acids, growth factors, and retinoids), pH, and osmolarity all affect the ability of mES cells to form muscle cells. The starting number of mES cells in the forming cell aggregate is critical: inappropriate numbers at this initial step may preclude muscle differentiation. And either premature plating of EBs (producing insufficient development) or delayed plating (causing cell death) may result in heterogeneous cell populations containing very few to no muscle derivatives. The plating of EBs on some proteins of the extracellular matrix (ECM) alters their potential to generate muscle cells, and therefore can be used to advantage (e.g. see footnote, *Protocol 5*). Lastly, as bacterial (or other) contaminations affect the differentiation potential of ES cells in general, good aseptic techniques should be used at all times. All incubations described below are performed at 37 °C in a humidified tissue-culture incubator, gassed with 5% CO_2 in air.

Protocol 1
Preparation of mES cell cultures for *in vitro* differentiation

Reagents and equipment

- Mouse ES cells, e.g. lines R1 (A. Nagy, Toronto) or D3 (Cat. No. CRL-1934, ATCC), cryopreserved in aliquots of 1.0×10^6 cells/ml
- Culture Medium, for mES cells: 1 × liquid Dulbecco's modified Eagle's medium with 4.5 g/L glucose (DMEM; Cat. No. 11995-040, Invitrogen) supplemented with 15% FCS (selected batches only), 2 mM L-glutamine (from 200 mM stock solution; Cat. No. 25030-081, Invitrogen), 10 μM β-mercaptoethanol (Cat. No. M7522, Sigma), 0.1 mM non-essential amino acids (100 × stock solution, Cat. No. 11140-050, Invitrogen), 50 U/ml penicillin, and 50 μg/ml streptomycin (Cat. No. 15070-063, Invitrogen)[a]
- Differentiation Medium I: DMEM (4.5 g/L glucose) supplemented with 20% FCS, 2 mM L-glutamine, 10 μM β-mercaptoethanol, 0.1 mM non-essential amino acids, 50 U/ml penicillin, and 50 μg/ml streptomycin[b]
- Feeder layers, consisting of mitomycin C-inactivated mouse embryonic fibroblasts (MEF) prepared as described in Chapter 2, *Protocols 2 and 3*; but seeded into 60 mm tissue-culture dishes precoated with gelatin[c]
- LIF (Chemicon International, via Sigma, Cat. No. L5158)
- PBS, without Ca^{2+} or Mg^{2+} (see *Protocol 1*, Chapter 2)
- Trypsin/EDTA solution: 0.25% trypsin, 1 mM EDTA (Cat. No. 25200-056, Invitrogen)
- Sterile, glass Pasteur pipettes (23 cm length)
- Bench-top centrifuge
- 15 ml centrifuge tube
- Haemocytometer

Methods

Thawing:

1 Thaw the vial of frozen mES cells quickly at 37 °C.

2 Transfer aseptically the cell suspension into a tube containing 9 ml prewarmed Culture Medium, and centrifuge (300 g or 1000 rpm, for 3-5 min). Resuspend the cell pellet in 1 ml Culture Medium.

3 Transfer the cell suspension onto a feeder layer in a 60 mm dish containing 2.5 ml Culture Medium ± 10 ng/ml (or equivalent to 1000 U/ml) LIF, as appropriate. Ensure that the cells are evenly distributed over the feeder layer by gently agitating, and place the dish in an incubator. Change medium the following morning and

Protocol 1 continued

continue incubation; and when mES cell colonies cover ~60% of the surface of the feeder layer, passage the cells.^{d}

Passaging:

2. Change medium 1–2 h before passaging.

3. Aspirate medium and wash cells with PBS. Add 1 ml trypsin/EDTA solution (pre-warmed to 37 °C) per dish, and incubate at r.t. for 45–60 s. Add 0.5–0.75 ml Culture Medium, and draw the cells up and down a Pasteur pipette (10 to 15 times) to achieve a single-cell suspension. Transfer the suspension to a 15 ml centrifuge tube containing 8.5 ml Culture Medium. Centrifuge (300 g, 3–5 min) and resuspend the cells either in Culture Medium ± LIF for passaging on fresh feeder layers (at a split ratio of 1:5 to 1:10), or in 5 ml Differentiation Medium I in preparation for EB formation, as follows.^{e}

Cell suspension for differentiation:

4. Count the number of viable mES cells using a haemocytometer: mES cells are distinguished by being small and rounded, whereas contaminating feeder cells are generally much larger and non-uniform in shape.^{f}

5. Adjust the mES cell suspension in Differentiation Medium I to give (a) 20–40 × 10^3 cells/ml, i.e. 400–800 viable cells per 20 µl, for hanging drops, or (b) 10–50 × 10^4 cells/ml for mass culture (see *Protocols 2–5*).

^{a} Iscove's modification of DMEM (IMDM) can be used in place of DMEM, but monothioglycerol (MTG; Cat. No. M6145, Sigma) should then be used, at a final concentration of 450 µM, instead of 10 µM β-mercaptoethanol. Note that MTG should be added to the medium just prior to use for cell culture.

^{b} For Differentiation Medium also, IMDM with 450 µM monothioglycerol can be used instead of DMEM with 10 µM β-mercaptoethanol.

^{c} For gelatin-coating of tissue-culture dishes or multiwells prior to seeding of feeder cells, add sufficient of an autoclaved solution of 0.1% gelatin (Cat. No. G-1890, Sigma) in PBS to cover the bases, incubate the dishes or wells at for least 0.5–1 h (overnight is optimal) at 4 °C, and aspirate residual gelatin solution.

^{d} After thawing, mES cells should be passaged at least once on feeder layers to promote cell recovery and resumption of proper growth characteristics.

^{e} Mouse ES cells should be passaged prior to EB production, on gelatin-coated dishes either with or without feeder layers. Passaging without feeder layers may be required to deplete the culture of feeder cells, which when present in high numbers in the forming EB can inhibit differentiation into mesodermal lineages.

^{f} Cell viability is assessed by the dye-exclusion test: viable cells exclude vital dyes, whereas dead cells are permeable. Trypan Blue (Cat. No. T8154, Sigma) is a dye commonly used for this purpose, but its interaction with serum proteins may require that counting be conducted in serum-free medium or PBS. Other dyes, such as Erythrosin B (Cat. No. E9259, Sigma), also can be used.

3 Differentiation of mES cells into muscle cell lineages

When mES cells are induced to form aggregates *in vitro*, in the absence of those self-renewal signals normally provided by mitotically inactivated feeder layers or LIF, cell differentiation is initiated (see Chapter 5). If aggregates are allowed to form spontaneously in suspension by seeding a culture onto a non-adhesive substratum (typically, a bacteriological-grade Petri dish), the number of cells incorporated into each aggregate is variable. But this number can be controlled through the use of a 'hanging-drop' technique, in which a defined number of cells are suspended in droplets and thereby induced to coalesce. As aggregates differentiate and develop into EBs, primary germ-like layers are produced spontaneously. Initially, an outer layer of endoderm-like cells forms around the EB, followed over a period of a few days by the development of an ectodermal 'rim' and subsequent specification of mesodermal cells. For differentiation into muscle cell lineages, EBs are most efficiently generated by the hanging-drop technique (*Figure 1*) but mass culture techniques also can be used. Differentiation into muscle cell lineages may be promoted by addition of drugs and growth factors. Several morphogens have proven useful in this respect, including the retinoids, all-*trans* retinoic acid (RA) and 9-*cis* RA, and bipolar compounds such as dimethyl sulphoxide (DMSO). Furthermore, κ opioid receptor agonists have been shown specifically to promote cardiomyogenesis (11), whilst induction of vascular smooth muscle (VSM) may require use of dibutyryl-cyclic AMP (db-cAMP) and transforming growth factor β_1 (TGF β_1).

3.1 Differentiation of mES cells into cardiac muscle cells (*Figure 2A, B*)

Cardiomyocytes differentiated from mES cells display properties similar to those observed both *in vivo* and in primary culture. Depending on their stage of differentiation or maturation, they: (i) express cardiac gene products in a developmentally controlled manner; (ii) display characteristic sarcomeric structures; and (iii) undergo spontaneous contractions triggered by cardiac-specific ion currents and membrane-bound ion channels (8). Electrophysiological analyses have also revealed in differentiated cultures those phenotypes typical of primary myocardium and, with continued differentiation, atrial-, ventricular-, Purkinje-, and pacemaker-like cell types (12).

Experiments with serum-free medium supplemented with non-protein additives, hormones, and growth factors show significant promise for the generation of cardiomyocytes under defined conditions (see Chapter 5). In mES cell cultures differentiated in serum-replacement medium containing insulin and transferrin, for example, the expression of cardiac-restricted genes (e.g. $\alpha-$ and $\beta-MHC$) is increased, and the addition of either platelet-derived growth factor BB (PDGF-BB) or sphingosine 1 phosphate enhances cardiogenesis (15).

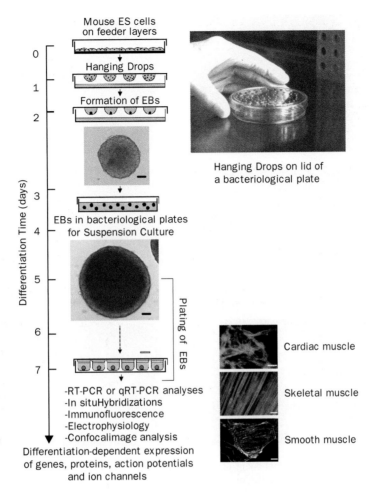

Figure 1 'Hanging-drop' method for the generation EBs for muscle-cell differentiation. The 'hanging-drop' method generates aggregates of defined cell number and size. Main stages include: the aggregation of mES cells, induced by their culture in hanging drops for 2 d; continued culture of EBs in suspension for 5 d; and plating of EBs on an adhesive substratum on day 7 postaggregation, followed by incubation for ≥ 30 d further. (For example of EB, scale bar = 50 mm.) Depending on the precise culture conditions, cardiac-, striated-, and smooth-muscle cells differentiate from the mES cells (scale bars = 10 µm). These differentiated derivatives, either within EB outgrowths or following their isolation, can be analysed for muscle-restricted gene expression, or for the presence and function of muscle-related proteins.

3.2 Differentiation of mES cells into skeletal myocytes (*Figure 2C, D*)

Despite the availability of skeletal-muscle cell lines (e.g. C2, L6, L8), and the ease of isolation of the myogenic stem-cell pool (i.e. satellite cells) from muscle, the mES-cell differentiation system has proven especially valuable for the study of skeletal muscle owing to the recapitulation of developmental processes. For

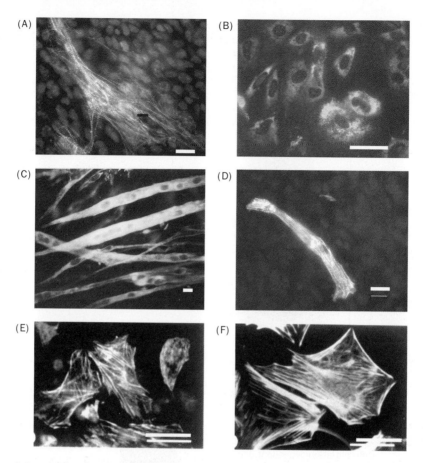

Figure 2 Immunofluorescence staining of muscle-cell derivatives from plated EBs. (A, B) Cardiac differentiation in the R1 line of mES cells. (A) Staining of cardiomyocytes for α-actinin, 10 d after EB plating at day 7, i.e. 7 + 10 d postaggregation (scale bar = 20 μm). (B) Staining of cardiomyocytes for ANP (a cardiac cell-restricted protein), 6 d after EB plating at day 7 (scale bar = 20 μm). (C) Staining of myotubes, obtained using the BLC6 line of mES cells, for Titin (a Z-band epitope) 31 d after EB plating at day 5 (scale bar = 10 μm). (D) Staining of skeletal myoblast/myotube, obtained using R1 cells, for α-actinin 6 d after EB plating at day 7 (scale bar = 20 μm). (E, F) Smooth-muscle differentiation in the D3 line of mES cells, induced by treatment of plated EBs with RA plus db-cAMP between 7 and 11 d after plating at day 7 (for methods see *Protocol 5*, or (18)). (E) Staining for smooth muscle α-actin, 7 d after plating (scale bar = 1 μm). (F) Staining for smooth-muscle MHC, 14 d after plating (scale bar = 1 μm).

example, when EBs are plated onto an adhesive substratum and cultured for 1–2 weeks, myoblasts are produced that subsequently fuse and form recognizable myocytes, which are characteristic of developing skeletal muscle. Moreover, mES cell-derived skeletal-muscle cells display many properties similar to those observed in established skeletal-muscle cell lines, including the expression of skeletal-muscle gene transcripts (e.g. Myf-5, Myogenin, and MyoD) in a manner that mimics the temporal activation seen in normal embryogenesis (16).

Protocol 2

Hanging-drop technique: production of cardiomyocytes

Reagents and equipment

- Mouse ES-cell suspension at 20–40 × 10^3 cells/ml in Differentiation Medium I (*Protocol 1*, step 7)
- Culture Medium (*Protocol 1*)
- PBS (*Protocol 1*)
- All-*trans*-retinoic acid (RA; Cat. No. R2625, Sigma): 10^{-3} M stock solution in 96% ethanol or DMSO, protected from light and stored at −20 °C
- 10 cm bacteriological-grade Petri dishes (e.g. Falcon®, Cat. No. 351029, Bacto Laboratories Pty Ltd)
- 60 mm suspension-culture dishes (Cat. No. 901, Kord Products Inc.)
- 60 mm dishes and 24-well plates, coated with 0.1% gelatin solution (footnote, *Protocol 1*) for 1–24 h at 4 °C before use
- MatTek® dishes (35 mm glass-bottom culture dishes supplied by MatTek® Corporation): these are supplied uncoated (Cat. No. P35G-0-14-C) and should be coated with 0.1% gelatin solution (footnote, *Protocol 1*) for 1–24 h at 4 °C before use; alternatively, collagen-coated dishes (Cat. No. P35GCol-1.5-14-C) may be used
- Glass coverslips, 10 × 10 mm

Method

1. Place 20 μl drops (n = 50–60) of mES cell suspension onto the undersides of inverted lids of bacteriological-grade Petri dishes, and replace the lids over their bases containing 5–10 ml PBS (*Figure 1*). The range of 400–800 mES cells (from lines D3, R1, or CCE) per hanging drop for the preparation of EBs is optimal for cardiac differentiation.[a]

2. Incubate the cells in hanging drops at 37 °C for 2 d, and *do not disturb the dishes during this period* when mES cells coalesce and aggregates form. The 24 h period following initiation of culture in hanging drops is designated 'day 1' (*Figure 1*).

3. On day 2, rinse and collect cell aggregates (i.e. nascent EBs) carefully from the lids with 2 ml Culture Medium, and transfer into a 60 mm suspension-culture dish containing 2–3 ml Culture Medium. Continue incubation.

4. At days 5 to 7, plate EBs onto gelatin-coated tissue-culture plates for adherent culture: for morphological analysis plate a single EB into each well of 24-well plates with 0.5 ml Culture Medium; for immunofluorescence analysis plate 20–40 EBs per 60 mm tissue-culture dish with 3 ml Culture Medium and containing 4 coverslips, or per MatTek® dish; and for RT-PCR analysis, or for dissections of EB outgrowths, plate 15–20 EBs per 60 mm tissue-culture dish with 3 ml Culture Medium. The first beating clusters in EBs can be observed at day 7, but maximal cardiac differentiation is achieved several days after plating. To investigate early cardiac stages, plate EBs at day 5.[b]

5. Change medium every second or third day. Count the number of EBs with foci of contraction (from at least 24 separate wells) to calculate the overall percentage of EBs containing cardiomyocytes. Routinely, >90% of EBs should contain beating areas.[c]

> **Protocol 2** continued
>
> [a] EBs can be initiated and cultured *en masse* to generate large numbers of differentiated muscle cells (Protocol 3).
>
> [b] To enhance the development of ventricular-like cardiomyocytes, add RA (10^{-9} M) between days 5 and 15 of culture (13).
>
> [c] Cardiomycocytes and skeletal-muscle cells may be stained histochemically by the Periodic Acid–Schiff (PAS) reaction (PAS diagnostic kit; Cat. No. 395-B, Sigma), which demonstrates glycogen and collagen by producing a purple or magenta colour (Figure 3C).

Protocol 3

Mass culture of EBs: production of cardiomyocytes

This mass-culture technique is less efficient (with respect to the overall percentage of EBs with muscle cells) than the hanging-drop method, but is less labour intensive and generates large numbers of EBs rapidly.

Reagents and equipment

- R1 line of mES cells, in suspension at $10–50 \times 10^4$ cells/ml in Differentiation Medium I (see step 7, Protocol 1)[a]
- Culture Medium (Protocol 1)
- Differentiation Medium I (Protocol 1)
- Suspension-culture dishes (Protocol 2)
- Gelatin-coated, 60 mm tissue-culture dishes (Protocol 2)

Method

1. Add 5×10^5 mES cells to a 10 cm suspension-culture dish containing 10 ml Differentiation Medium I, and incubate at 37 °C.[b]

2. After 2 d, tilt the plate to one side and allow cell aggregates to settle. Carefully remove medium and replace with 10 ml Culture Medium.

3. Change medium every second day, supplementing the cell suspension with 5 ml Culture Medium on intervening days.

4. On day 7, plate EBs onto gelatin-coated tissue-culture dishes (at densities given in Protocol 2) in Culture Medium, and incubate. Spontaneously beating cardiomyocytes should become visible 1–2 d after plating. Because of the large number of EBs that can be plated in this way it is essential to change medium regularly, otherwise changes in pH will adversely affect the longevity of the mES cell-derived cardiomyocytes.

[a] Generally, 5×10^5 to 2×10^6 mES cells (depending on the cell line) are seeded into suspension-culture dishes in the appropriate medium.

[b] Spinner flasks also can be employed for the generation of very large numbers of differentiating cells (14; and see Chapter 11 for human ES cells).

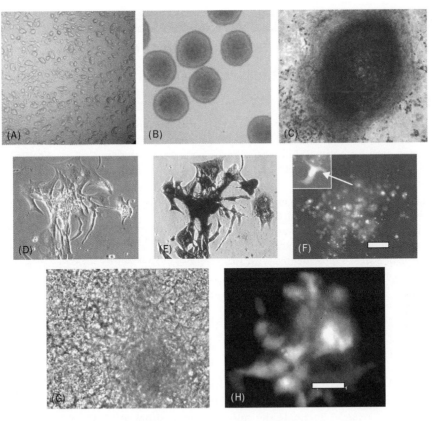

Figure 3 Cardiomyocyte differentiation in genetically modified mES cells. (A) to (C) Stages in the modification and differentiation of a mES cell line transformed with a puromycin-selection cassette controlled by the rat *MLC2v* promoter (*MLC2v-pac^r*/PGK-*neo^r*). After introduction of the construct into mES cells by electroporation, cells were treated with G418 to select for transformants. Following selection and expansion of undifferentiated, G418-resistant colonies (A), cells from a clonal line were induced to differentiate via the hanging-drop method into cardiomyocytes. Day 7 EBs in suspension culture are shown in (B). After EB plating at day 7 and following 3 d of culture, puromycin was added to the medium for a period of 4 d, and cells were stained by the periodic acid–Schiff (PAS) reaction to visualize areas of cardiogenic induction (C) (31). (D) to (H) Cardiomyocyte differentiation in clonal mES cell lines transformed with a neomycin-resistance cassette (*NCX1-neo^r*; D, E) or a puromycin-IRES-EGFP cassette (*NCX1-pac^r-IRES-EGFP*; F– H) under the control of the rat *NCX1* promoter (35). (D), (E) A cluster of differentiated *NCX1-neo^r* mES cells selected in G418 for 8 d to eliminate non-cardiomyocytes; (D) phase-contrast image, and (E) fluorescent-antibody staining with the anti-MHC antibody, MF20 (adapted from (24)). (F) Area of beating cardiomyocytes expressing EGFP within a mixed population of differentiated *NCX1-pac^r-IRES-EGFP* mES cells. After dissection (*Protocol 6*), individual cardiomyocytes continue to express EGFP (inset). (G) Phase-contrast image of an *NCX1-pac^r-IRES-EGFP* mES cell line following differentiation into cardiomyocytes and selection with puromycin: these cells express EGFP (H). Scale bars = 20 µm.

Protocol 4
Hanging-drop technique: production of skeletal muscle cells

Reagents and equipment

- Suspension of mES cells (*Protocol 1*, step 7): for lines D3 and R1, use 800 cells/20 μl hanging drop for optimal differentiation; and for line BLC6, 600 cells/20 μl hanging drop
- Differentiation media for hanging drop culture: for lines D3 and R1, use Differentiation Medium I (*Protocol 2*); and for line BLC6, Differentiation Medium II which consists of DMEM supplemented with 15% dextran-coated-charcoal-treated FCS (9), 2 mM L-glutamine, 10 μM β-mercaptoethanol, 0.1 mM non-essential amino acids, 50 U/ml penicillin, 50 μg/ml streptomycin, 10 μg/ml human transferrin (Cat. No. 11107, Invitrogen), and 0.2% bovine serum albumin Fraction V (Cat. No. 738 325, Roche Applied Science)
- PBS (*Protocol 1*)
- Culture Medium (*Protocol 1*)
- 10 cm bacteriological-grade Petri dishes (*Protocol 2*)
- Suspension-culture dishes (*Protocol 2*)
- Gelatin-coated tissue-culture dishes (*Protocol 2*)

Method

1. Place 20 μl drops (n = 60) of mES cell suspension in Differentiation Medium I or II (as appropriate) on the undersides of lids of bacteriological-grade Petri dishes containing PBS, as in *Protocol 2*. Incubate the cells in hanging drops at 37 °C for 2 d.

2. Wash and collect the aggregates carefully from the lids with 2 ml appropriate Differentiation Medium, transfer into a 60 mm suspension-culture dish containing 2–3 ml Differentiation Medium, and continue incubation.

3. At day 5 (i.e. 5 d after initiation of hanging drops at time '0') plate EBs in Culture Medium (as in *Protocol 2*). The first myoblasts appear 4 d for line BLC6, or 5–6 d for lines D3 and R1, after EB plating; and myoblasts begin to fuse into myotubes 1–2 d later. These elongated, multinucleated cells are readily visible and are identifiable either morphologically or (in some instances) following field stimulation, which induces their contraction.[a,b]

[a] Differentiation of skeletal-muscle cells can be enhanced by addition of 10^{-8} M RA (*Protocol 2*) to the medium (17).

[b] We also observe a large number of myotubes appearing approximately 2–3 weeks after plating EBs (generated using lines R1 or D3) for the production of cardiomyocytes, as described in *Protocol 2*.

3.3 Differentiation of mES cells into smooth-muscle cells (*Figure 2E, F*)

Smooth-muscle cells (SMC) can be induced to differentiate at high efficiency from progenitor cells in EBs using additives, or may be selectively purified from differentiated derivatives. Vascular smooth-muscle (VSM) cells so derived express a number of gene transcripts in common with their *in vivo* counterparts, for example VSM myosin heavy chain (*MHC*) and smooth muscle α-actin (*SMA*); and the *in vitro* differentiated cells show three distinct voltage-sensitive ion channels, which also are characteristic of VSM *in vivo* (16). Slow, peristaltic-like contractions are regarded as a defining feature of SMC differentiated from mES cells *in vitro*.

Protocol 5
Hanging-drop technique: production of vascular smooth-muscle cells

Reagents

- EBs generated using mES cell lines D3 or R1 by incubation in hanging drops (800 cells/20 µl in Differentiation Medium I) for 2 d, followed by incubation in suspension for 5 d; *or* using mES cell lines AB1 or AB2.1 by incubation in hanging drops (400 cells/20 µl in M15 medium) for 2 d, followed by incubation in suspension for 2.5 d (see *Protocol 2*)
- M15 Medium: DMEM supplemented with 15% FCS, 2 mM L-glutamine, 0.1 mM β-mercaptoethanol, 0.05 mg/ml streptomycin, and 0.03 mg/ml penicillin
- RA (*Protocol 2*)
- Dibutyryl cyclic AMP (db-cAMP; Cat. No. D0627, Sigma), made up as a 100 mM stock in distilled water
- TGF β_1 (Cat. No. TGFβ1-2, Strathmann Biotech), resuspended in 0.1 M acetic acid or 0.05 M HCl, and stored as aliquots at $-20\,°C$ in precoated (e.g. siliconized) glass tubes to prevent adsorption

Methods[a]

(i) For lines D3 and R1, plate EBs at day 7, and induce differentiation of VSM cells by treatment with 10^{-8} MRA and 0.5×10^{-3} M db-cAMP between 7 and 11 d after plating. The exact timing of treatment for these additives must be optimized for each cell line. It is critical that the medium be changed every 1–2 d. The first spontaneously contracting VSM cells, which express the vascular-specific splice variant of the VSM myosin heavy chain (*MHC*) gene, normally appear in the EBs ~1 week after plating (18).

(ii) Alternatively for D3 cells, culture EBs in Differentiation Medium II containing 2 ng/ml TGF β_1 from day 0 to 5, and plate at day 5. The first spontaneously contracting VSM cells appear in EBs 10 d later, and maximal VSM-cell differentiation (evident in ~60% of EBs) is achieved at days 5 + 24 to 5 + 28.

(iii) For lines AB1 and AB2.1, plate EBs at day 4.5. Partially exchange the medium (half of the volume) with fresh medium every third day. Maximal VSM-cell differentiation (in 30% of EBs) normally is achieved for these cell lines at days 4.5 + 17 to 4.5 + 19 (19).

Protocol 5 continued

[a] Alternative protocols for the generation of VSM cells from mES cell-derived progenitor cells have recently been described (20, 21). EBs produced from lines CCE or D1 contain Flk1$^+$ (or VEGFR2-expressing) cells, which can give rise to cells expressing smooth muscle α-actin (SMA), desmin, and smooth muscle tropomyosin. Flk1$^+$/E-cadherin$^-$ progenitor cells, isolated by sorting using anti-Flk1 and anti-E-cadherin monoclonal antibodies, are cultured as a monolayer on collagen IV-coated dishes for 2–3 d in serum-free culture medium (SFO3) in the presence of PDGF-BB (21), and induction of SMA$^+$ cells from Flk1$^+$ cells occurs 1.5–2.5 d later.

4 Isolation techniques for differentiated muscle-cell derivatives

Within differentiating cell populations, smooth-muscle or cardiac-muscle cells are generally observable in clusters and are identifiable by their typical, spontaneous contractile activity; and skeletal-muscle cells are readily distinguished morphologically by their characteristic multinuclear and bipolar structure. Isolation of a specific muscle-cell type from the heterogeneous cell populations found in EBs and their plated outgrowths can be achieved either by mechanical harvesting and disaggregation of clusters, or by positive selection for genetic markers.

The disaggregation of EB cultures usually relies on enzymes that are active on fibrous structures, such as collagenase B; but other proteases (e.g. trypsin, pronase, or dispase) may prove useful, particularly for fluorescence-activated cell sorting (FACS®), which requires single cells for effective isolation. Factors to be considered when performing enzymatic digestions include medium components, time of digestion, temperature, pH, and ions. In particular, batches of enzyme should be tested as enzyme activity, duration of digestion, and the Ca^{2+}-ion concentration in the digestion solution can all adversely affect the cell isolation process. Increasing any of these three parameters promotes digestion, but potentially leads to cell membrane damage and necrosis.

Protocol 6

Mechanical dissection and recovery of muscle-cell derivatives

Reagents and equipment

(Standard reagents are from Sigma, unless otherwise stated.)

- Cultures of EBs differentiated according to *Protocols 2–6* for cardiomyocytes, skeletal-muscle, or VSM cells, plated on gelatin-coated 60 mm dishes as described in *Protocol 2*, and incubated for 2–30 d according to the specific differentiation protocol
- Ca^{2+}-free (modified) Hanks' solution: 120 mM NaCl, 5.4 mM KCl, 5 mM $MgSO_4$,

Protocol 6 continued

  5 mM sodium pyruvate, 20 mM glucose, 20 mM taurine, and 10 mM Hepes, pH 7.2 (NaOH-adjusted), made up in HPLC-grade water (i.e. Ca^{2+}-free)
- Digestion Solution: 120 mM NaCl, 5.4 mM KCl, 5 mM $MgSO_4$, 5 mM sodium pyruvate, 20 mM glucose, 20 mM taurine, and 10 mM Hepes, pH 7.2 (NaOH-adjusted), adjusted to final pH 6.9 at 24 °C, and supplemented with enzymes and $CaCl_2$ (see below)
- Digestion supplements for cardiomyocytes and skeletal-muscle cells: 1 mg/ml collagenase B (Cat. No. 1088807, Roche Applied Science) and 30 μM $CaCl_2$[a]
- Digestion supplements for SMC: 2 mg/ml collagenase IA (Cat. No. C9891 Sigma), 0.5 mg/ml elastase Type IIA (Cat. No. 0100905, Roche Applied Science), and 1.6 mM $CaCl_2$
- KB medium: 85 mM KCl, 30 mM K_2HPO_4, 5 mM $MgSO_4$, 1 mM EGTA, 5 mM sodium pyruvate, 5 mM creatine, 20 mM taurine, 20 mM glucose, and freshly added 2 mM Na_2ATP, adjusted to pH 7.2 at 24 °C
- Culture Medium (*Protocol 1*)
- DMEM supplemented with 20% FCS
- PBS
- Sterile Pasteur pipettes, pulled over a spirit burner, etched using a diamond-tipped pen, and broken to have a bevelled end
- Aspirator tube assembly, consisting of latex tubing, nosepiece with cotton plug, and plastic mouthpiece (e.g. Cat. No. A5177, Sigma)
- Inverted phase-contrast microscope
- 15 ml polypropylene centrifuge tube
- Benchtop centrifuge
- Gelatin-coated glass slides or MatTek® dishes (*Protocol 2*)

Methods

1. Observe a culture of 15–20 plated EBs through an inverted microscope, and locate spontaneously beating areas or (under field stimulation) induced-beating areas. Harvest these areas of cells mechanically using a pulled Pasteur pipette connected to an aspirator tube assembly: with the tip of the pipette dissect as closely to the outer edges of the clusters of beating cells as possible, to minimize non-muscle-cell contamination.

2. Transfer the dissected areas of cells into a centrifuge tube containing ∼12 ml of either Ca^{2+}-free Hanks' solution or PBS at r.t. Leave the cells in the Ca^{2+}-free solution for at least 20 min (or up to 60 min for terminally differentiated cells) to minimize Ca^{2+} extrusion into the medium from damaged cells, and centrifuge at 300 g for 2 min. Aspirate the supernatant.

Isolation of cardiomyocytes:[a]

3. Add to the dissected areas of cells 1 ml Digestion Solution with supplements for cardiomyocytes. Incubate the cells at 37 °C for 25–45 min, the exact duration depending on collagenase B activity. For the isolation of cardiac clusters, shorten the incubation time to 10–20 min. Centrifuge the dissociated cells at 300 g for 2 min, and aspirate the supernatant. Resuspend the cell pellet in 0.5–1.0 ml KB medium, and incubate the tube for 30–60 min at 37 °C (or 24 °C) with gentle, periodic shaking.

Protocol 6 continued

4 Transfer the cell suspension into tissue-culture plates containing gelatin-coated slides (or into gelatin-coated MatTek® dishes), dilute the KB medium at least 1:10 with Culture Medium, and incubate at 37 °C. The next morning, change medium. Cardiomyocytes start beating within a few hours of plating and can be employed for rate studies, morphological analyses, or immunofluorescence up to 72 h after plating. For electrophysiological measurements it is advisable to incubate the cells for 18–24 h prior to experimentation. These cardiomyocytes can be maintained in culture for 5–10 d, but some of their physiological and morphological properties may change with increasing time of incubation.

Isolation of skeletal muscle cells:[a]

3 Add to the dissected areas of cells 1 ml Digestion Solution with supplements for skeletal-muscle cells. Incubate the cells at 37 °C for 20–40 min, the exact duration depending on collagenase B activity. For the isolation of clusters of skeletal-muscle cells, shorten the incubation time to 10–20 min. Halt the enzymatic digestion by addition of an equal volume of DMEM containing 20% FCS. Centrifuge the suspension at 300 g for 2 min, resuspend cells and replate in either Culture Medium or DMEM into tissue-culture plates containing gelatin-coated slides (or into gelatin-coated MatTek® dishes), and incubate at 37 °C.

Isolation of SMC:[a]

3 Add to dissected areas of cells 1 ml Digestion Solution with supplements for SMC. Digest for 30 min at 37 °C, mixing periodically. Centrifuge at 300 g for 2 min, resuspend cells and replate in either Culture Medium or DMEM into tissue-culture plates containing gelatin-coated slides (or into gelatin-coated MatTek® dishes), and incubate at 37 °C.

[a] Supplemented Digestion Solutions for muscle cells can be prepared in advance, but should not be stored longer than 3–4 d at 4 °C.

4.1 Genetically modified mES cells for enrichment of muscle-cell derivatives (*Figure 3*)

Mouse ES cells are amenable to genetic modification to facilitate the purification of specific cell types from other derivatives following *in vitro* differentiation protocols. For muscle-cell differentiation, a number of suitably modified mES cell lines are already available in which an antibiotic resistance cassette [e.g. puromycin (pac^r), hygromycin ($hygro^r$), or aminoglycoside phosphotransferase (neo^r)] is controlled by a muscle-restricted promoter sequence, thereby permitting positive selection of muscle-specific lineages (11, 22–24): once these stably transformed lines have been induced to differentiate and the specific muscle-cell type is observed, cultures are treated with antibiotic to effectively kill the non-myocytes, leaving the antibiotic-resistant muscle cells of choice. Here (Protocol 7) we provide an overall strategy for (a) obtaining transformed mES-cell clones, and (b) selectively enriching for cardiomyocytes following EB-mediated differentiation; but similar protocols should prove effective for other muscle derivatives, provided that expression of the selectable marker is controlled by a suitable tissue-restricted promoter sequence.

Protocol 7

Transformation of mES cells for selective enrichment of EB-derived cardiomyocytes

The requisite mES-cell modification involves transfection with a plasmid containing: (i) a gene allowing positive selection of transformed, undifferentiated mES-cell clones, and (ii) a gene conferring antibiotic resistance (or, alternatively, a fluorescence cell marker such as EGFP) controlled by a muscle-restricted promoter that is active in the differentiated cells. For (i), the relatively position-independent, constitutive promoters for phosphoglycerate kinase (*PGK-1*), RNA polymerase II, β-actin, and elongation factor 1 alpha (*EF-1α*) are most commonly used (30). For (ii), several tissue-restricted promoters are effective for isolation of specific muscle-cell types/lineages, e.g. the cardiac-restricted *NCX1* and α*Mhc* promoters, and the ventricular-associated *Mlc2v* promoter. For positive selection in (i) and (ii), the most commonly used genes are those conferring resistance to the antibiotics, neomycin (*neo*), puromycin (*pac*), and hygromycin (*hph*), for which the *in vitro* selective agents are geneticin (antibiotic G418), puromyocin, and hygromycin B, respectively. (Clearly, for (i) and (ii) the selectable marker genes must differ.) When mES cells are transformed to carry fluorescence markers (such as EGFP), selection is performed by FACS® analysis on single-cell suspensions derived from EB cultures.

(a) Transfection of undifferentiated mES cells:

Reagents and equipment

- Minimum of 25 mg linearized plasmid DNA containing an appropriate tissue-restricted promoter and selection cassette (e.g. *NCX1-pac^r/PGK-neo^r*)[a]
- 3 M sodium acetate
- Absolute ethanol
- Mouse ES cells, 80% confluent culture: one 60 mm dish is required per electroporation
- LIF-supplemented Culture Medium (Protocol 1)
- Trypsin/EDTA solution (Protocol 1)
- PBS (Protocol 1)
- Gene Pulser II with capacitance extender (BioRad)

- 0.4 cm electrode-gap electroporation cuvettes (Cat. No. 165–2081EDU, BioRad)
- Antibiotic: e.g. G418 (Cat. No. 11811–031, Invitrogen) or puromycin (Cat. No. P8833, Sigma)
- 96-well tissue-culture plates
- Feeder layers (Protocol 3, Chapter 2), antibiotic-resistant if available, seeded into gelatin-coated (Protocol 1) 100 mm tissue-culture dishes, 96-well and 24-well plates
- Gilson P20 pipette
- 2 × freezing mixture (Protocol 3, Chapter 2)
- Benchtop centrifuge
- Microcentrifuge
- Haemocytometer

Protocol 7 continued

Method

1. Prepare vector DNA for electroporation with two rounds of phenol/chloroform extraction (optional) and by ethanol precipitation, using 1/10 volume of 3 M sodium acetate and 2.2 volumes of ethanol. Precipitate overnight at $-20\,°C$, or for 1 h at $-70\,°C$. Pellet DNA in Eppendorf tubes by microcentrifugation at 12 000 rpm for 15 min at $4\,°C$. Discard supernatant and wash pellet twice with 0.5 ml 70% ethanol; after each wash, centrifuge at 12 000 rpm for 5 min at $4\,°C$. Remove supernatant and air-dry the DNA pellet. Dissolve pellet completely in PBS at a concentration of 1-5 mg/ml DNA.

2. Change medium on the mES cell culture 1-2 h before harvesting. Trypsinize the culture to achieve a single-cell suspension (*Protocol 1*). Centrifuge cells at 500 g, aspirate the supernatant, and resuspend the cells in 10 ml (ice-cold) PBS; recentrifuge and resuspend the cells in 10 ml PBS. Count the cells: from a 60 mm dish, $3-5 \times 10^6$ mES cells should be harvested. Recentrifuge and resuspend the cells to give $4.0-5.0 \times 10^6$ mES cells/ml in 0.8 ml PBS (or Culture Medium), and transfer the suspension to an electroporation cuvette on ice.

3. Add 20-40 µg purified plasmid DNA to the cell suspension in the recooled cuvette, and gently mix by inversion. Leave the suspension on ice for 15 min.

4. Electroporate the cells by delivering a pulse of 250 V at 500 µF. Leave the cuvette on ice for 15-20 min further (optional), and transfer the suspension to a tube containing 15 ml Culture Medium. Plate half of the tube's contents onto each of two feeder layers in 100 mm dishes with 15 ml LIF-supplemented medium. After overnight incubation, change medium; but incubate the cells for a total of 24-36 h before applying antibiotic selection.

5. Switch the cells to LIF-supplemented Culture Medium containing the appropriate selective agent (e.g. $\sim$200 µg/ml G418, or $\sim$1 µg/ml puromycin). Change medium every 1-2 d during a 7-10 d selection period, during which time resistant, clonal colonies of transformed mES cells should become visible to the naked eye.

6. Wash the cells with PBS twice, and add 6-8 ml PBS to cover the dish. Harvest individual colonies using a Gilson P20 pipet set at 7-10 µl, and transfer singly to 96-well plates containing 20 µl trypsin/EDTA solution at r.t. Once 20-50 individual colonies have been picked and transferred in this way (within 30 min), incubate the plate for 5 min at $37\,°C$. Add 50-70 µl Culture Medium to each well, gently triturate several times to disaggregate cells, and transfer cell suspensions into 96-well plates containing feeder layers, fresh medium and the selective agent.

7. Individual wells should be monitored daily, changing media when necessary to avoid acidification. Expand clones by passaging (usually after 3-4 d) into 24-well plates with feeder layers (or alternatively into two 96-well plates, a master and a replica). Incubate for 24-48 h.

Protocol 7 continued

8   Trypsinize the cells with 0.2 ml trypsin/EDTA solution at 37 °C for 5–10 min, and halt trypsinization by addition of 0.3 ml Culture Medium. Pipette suspensions up and down 10 times to mix well and fully dissociate colonies (using a fresh pipette tip for each well).

9   Remove half of the cell suspension from each well for further expansion or for DNA extraction; and freeze the remaining half *in situ*, by adding an equal volume of 2 × freezing mixture to each well and storing the multiwell plate at −80 °C.[b]

10 Once expanded, differentiate the cells as described under *Protocols 2–5*, and select muscle-restricted lineages as follows.

[a] The super-coiled plasmid DNA should be prepared either by caesium chloride centrifugation or ion-exchange chromatography.
[b] Extracted DNA is used in either PCR reactions or Southern analysis to determine the presence and integration of the selectable marker genes.

(b) Positive (antibiotic) selection for cardiomyocytes:

Reagents

- EBs derived using a transformed mES cell line carrying an antibiotic-resistance cassette controlled by a cardiomyocyte-restricted promoter (e.g. *NCX1* promoter), either plated or in suspension culture (*Protocols 2 or 3*)
- Digestion Solution for cardiomyocytes (*Protocol 6*)
- Differentiation Medium I (*Protocol 1*) supplemented with the appropriate antibiotic: for *neo*^{r}-transformed (G418-resistant) cells, use 200–400 µg/ml G418 (Cat. No. 11811-031, Invitrogen); and for *pac*^{r}-transformed (puromycin-resistant) cells, 1–6 µg/ml puromycin (Cat. No. P8833, Sigma).[a]
- Culture Medium (*Protocol 1*)

Method

1   When EB cultures (either plated or in suspension) display spontaneous contractile activity, incubate in antibiotic-supplemented Differentiation Medium I. Alternatively, EBs or outgrowths may be disaggregated using Digestion Solution (as described in *Protocol 6*, step 3, for the isolation of cardiomyocytes, but *en masse*) and replated onto gelatin-coated tissue-culture dishes prior to antibiotic treatment; under these conditions selection occurs relatively faster, but the loss of cells due to membrane damage and antibiotic-related cytotoxicity may decrease overall yields.[b]

2   Once selection is complete, as judged by morphological assessment of the complete loss of non-muscle cells, replace Differentiation Medium I with Culture Medium. For *neo*^{r}-transformed mES cells, EB cultures can be maintained in the presence of G418 for as long as required to eliminate non-cardiomycoytes (usually 1 week). For *pac*^{r}-transformed mES cells, puromycin selection (typically at 2.5–5.0 µg/ml) is maintained for at least 48 h and, depending on the dose, up to 96 h. Comparative and functional assessments may be conducted on the purified cardiomyocytes obtained

> **Protocol 7 continued**
>
> by G418 or puromycin selection, to reveal the presence of atrial- and ventricular-like lineages (see sections below).
>
> [a] If the genetic modification was achieved via random integration events (as opposed to gene targeting), mES-cell clones will contain different copy numbers of promoter-driven antibiotic-resistance cassettes; and therefore the appropriate concentration of antibiotic and duration of treatment must be determined empirically for each subline.
>
> [b] Regarding the initiation of selection, in general the highest yields of cardiomyocytes are obtained when selection is applied just after the appearance of beating cardiomyocytes; however, the exact timing is promoter dependent. For example, the cardiac-restricted *NCX1* promoter is relatively 'strong' in the early embryo, and in differentiating EBs becomes active prior to – or concomitant with – the first appearance of spontaneous contractions, when selection may be initiated. In contrast, the rat ventricular-associated *Mlc2v* promoter is relatively 'weak' during early stages of *in vitro* differentiation, and therefore may require more time (up to 5–15 d) to confer antibiotic resistance to cardiomyocytes before selection is applied.

5 Assessing muscle-gene expression during *in vitro* differentiation of mES cells

5.1 RNA isolation from cell extracts

The expression of muscle-associated genes during EB culture and ongoing differentiation is most commonly analysed by RT-PCR using isolated RNA. As a general guide, we recommend that at specified times EBs should be washed with PBS and, following cell lysis in an appropriate buffer, total RNA extracted according to established methods. We routinely purify RNA by ultracentrifugation through a caesium chloride gradient and employ guanidinium isothiocyanate, a chaotropic agent that inactivates ribonucleases (25). However, RNeasy Mini Kits (from Qiagen) or other similar kits also yield good-quality RNA. For detailed methods for RNA preparation, the reader is referred to Wobus *et al.* (9) or Chirgwin *et al.* (25).

5.2 Reverse-transcription PCR, and quantitative reverse-transcription PCR (*Figure 4*)

To determine the presence of muscle-associated gene transcripts, semiquantitative reverse-transcription PCR (or RT-PCR) may be employed (*Figure 4A*). Typically, total RNA (0.5–1.0 µg) is reverse transcribed (with MuLV reverse transcriptase) using as primers either oligo d(T)$_{16}$ or random hexamers; RT reactions are carried out at 42 °C for 0.5–1 h, and then heated at ≥95 °C for 5 min to terminate the reaction (see *Protocol 8*). Complementary DNA sequences (cDNAs) to the reverse DNA transcripts are then amplified from (~1 µl) aliquots of the RT-reactions with DNA polymerase (e.g. Taq, AmpliTaq®) during PCR using primers specific for the genes to be analysed (*Table 1*). Mouse RNA extracted from heart, soleus muscle, or aorta should be used as positive controls in RT-PCR; and a distilled water sample should always be included as a negative, no-template control (NTC). Resulting

PCR fragments are separated electrophoretically on 2% agarose gels, and analysed by computer-assisted densitometry in relation to HPRT, β-tubulin, or GAPD (household genes, or internal controls). To collect reliable data, we recommend that RT-PCR be performed two to three times on a set of three to four independent RNA preparations. Data may be evaluated using statistical analyses, as described in 'User Bulletin #2' for the ABI PRISM7700 Sequence Detection System (which is accessible at the Applied Biosystems website, http://home.appliedbiosystems.com/).

By using quantitative (q)RT-PCR methods incorporating real-time detection systems, it is possible to determine the number of amplification cycles (defined as the 'threshold cycle') necessary to detect a signal, which is indicative of the abundance of a given transcript in the original RNA preparation (*Figure 4B*). There are two forms of quantification of RNA – absolute and relative. Absolute quantification assesses the number of template copies (as moles, grams, or molecules) in the sample, by reference to a defined, standard signal. To this end, standards corresponding to the target transcript usually are synthesized from reference DNA sequence with RNA polymerase, quantified spectrophotometrically, and added to the RT reactions in precise amounts (see User Guide for 7900 Sequence detection system, PE Applied Biosystems website). Relative quantification is simpler, giving the proportion of target to reference sequence as well as signal ratios between samples. Two fluorescence-based formats are available for detecting the threshold cycle, employing: (i) fluorescent-labelled, sequence-specific hybridization probes (e.g. TaqMan®, Amplifluor®); or (ii) the DNA-binding dye, SYBR Green I, whose fluorescence increases over 100-fold on binding to the minor groove of double-stranded DNA. In general, and when multiple gene targets are to be measured from limited sample numbers, the SYBR Green 1 protocol is the method of choice; but, when a limited number of gene targets are to be quantified from numerous samples, a TaqMan®-like protocol may be considered preferable. We perform real-time quantitative PCR (or Q-PCR) reactions (*Protocol 9*) with an ABI PRISM 7900 Sequence Detection System and a SYBR Green 1 protocol, in a 384-well plate format, after determining the optimal amplification programme for each primer set. More technical details are provided in the 'SYBR Green® PCR Master Mix and RT-PCR protocol' on the PE Applied Biosystems website, as are details of the TaqMan®-based methods. (Other systems also may give reliable data.) In addition to determining the dissociation curve for each primer set, we recommend separating electrophoretically the products of Q-PCR amplifications on an agarose gel after each run. For the generation of novel primer sets we use Primer Express 2.0 software (PE Applied Biosystems), but a list of useful primers for analysing muscle-restricted gene expression by qRT-PCR is provided in *Table 2*. All procedures should be performed under conditions that minimize DNA contamination.

5.3 Non-radioactive *in situ* detection of transcripts in EBs

In situ hybridization (ISH) is a technique that permits detection of RNA (or DNA) sequences in tissues, cells, or chromosomal spreads, based on the formation

Table 1 Primers for RT-PCR amplification of muscle-specific gene transcripts

Cell type	Gene transcript	Primer sequences (5' to 3', F_w/R_v)	Product size (bp)	Annealing temp. (°C)	References
Cardiac muscle	Cardiac α-MHC	CTGCTGGAGAGGTTATTCCTCG GGAAGAGTGAGCGGCGCATCAAGG	301	64	(9)
	Cardiac β-MHC	TGCAAAGGCTCCAGGTCTGAGGGC GCCAACACCAACCTGTCCAAGTTC	205	64	(9)
	Myosin light chain-2V (MLC-2V)	TGTGGGTCACCTGAGGCTGTGGTTCAG GAAGGCTGACTATGTCCGGGAGATGC	189	60	(9)
	Myosin light chain-2A (MLC-2A)	CAGACCTGAAGGAGACCT GTCAGCGTAAAACAGTTGC	286	52	(32)
	Atrial natriuretic factor (ANF)	TGATAGATGAAGGCAGGAAGCCGC AGGATTGGAGCCCAGAGTGGACTAGG	203	64	(9)
	Nkx2.5	CGACGGAAGCCACGCGTGCT CCGCTGTCGCTTGCACTTG	181	60	(9)
Skeletal muscle	Myf5	TGCTGTTCTTTCGGGACCAGACAGG GGAGATCCTCAGGAATGCCATCCGC	132	65	(9)
	Myogenin	CAACCAGGAGGAGCGCGATCTCCG AGGCGCTGTGGGAGTTGCATTCACT	85	60	(9)
	MyoD	ATGCTGGACAGGCAGTCGAGGC GCTCTGATGGCATGATGGATTACAGCG	144	65	(9)
	Myf6	GAGGGTGCGGATTTCCTGCGCACC GGAGGCTGAGGCATCCACGTTTGC	117	60	(9)
	M-cadherin	AACTGGAGCGTCAGCCAGATTAACG GCGCGGCAAACAGGATGAGAAC	386	56	(9)
Smooth muscle	Smooth muscle myosin heavy chain (SM-MHC)	GGATGCCACCACAGCCAAGTA TGGTGTGGGTCCCTTCAGAGA	497	60	(9)
Muscle	Sarcoplasmic reticulum Ca^{2+}-ATPase	TGTGTGATGTGGAGGAAATGTGTA TACAACTGAAGGCATGCATTACAA	224	65	Unpublished
	Phospholamban (PLB)	GCTAAGCTCCCATAAGACTT CCAGACTGGAGCATATAAAGTG	695	58	(33)
	Na^+/Ca^{2+} exchanger (NCX1)	TCAAGGTAATCGATGACGAGGA TCTCTAGCATGGACCTTCCTGA	181	65	Unpublished
	L-type calcium channel	GGTAATCCACCCACGGAGAAGC GAGGCTCAAGGTCACAGCCA	295	65	Unpublished
Internal controls	β-tubulin	GGAACATAGCCGTAAACTGC TCACTGTGCCTGAACTTACC	317	60	(9)
	HPRT M1/P1	CGCTCATCTTAGGCTTTGTATTTGGC AGTTCTTTGCTGACCTGCTGGATTAC	447	60	(9)
	HPRT M2/P19	GCCTGTATCCAACACTTCG AGCGTCGTGATTAGCGATG	502	64	(9)

DIFFERENTIATION OF mES CELLS INTO MUSCLE CELLS

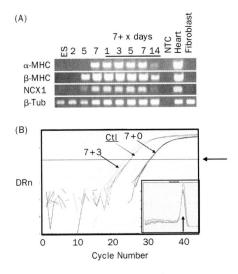

Figure 4 Analysis of muscle-restricted gene expression in R1 mES cell line-derived EBs. (A) Semiquantitative RT-PCR amplification of muscle-related gene transcripts. RNA was prepared from undifferentiated mES cells, and from EBs at the indicated time points after plating at day 7. A no-template control (NTC) was included. Primers for the housekeeping gene, β-tubulin (β-Tub), were used as an internal standard for RT-PCR reactions. (MHC = myosin heavy chain; NCX1 = Na^+/Ca^{2+} exchanger.) (B) Differentiation-regulated transcript abundance as determined by quantitative (q) RT-PCR with SYBR Green 1. In this plot of signal against PCR cycle number, smooth muscle α-actin (SMA) amplifications are shown for RNA from EBs at days 7 + 0, and 7 + 3. A shift of the curve to the left with increasing time of culture indicates increased transcript abundance. $\Delta Rn = (Rn^+) - (Rn^-)$, where Rn is the ratio of SYBR Green 1 Emission intensity to the Passive Reference (ROX) for the sample (R^+), or for the NTC/early cycles of amplification (R^-). The horizontal line (see horizontal arrow) indicates the threshold points through the amplification curve. 'Ctl' indicates the amplification plot for a control sample of known concentration. Insert: the dissociation plot for SMA amplicons shows a single spike at ~81.5 °C, which corresponds to the predicted melting temperature of this product.

of double-stranded hybrid molecules between those target sequences and complementary, single-stranded, labelled probes. For gene expression analysis in differentiating cell populations, ISH is particularly informative when combined with qRT-PCR based approaches, which cannot independently

Protocol 8
Reverse transcription

Reagents and equipment

- EB-derived RNA preparation (see Section 5.1)
- DNA-free kit (Cat. No. 1906, Ambion)
- TaqMan® Reverse Transcription Reagents (Cat. No. 8080234, PE Applied Biosystems)
- DNase-free microfuge tubes (Eppendorf)

Protocol 8 continued

- Benchtop microfuge
- Thermal cycler, e.g. Gene Amp PCR System 9700 (Applied Biosystems)
- Spectrophotometer, e.g. GeneQuant pro (Pharmacia Biotech)

Method

1. Prepare EB-derived RNA as a solution in diethyl pyrocarbonate (DEPC)-treated water (at neutral pH), quantify spectrophotometrically from the absorption at 260 nm, and adjust the concentration to $\sim 0.1\,\mu g/\mu l$. Typically the yield of RNA from EBs, where n = 20, is in the range 20–100 μg.[a]

2. Pretreat the RNA samples with DNAse as supplied in the DNA-free kit, according to the manufacturer's recommendations.

3. Using the TaqMan® system, mix reagents in the following order in Eppendorf tubes on ice.

10 × TaqMan® RT buffer	2.5 μl
DeoxyNTPs mixture (2.5 mM)	5 μl
Random hexamers (50 μM)	1.25 μl
RNase inhibitor (20 U/μl)	1 μl
Total RNA	1 μg
MultiScribe™ Reverse Transcriptase (50 U/μl)	0.725 μl
Mg^{2+} (25 mM)	5.5 μl

   Mix reagents well, either by pipetting or a quick vortex, and pulse-spin in a microfuge to bring to the bottom of the tube.

4. As a negative control for RT reactions (RTNC), prepare one tube as in Step 3 but without reverse transcriptase. *The RTNC is crucial to determining the potential presence of contaminating genomic DNA in RNA preparations.*

5. Incubate samples as follows: 25 °C, 10 min; 48 °C, 30 min; and 95 °C, 5 min.

6. Proceed to Q-PCR step immediately (Protocol 9).[b]

[a] RNA samples are stored either at −80 °C as an aqueous solution, or at −20 °C as an ethanol precipitate.

[b] Alternatively, cDNA samples can be stored either for up to 48 h at 4 °C, or indefinitely at 7–20 °C; but avoid multiple freeze/thawing of samples, which leads to degradation of cDNA.

discriminate differences in mRNA levels arising from cellular heterogeneity. For successful hybridizations a number of parameters should be considered, including: (i) the choice of probe (RNA or DNA) and method of labelling (radioactive *versus* nonradioactive); (ii) fixation protocols; (iii) hybridization conditions; (iv) washing conditions: and (v) signal detection. A thorough

Table 2 Primers for qRT-PCR amplification of muscle-specific gene transcripts

Cell type	Gene transcripts	Primer sequences (5′ to 3′, F_w/R_v)	Product size (bp)	Annealing temp. (°C)	References
Cardiac muscle	GATA-4*	GAGTGTGTCAATTGTGGGGCCATGT CCGCAGGCATTACATACAGGCTCA	242	65	(34)
	GATA-5*	CCCCACAACCTACCCAGCATACAT AGGGTTCTCTGGCTTCCGTTTTCT	499	65	(34)
	GATA-6	GCGGTCTCTACAGCAAGATGAA ACAGGCTCACCCTCAGCATT	151	65	(34)
	Nkx 2.5*	CGACGGAAGCCACGCGTGCT CCGCTGTCGCTTGCACTTG	181	60	(34)
	Hand 1	GGGTTAAACCCGGTCTTTGG GAGCAAGGCTGGAGATGACAC	101	60	(34)
	Calsequestrin*	TGAACTTCCCCACGTACGATG AAACTCAATCGTGCGGTCACC	307	60–65	(35)
	Cardiac troponin I (cTnI)	AGGGCCCACCTCAAGCA GGCCTTCCATGCCACTC	103	58	(24)
	Na^+/Ca^{2+} exchanger (NCX1)	CAGCTTCCAAAACTGAAATCGA GTCCCTCTCATCAACTTCCAAAA	100	58	(35)
Skeletal muscle	Myf 5	CTAGGAGGGCGTCCTTCATG CACGTATTCTGCCCAGCTTTT	72	57–60	(36)
	MyoD	GGACAGCCGGTGTGCATT CACTCCGGAACCCCAACAG	96	57–60	(36)
	Myogenin	GGAGAAGCGCAGGCTCAAG TTGAGCAGGGTGCTCCTCTT	47	57–60	(36)
Smooth/ cardiac muscle	Vasular smooth muscle α-Actin	TCCTGACGCTGAAGTATCCGA GGCCACACGAAGCTCGTTAT	82	65	(34)
Internal controls	M-5-β-tubulin*	GAAGAGGAGGCCTAACGGCAGAGAGCCCT GAGTGCCTGCCATGTGCCAGGCACCATTT	174	65	(34)
	GAPD*	TCCCGTAGACAAAATGGTGAAGG TTTGATGTTAGTGGGGTCTCGCT	265	65	(37)
	H1 (Histone 1)	AGATCGCGAGTCAGGTTCTG GTGGAGTTCTCGGTCATGGT	127	60	(41)

*Primers can be used for both RT and qRT-PCR.

discussion of these parameters is beyond the scope of this chapter, and the reader is instead referred to excellent reviews (26, 27). Here we present a low-throughput, non-radioactive, ISH-based technique that provides high sensitivity and high resolution in the cellular context of EBs, based on modified whole-mount procedures that were recently described (24, 28). Throughout the hybridization steps all solutions are DEPC-treated and all equipment is sterilized, unless otherwise noted. In addition, Tyramide amplification kits (Cat. No. NEL700, Perkin Elmer) may be used to enhance signal detection of hybridization on sections of EBs.

Protocol 9
Real-time quantitative PCR (Q-PCR)

Reagents and equipment

(all from PE Applied Biosystems)

- Template cDNA (*Protocol 8*)
- Sybr® Green PCR Core Reagents (Cat. No. 4306736)
- ABI PRISM 7900 Sequence Detection System, or similar
- 96-well or 384-well optical plates (Cat. Nos. 4326270 and 4306737)
- Optical adhesive covers (Cat. No. 4311971)

Method

1. Set up Q-PCR reactions in a 10 µl volume, in optical plates with covers, as follows:

H_2O	5.45 µl
cDNA	1 µl
10 × SybrGreen® PCR buffer	1 µl
DeoxyNTPs mixture (2.5 mM)	0.8 µl
Forward primer (F_w; 10 pmoles/µl)	0.2 µl
Reverse primer (R_v; 10 pmoles/µl)	0.2 µl
AmpErase® uracyl N-glycosylase (UNG; 1 U/µl)	0.1 µl
AmpliTaq Gold® DNA polymerase (5 U/µl)	0.05 µl
$MgCl_2$ (25 mM)	1.2 µl

2. Conduct an amplification program consisting of 1 cycle at 95 °C (for 10 min), followed by 40 cycles of: dissociation at 95 °C (30 s), annealing at 60–65 °C (*Table 2*) (30 s), and elongation at 72 °C (30 s).

3. To determine melting curves, and to eliminate nonspecific amplifications, run dissociation stages as follows: 95 °C (15 s), 100% ramp rate; 60 °C (15 s), 100% ramp rate; 95 °C (15 s), 2.0% ramp rate.

4. For each set of analyses include a no-template control (NTC; reaction mix with distilled water instead of cDNA), a RTNC (*Protocol 8*, step 4), and a series of dilutions of positive-control samples to generate a standard curve. Run all samples in quadruplicate.

5. Perform data analyses with SDS 2.0 software (PE Applied Biosystems), from cDNA dilutions (1:1, 1:10, 1:100, 1:1000, 1:10 000). For each primer set, the determined melting temperatures must correspond to those predicted, and a single peak should be observed.[a]

6. Calculate the amplification efficiency (E) according to the equation: $E = 10^{-1/\text{slope value}} - 1$. An ideal amplification should range from 90–100%.

7. Generate standard curves and determine relative quantities of gene expression according to the protocols given in User Bulletin #2, ABI PRISM7700 Sequence Detection System, on http://home.appliedbiosystems.com/.

Protocol 9 continued

^a If determined and predicted melting temperatures are dissimilar or multiple peaks are observed, the data are considered unreliable and new primer sets must be generated.

Protocol 10
Digoxigenin-labelled RNA probes

The RNA probe-labelling system we routinely employ is based on digoxigenin (DIG), and has the advantage of a high labelling efficiency. The detection system uses alkaline phosphatase-conjugated anti-DIG antibodies to provide a colourimetric assay: here, nitroblue tetrazolium/5-bromo-4-chloro-3-indolyl-phosphate (NBT/BCIP) is the substrate for alkaline phosphatase, giving a blue precipitating product.

Reagents and equipment

- Purified template DNA (cDNA), cloned into the multiple cloning site of a plasmid containing promoters for SP6, T7, or T3 RNA polymerases – the orientation of the insert must be determined
- NBT/BCIP colour development substrate (Cat No. S3771, Promega)
- Anti-DIG antibody solution: sheep anti-DIG, alkaline phosphatase-coupled antibodies (Cat. No. 11093272001, Roche Applied Science)
- RNase-free microfuge tubes (Eppendorf)
- Hybond N$^+$ membranes (Amersham)
- DIG RNA Labelling Kit (SP6/T7) (Cat. No. 11175025001, Roche Applied Science)

Method

1. Linearize the cloned plasmid with an appropriate restriction enzyme to give a synthesized transcript of 500–600 bp, an optimal length for DIG-labelled RNA probes.

2. Prepare DIG-labelled probes according to the manufacturer's instructions using DIG-UTP. Typically, 1 µg template DNA provides ~10 µg DIG-labelled probe. The labelling efficiency can be determined by dot blot, using anti-DIG-alkaline phosphatase and NBT/BCIP colour development substrate, according to the manufacturer's protocol. Labelled probes are stable for 1 year when stored at -20 °C. Avoid repeated freezing and thawing of probes.

5.3.1 Probe preparation

Probes for ISH can be either RNA or DNA (single- or double-stranded). RNA probes are particularly effective, and are generated from plasmids containing SP6, T7, or T3 RNA polymerase recognition sequences. Ideally, the plasmid should contain two such recognition sites, placed at either end of the DNA insert, so that

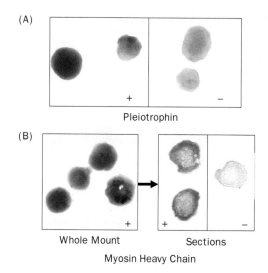

Figure 5 Examples of ISH performed on EBs generated from the R1 line of mES cells, for the localization of muscle-associated transcripts. (A) Whole-mount ISH with DIG-labelled RNA probes generated against pleiotrophin. (B) Whole-mount ISH, and ISH on sections of EBs, with probes against myosin heavy chain (MHC). Dark staining indicates the location of pleiotrophin or MHC transcripts in these preparations. Probe orientations are indicated by '+' for the antisense, and '−' for the sense probe.

both the identical (negative control) and complementary (experimental) probes to the target species can be synthesized by RNA polymerase. An important consideration is that hybrid formation in solution is proportional to the single-stranded probe's length; this however must be counterbalanced, in the context of cellular aggregates, with the relative ease of diffusion of smaller probes into the dense matrix surrounding the cell.

5.3.2 Whole-mount *in situ* hybridization of EBs (*Figure 5*)

ISH is a cornerstone of molecular analysis in developmental biology, having proven fundamental to the delineation of patterns of gene expression during development. It also represents the method of choice for visualizing mRNA molecules in circumstances where they are distributed in a non-random, topographical manner, as in EBs.

6 Assessing expression and function of muscle-specific proteins

6.1 Immunofluorescence (*Figure 2*)

Due to the inherent heterogeneity of EBs, differentiated derivatives are commonly identified and discriminated by immunohistochemical techniques

Protocol 11
Whole-mount ISH of EBs

Reagents and equipment

- EBs (n = 20) in suspension culture (*Protocols 2–5*)
- PBS (*Protocol 1*)
- Paraformaldehyde/PBS solution, freshly prepared: 4% paraformaldehyde (Cat. No. P-6148, Sigma) in PBS (see Chapter 9, *Protocol 6*)
- PBST: 0.1% Tween-20 (Cat. No. P-1379, Sigma) in PBS
- Paraformaldehyde/PBST solution, freshly prepared: 4% paraformaldehyde in DEPC-treated PBST
- Methanol solutions: 25%, 50%, and 75% methanol in PBST
- Proteinase K solution: 10–20 µg/ml proteinase K (Cat. No. 25530-015, Invitrogen) in DEPC-treated PBST
- Glycine solution: 2 mg/ml glycine (Cat. No. G-7403, Sigma) in DEPC-treated PBST
- Hybridization mix: 50% formamide, 5 × SSC pH 7.0, 1% blocking solution (Cat. No. 1175041, Roche Applied Science), 5 mM EDTA, 0.1% Tween-20, 0.1% CHAPS (Cat. No. 220201, Calbiochem), 0.1 mg/ml heparin (Cat. No. 4-4784, Sigma), and 1 mg/ml yeast total RNA (Cat. No. 0109223, Roche Applied Science)
- Wash Solution I: 50% formamide, 5 × SSC (adjusted to pH 4.5 with citric acid), and 0.1% Tween-20
- Wash Solution II: 50% formamide, 2 × SSC (pH 4.5, as above)
- Blocking buffer: 1% blocking solution and 10% goat serum (Cat. No. G-9023, Sigma) in PBST
- Anti-DIG antibody solution (*Protocol 10*)
- NBT/BCIP colour development substrate (*Protocol 10*)
- NTM-T solution: 100 mM NaCl, 100 mM Tris pH 9.5, 50 mM $MgCl_2$, and 0.1% Tween-20
- NTM-T solution with NBT/BCIP: 4.5 µl NBT and 3.5 µl BCIP per ml NTM-T solution
- NTM-T/levamisole solution: NTM-T solution supplemented with 5 mM levamisole (Cat. No. L 9756, Sigma)
- 5 ml culture tubes (Cat. No. 35-2058, Falcon) for fixation and dehydration
- 3 ml transfer pipette, with large-bore tip
- 1.8 ml round-bottom cryotubes (Cat. No. 368632, Nunc)
- Sterile 50 ml centrifuge tubes
- Glass hybridization tubes
- Hybridization oven

Method

1 Transfer the suspension of EBs into a 5 ml culture tube using a wide-bore pipette (cut the tip if necessary). Once EBs have settled to the bottom, aspirate medium and wash EBs with PBS. Allow the EBs to re-settle and aspirate PBS.[a]

2 Fix EBs with paraformaldehyde/PBS solution either for 1 h at r.t., or overnight at 4 °C.

3 Wash fixed EBs twice with PBS for 5 min at r.t.[b]

Protocol 11 continued

4. Treat EBs with proteinase K solution at r.t. with rocking: typically, smaller EBs at day 2–4 need only 4–5 min digestion, whereas day 6–8 EBs require ∼10 min. Halt digestion by removing proteinase K solution and washing EBs with freshly prepared glycine solution for 5 min at r.t. Rinse twice in PBST for 5 min at r.t.

5. Post-fix samples with paraformaldehyde-PBST solution for 20 min at r.t. Discard post-fixative, and wash EBs twice with PBST for 5 min at r.t.

6. Transfer ≥10 EBs per probe into individual cryotubes; add 0.9 ml hybridization mixture per cryotube, secure the tops, and place the tubes inside 50 ml centrifuge tubes. Place these in turn inside a hybridization tube and incubate in a hybridization oven for 1 h at 70 °C with rotation, for prehybridization.

7. Add DIG-labelled RNA probe to 100 µl hybridization mixture, and heat for 5 min at 95 °C. Quench on ice, and add probe to the prehybridization solution containing the EBs. The final probe concentration should be determined empirically, but typically 0.5–2.0 ng/µl is sufficient. Incubate overnight (i.e. >8 h) at 70 °C with rotation, for hybridization.

8. From this point on, solutions no longer need to be DEPC-treated. Remove hybridization solution from EBs in cryotubes; transfer the hybridization solution to a clean Eppendorf tube and freeze.[c]

9. Rinse EBs briefly with Wash Solution I, twice at r.t.; then wash three times with Wash Solution I for 30 min with rotation at 70 °C.

10. Wash EBs with Wash Solution II, three times for 30 min with rotation at 65 °C; transfer EBs to 5 ml culture tubes, and wash with PBST three times for 5 min at r.t.

11. Block non-specific antibody binding by incubating EBs in 1.5 ml blocking buffer for 1.5–2 h at 4 °C. Meanwhile, preabsorb the anti-DIG antibody solution by diluting in blocking buffer to a concentration of 1:2000 to 1:3000, and incubating at 4 °C. This and all subsequent incubations and washing of EBs in Steps 12 and 13 are performed with rocking.

12. Add 1.5 ml of the antibody solution to the tubes containing EBs. Depending on the strength (i.e. efficiency of labelling) of the probe, the final concentration of antibody should be between 1:4000 and 1:6000. Incubate tubes overnight at 4 °C.[d]

13. Wash EB with PBST three times for 5 min at r.t., six to eight times for 30 min at r.t., and finally overnight at 4 °C.[e]

14. Wash EBs once in fresh NTM-T, and a second time in NTM-T/levamisole, for 10 min at r.t.[f]

15. Incubate EBs in NTM-T with NBT/BCIP colour development substrate until the desired degree of colouration is reached or when the negative controls, hybridized with the sense probe, begin to develop colour (i.e. background appears). Staining can take 1–18 h at r.t. Halt colour development by washing samples briefly in NTM-T, and then three times with PBST for 15 min at r.t.[g]

Protocol 11 continued

16 Fix the stain by incubating samples in paraformaldehyde/PBST overnight at 4, then wash EBs three times with PBST for 5 min at r.t.

17 Photograph the stained EBs as soon as possible, as the signal may fade or turn blue with storage. For extended storage at 4 °C, EBs should be taken through a series of increasing concentrations of glycerol (up to 80%) in PBST. *This procedure does not remove the staining.*

[a] Unless otherwise specified, volumes of ~1.5 ml are used for processing and washing EBs.

[b] If EBs are to be processed immediately, proceed to Step 4. For intermediate storage, dehydrate EBs through steps of 25%, 50%, and 75% methanol in PBST, for ~7 min at each step. Finally, store EBs in 100% methanol at −20 °C for up to 1 year. Immediately prior to use, rehydrate EBs through steps of 75%, 50%, and 25% methanol in PBST, then twice in PBST, with 3–5 ml volumes and rocking at r.t. for 7 min at each step.

[c] Probes frozen in this manner can be reused several (at least four) times, effectively increasing the signal-to-noise ratio for future hybridizations.

[d] The antibody mixture also can be reused several times within 1 week.

[e] Extensive washing is particularly important for the larger EBs.

[f] Levamisole inhibits endogenous alkaline phosphatase activity.

[g] Washing in PBST will turn the stain blue and decrease the background; and so this step can be shortened or lengthened as required, depending on the strength of the signal.

Protocol 12
Cryosectioning of stained EBs

Reagents and equipment

- Stained EBs, in either PBST or 80% glycerol in PBST (*Protocol 11*)
- 3 ml transfer pipettes, large bore
- Tissue-Tek® Cryomold® #4565, 10 mm × 10 mm × 5 mm (Cat. No. 62534-10, Electron Microscopy Sciences)
- Esco Superfrost® Plus glass slides (Cat. No. 71869-10, Electron Microscopy Sciences)
- Coverslips
- Cryostat
- Steel or disposable microtome knives (Delaware Diamond Knives)
- Tissue-Tek® O.C.T. freezing compound 4583 (Cat. No. 62550-01, Electron Microscopy Sciences)
- Paraformaldehyde/glutaraldehyde solution: 4% paraformaldehyde (*Protocol 11*) and 0.2% glutaraldehyde (Cat. No. G7651, Sigma) in PBS
- Distilled water
- Graded series of ethanol solutions in distilled water: 25%, 50%, 75%, 95%, and 100% ethanol
- Xylene (Cat. No. X-2377, Sigma)
- Non-aqueous mounting medium, such as Entellan® (Cat. No. 14802, Electron Microscopy Sciences)
- Isopentane, precooled in liquid nitrogen
- 0.1% Nuclear fast red (Cat. No. N 8002, Sigma) in distilled water

Protocol 12 continued

Method

1. Transfer EBs from 80% glycerol/PBST into freezing compound, and incubate for 30 min at r.t.
2. Transfer EBs into a cryomold containing freezing compound, and either transfer to −80 °C or quick-freeze in isopentane precooled in liquid nitrogen.[a]
3. When ready to section bring samples to −20 °C, and cut sections to desired thickness using a cryostat. Sections of 10–15 microns (μm) usually give good results.
4. Collect sections on microscope slides, and dry for no more than 1 h at r.t.
5. Fix sections in paraformaldehyde/glutaraldehyde solution for 5 min, and rinse with distilled water.
6. Counterstain sections, if desired, using nuclear fast red or other compatible stains (depending on the researcher's experimental requirements); dehydrate through a graded ethanol series (50%, 75%, 95%, 95%, 100% ethanol) for 1 min at each step, followed by xylene for 2 min; and mount coverslips using a non-aqueous mounting medium, such as Entellan®.

[a] Samples can be stored frozen for an extended period at −80 °C.

---

based on cell-specific proteins. In particular, the presence of muscle-associated proteins within cells, either in outgrowths or isolated from EBs, may be detected readily by immunofluorescence (using either a fluorescence or a confocal microscope equipped with epifluroescence optics, and with appropriate filter systems for excitation and visualization of fluorochromes). Monoclonal, and some polyclonal, antibodies against tissue-specific intermediate-filament proteins or sarcomeric proteins are suitable for the characterization of mES cell-derived muscle phenotypes; however, it should be noted that not all antibodies are tissue-restricted. For example the presence of desmin may indicate skeletal muscle, cardiac muscle, and also some types of smooth-muscle cells, whilst α-actinin is typical of both forms of striated muscle. Analysis of muscle-restricted isoforms of actins and myosins, however, allows muscle cell types to be distinguished (Table 3).

The method for fluorescence analysis of EB-derivatives described here features the fixation of cells to retain the intracellular distribution of antigens whilst preserving overall cellular morphology. This is followed by the permeabilization of cells to ensure subsequent access of the primary antibody to the epitope. The epitope-bound primary antibody is subsequently identified by incubation with a fluorescence-tagged, secondary antibody. (Useful information with respect to fluorescence techniques and fluorescent probes can be found on the website http://www.probes.com/handbook/.)

Table 3 Antibodies for immunostaining of muscle-restricted proteins in differentiated EB derivatives

Muscle cell type	Cell-specific antigens	Antibody	Antibody type	Company	References
Striated	Titin (Z-disk)	T11, T12	m	United States Biology	(9)
	Titin (M-band)	T51	m	United States Biology	(9)
	α-Actinin	653	m	BD Biosciences Pharmingen,	(9)
	Myomesin	MyBB78	m	–	(9)
	Sarcomeric MHC	MF20	m	Hybridomabank	(9)
	Sarcomeric α-actin	5C5	m	Abcam	(9)
	M-Protein	MpAA241	m	Alpha Diagnostic International	(9)
	Cardiac α-MHC	BA-G5	m	–	(9)
	Desmin	Desmin	m	Abcam,	(38)
	SERCA2A	SERCA2a	p	–	(39)
Cardiac	Connexin 45	Cx45	p	Alpha Diagnostic International	(40)
	ANP	ANP (FL-153)	p	Biodesign International	Unpublished
	Troponin T	TnT	m	Biodesign.International	(9)
Skeletal	Nebulin	Nb2	m	–	(9)
	Slow myosin-binding protein C	L-sMyBP-C	m	–	(9)
	M-Cadherin	M-Cadherin	m	BD Biosciences/ Pharmingen,	(9)
	Myogenin	Myogenin	m	–	(38)
Smooth	SM-MHC (fast)	MY-32	m	Abcam	(9)
Smooth and cardiac	Smooth muscle α-actin	1A4	m	Abcam	(9)

'm' = monoclonal; 'p' = polyclonal.

Protocol 13
Immunofluorescence analysis of EB outgrowths

Reagents and equipment

- Undifferentiated mES cells or day 5–30 EB cultures, cultured on glass coverslips or MatTek® (35 mm) dishes (see Sections 3 and 4)
- 2% paraformaldehyde solution: prepared from a 1:1 mixture of 4% paraformaldehyde solution (*Protocol 11*) and PBS
- PBS pH 7.2, without Ca^{2+} or Mg^{2+} (Chapter 2, *Protocol 1*)
- Glycine/PBS solution: 100 mM glycine (Cat. No. G-7403, Sigma) in PBS
- Triton X-100 solution: 0.2% Triton X-100 (Cat. No. T 9284, Sigma) in PBS

Protocol 13 continued

- Goat serum (Cat. No. 50-062, Zymed Laboratories): dilute to 10% in PBS
- Primary antibodies (Table 3)
- Fluorescence-conjugated secondary antibodies (commercial preparations)
- Hoechst dye: 10 mg/ml Hoechst 33342 (Cat. No. B2261, Sigma) in distilled water
- TO-PRO solution: dilute TO-PRO-3-iodide (Cat. No. T3605, Molecular Probes) 1:5000 in PBS
- Mounting medium: Vectashield® Mounting Medium (Cat. No. H-1000, Vector Laboratories)
- Humidified chamber
- Fluorescence microscope with Zeiss objectives (e.g. Axiovert 35), or Confocal microscope (e.g. CLSM 410, Carl Zeiss)

Method

1. Wash undifferentiated mES cells or EB cultures twice with PBS at r.t. Fix the cells with 2% paraformaldehyde solution for 10 min at r.t., and wash twice with glycine/PBS solution for 15 min at r.t.

2. To permeabilize the fixed cells, incubate with Triton X-100 solution for 5 min at r.t. (the timing is crucial!); and wash twice with chilled glycine/PBS solution (at 4–10 °C) for 10 min.

3. Incubate the fixed cells in 10% goat serum for 1 h at 37 °C (or overnight at 4 °C). Remove goat serum by aspiration.

4. Apply to each sample (500 µl) primary antibody solution diluted in 10% goat serum (e.g. 1:200, but the optimum dilution must be determined for each antibody); and incubate in a humidified chamber at either r.t. or 37 °C (depending on the antibody) for 30–60 min, or preferably at 4 °C overnight. Rinse samples with PBS, five times for 10 min at r.t.

5. Incubate samples with secondary antibody solution diluted in 10% goat serum, in a humidified chamber for 45 min at r.t. (Depending on the primary antibody, either dilute goat anti-mouse IgG 1:1000, or prepare goat anti-rat IgG at 12 µg protein/ml final concentration.) Wash samples with PBS, five times for 10 min at r.t. After the fourth wash, cell nuclei can be either washed again or counterstained with Hoechst dye or TO-PRO solution, for 5 min at 4 °C.[a]

6. Embed coverslips in prewarmed (65 °C) mounting medium (1 drop), and analyse immunolabelled cells using either a conventional fluorescence microscope (equipped with appropriate filter cubes) or a confocal laser-scanning microscope.[b]

[a] The period of TO-PRO staining must not exceed 5 min; and the stained cells should be washed once more with PBS prior to visualization.

[b] Excitation/emission wavelengths for commonly used fluorochromes conjugated to antibodies can be found on the website http://micro.magnet.fsu.edu/primer/techniques/fluorescence/fluorotable2.html. Examples of common excitation/emission wavelengths include: 355/465 nm for Hoechst 33342; 490/520 nm for fluorescein isothiocyanate (FITC); and 578/603 nm for the rhodamine substitute, Alexa™ 568.

6.2 Functional analyses of muscle derivatives

6.2.1 Rate measurements in EBs

Cardiomyocytes differentiated from mES cells develop appropriate physiological properties, express muscle-specific receptors, and show differentiation-dependent signal transduction mechanisms. Measurements of beating frequencies (*Protocol 14*), as a function of differentiation time and after application of cardiotropic agents, can be useful for the analysis of the physiological responses of these mES-cell derivatives. Such data can be used to determine the degree of cellular maturation, and the correlation between ion channel expression and electrical activity, in the early embryonic-like cardiomyocytes.

Protocol 14
Rate measurements in EBs

EBs should be assayed when beating cardiomyocytes comprise at least 5%, and up to 30%, of the EB outgrowth's area.

Reagents and equipment

- Day 5–30 EBs, cultured singly after plating on gelatin-coated, 24-well plates (see Sections 3 and 4)
- Culture Medium (*Protocol 1*)
- Cardiotropic drugs
- Phase-contrast inverted microscope fitted with a 37 °C heating plate and a CO_2-gassed incubation chamber
- Video-based camera system, e.g. Video Edge Detector (Colorado Video), and TV camera

Methods

1. One day prior to beating-rate measurement, change medium adding *exactly* 1 ml fresh Culture Medium per well.
2. Place the microwell plate onto the microscope stage and localize independent, beating areas (n = 20). Observe and record the spontaneous beating frequency of these areas – if possible after relaying the image through a video-edge detector system. Rate data should be analysed in real-time using a threshold-crossing criterion, and averaged every 5 s. Measurements should be for 1–15 min.[a]
3. To test the effects of cardiotropic drugs on beating rates, add appropriate concentrations (e.g. 10^{-9}–10^{-4} M) of test substances and incubate for about 3 min. Depending on the pharmacological agent, optimal incubation times can vary considerably and should be based on the pharmacological ED_{50}. Once an optimal time has been determined, ascertain dose-dependent effects on the beating frequency after cumulative application (or with increasing concentrations) of the test substance.[b]

Protocol 14 continued

4 Calculate the mean values (± standard error of the mean) of beats per min for each data point from the beating rates of the mES cell-derived cardiomyocytes, with and without addition of drugs. Test for significance by the Mann–Whitney U-test, and calculate dose–response curves.

[a] The automated LUCIA 'HEART' Imaging System also can be used for data acquisition and processing (9), but this system must be modified for use within the US.

[b] We recommend addition of either positive chronotropic drugs for which Bay K8644 may be used as a control (Cat. No. B6174, Sigma), or negative chronotropic drugs with Diltiazem (D2521, Sigma) as a control, at concentrations in the range of 10^{-9}–10^{-5} M (depending on the test substances).

6.3 Electrophysiological measurements in mES cell-derived cardiac cells

Mouse ES cell-derived muscle cells are well suited for electrophysiological measurements, tolerating both whole-cell patch and high-resistance microelectrode recording techniques, the details of which are beyond the scope of this chapter. However, brief descriptions of electrophysiological measurements are given below. For such experiments, cardiomyoctes are generated from EBs, isolated, and cultured on coverslips or MatTek® dishes as described in Sections 3 and 4: cardiomyocytes cultured on a glass substratum are distinguishable either by their morphology or their contractile activity (e.g. as demonstrated in Ca^{2+}-containing Hanks' solution).

6.3.1 Voltage clamp and action potential recordings

It is possible to form 'giga-seals' on cardiomyocytes with standard patch pipettes; and, by varying the contents of the pipette, to study specific ion currents (such as the Na^+ and K^+ currents, and L- and T-type Ca^{2+} currents). During early stages of their differentiation or maturation, mES cell-derived cardiomyocytes have no T-tubular system and a small cellular capacitance (ranging from 10–25 pF), which features may be exploited to evaluate the role of specific structural elements within the cell. High-resistance microelectrodes can be used to evaluate the transmembrane potential, and to record both spontaneous and evoked action potentials. Voltage or current clamping to control the transmembrane potential is best accomplished with larger whole-cell patch electrodes, which provide more reliable control of membrane currents. ('Rundown' of the cells is, however, a possible problem, largely due to cell dialysis induced by the contents of the micropipette.) Several well-established systems for these experiments are commercially available. The P-Clamp system of software from Axon Instruments is widely employed for these purposes, in combination with one of their several patch clamp amplifiers, for example Axopatch-1D patch clamp amplifier.

Voltage clamp recordings are made on mES cell-derived cardiomyocytes in the voltage-clamp mode. For these experiments a microelectrode is inserted into a cell, and the cell's membrane potential is held at a predefined level. To achieve a giga-seal between the patch pipette and the plasma membrane, it is necessary to apply suction immediately after cell contact to avoid damaging the cell. The pressure in the pipette should be set to 0 before contact, and pipette resistance observed. Resistance should change only slightly after contact (e.g. from 3 mΩ to 3.5 mΩ). For successful patch clamping, choose the bulkiest part of the cell (usually in the geometric centre) where there is room along the vertical, Z-axis. Action potential (AP) parameters (amplitude, duration, maximum diastolic potential, upstroke velocity) can be acquired on isolated myocytes in the current-clamp mode using a perforated patch-clamp technique to record spontaneous APs. The pipette solution for perforated patch experiments may consist of the following: 120 mM K^+-gluconate, 5 mM NaCl, 5 mM Mg^{2+} ATP, 5 mM Hepes, and 20 mM KCl, adjusted to pH 7.2. For the evaluation of the resting membrane potential, the minimal diastolic potential can be taken and the AP duration estimated (29).

Microelectrode array (MEA) systems have been developed (by Multi Channel Systems MCS GmbH) to allow long-term, high-resolution, functional assessments of spontaneously contracting cardiomyocytes within EBs. Once EBs are plated on the MEA, field potentials are recorded over a period of several days from multiple non-invasive electrodes having an inter-electrode distance of 100 or 200 µm. This technique is useful for analysis of AP propagation, and for characterization of pacemaker activity.

6.3.2 Confocal imaging

The spatial and temporal distribution of intracellular ions (e.g. Ca^{2+}) and of specific enzyme markers (e.g. the ATPase family) in muscle cells can be gained utilizing fluorescent probes. A variety of useful probes are available from Molecular Probes (http://www.probes.com/), and the reader is referred to their catalogue for dye specifications and selection criteria. As the basis for measurement is the change in fluorescent characteristics of the dyes when bound, the selection of appropriate dyes requires careful consideration of the potential buffering effects of the dye on the process being studied. In addition, the dissociation constant, Kd, of the dye must be considered with respect to expected levels of ion or compound being detected. Finally, the selection of a particular dye may depend on the availability of instrumentation. Two types of ratiometric dyes deserve further discussion. (i) A dual-emission dye shifts its fluorescence emission peak between the bound and unbound forms, and thereby reports the amount of free target substance in the form of a ratio of bound to free. These dyes are excited by wavelengths shorter than their emission wavelengths; however, wavelengths much below 300 nm require more costly quartz optics. An example is Indo-1, an UV light–excitable (351–364 nm), ratiometric Ca^{2+} indicator, whose emission maximum shifts from $\sim$400 nm, when saturated with Ca^{2+}, to 475 nm in Ca^{2+}-free medium. (ii) A ratiometric dye of the second class is measured at a fixed wavelength, but its excitation maximum shifts when bound (e.g. Fura-2). While the former class requires (ideally) two detectors and one light source, allowing simultaneous measurement of bound and unbound forms, the latter requires only one detector but two different excitation wavelengths, which must be presented to the sample individually by scanning the excitation spectrum making simultaneous measurements impossible.

For confocal imaging of intracellular Ca^{2+}, mES cell-derived cardiomyocytes may be loaded with Fluo-4-acetoxymethyl ester (22 μM) or Fluo-3-AM (5 μM) (Molecular Probes) for 10 min at 25 °C. Intracellular Ca^{2+}, or $[Ca^{2+}]_I$, fluorescence is excited by 488 nm argon laser, and measured using an inverted confocal microscope (e.g. LSM 410, Carl Zeiss) working in line-scan mode (29). Data can be analysed with IDL software (version 5.2, Research Systems Inc.).

Acknowledgements

We would like to thank Dr Anna Wobus (Germany) for sharing information, expertise, and photographs, David G. Crider M.Sc. for technical assistance, Dr Antoon Moorman (Netherlands) for help in establishing the ISH protocols, and Dr Harold Spurgeon for comments pertaining to the electrophysiological presentation.

References

1. Evans, M.J. and Kaufman, M.H. (1981). *Nature*, **292**, 154-6.
2. Martin, G.R. (1981). *Proc. Natl. Acad. Sci. USA*, **78**, 7634-8.
3. Wobus, A.M., Holzhausen, H., Jakel, P., and Schoneich, J. (1984). *Exp. Cell Res.*, **152**, 212-9.
4. Doetschman, T.C., Eistetter, H., Katz, M., Schmidt, W., and Kemler, R. (1985). *J. Embryol. Exp. Morphol.*, **87**, 27-45.
5. Moorman, A.F. and Christoffels, V.M. (2003). *Physiol. Rev.*, **83**, 1223-67.
6. Buckingham, M., Bajard, L., Chang, T., Daubas, P., Hadchouel, J., Meilhac, S., Montarras, D., Rocancourt, D., and Relaix, F. (2003). *J. Anat.*, **202**, 59-68.
7. Gittenberger-de Groot, A.C., DeRuiter, M.C., Bergwerff, M., and Poelmann, R.E. (1999). *Arterioscler. Thromb. Vasc. Biol.*, **19**, 1589-94.
8. Boheler, K.R., Czyz, J., Tweedie, D., Yang, H.T., Anisimov, S.V., and Wobus, A.M. (2002). *Circ. Res.*, **91**, 189-201.
9. Wobus, A.M., Guan, K., Yang, H.-T., and Boheler, K.R. (2002). *Methods Mol. Biol.*, **185**, 127-56.
10. Boheler, K.R. (2003). *Methods Enzymol.*, **365**, 228-41.
11. Ventura, C., Zinellu, E., Maninchedda, E., Fadda, M., and Maioli, M. (2003). *Circ. Res.*, **92**, 617-22.
12. Hescheler, J., Fleischmann, B.K., Lentini, S., Maltsev, V.A., Rohwedel, J., Wobus, A.M., and Addicks, K. (1997). *Cardiovasc. Res.*, **36**, 149-162.
13. Wobus, A.M., Guan, K., Jin, S., Wellner, M.C., Rohwedel, J., Ji, G., Fleischmann, B., Katus, H.A., Hescheler, J., and Franz, W.M. (1997). *J. Mol. Cell. Cardiol.*, **29**, 1525-39.
14. Hescheler, J., Wartenberg, M., Fleischmann, B.K., Banach, K., Acker, H., and Sauer, H. (2002). *Methods Mol. Biol.*, **185**, 169-87.
15. Sachinidis, A., Gissel, C., Nierhoff, D., Hippler-Altenburg, R., Sauer, H., Wartenberg, M., and Hescheler, J. (2003). *Cell. Physiol. Biochem.*, **13**, 423-9.

16. Rohwedel, J., Maltsev, V., Bober, E., Arnold, H.H., Hescheler, J., and Wobus, A.M. (1994). *Dev. Biol.*, **164**, 87-101.
17. Wobus, A.M., Rohwedel, J., Maltsev, V., and Hescheler, J. (1994). *Roux's Arch. Dev. Biol.*, **204**, 36-45.
18. Drab, M., Haller, H., Bychkov, R., Erdmann, B., Lindschau, C., Haase, H., Morano, I., Luft, F.C., and Wobus, A.M. (1997). *FASEB J.*, **11**, 905-15.
19. Weitzer, G., Milner, D.J., Kim, J.U., Bradley, A., and Capetanaki, Y. (1995). *Dev. Biol.*, **172**, 422-39.
20. Yamashita, J., Itoh, H., Hirashima, M., Ogawa, M., Nishikawa, S., Yurugi, T., Naito, M., and Nakao, K. (2000). *Nature*, **408**, 92-6.
21. Fraser, S.T., Yamashita, J., Jakt, L.M., Okada, M., Ogawa, M., and Nishikawa, S. (2003). *Methods Enzymol.*, **365**, 59-72.
22. Klug, M.G., Soonpaa, M.H., Koh, G.Y., and Field, L.J. (1996). *J. Clin. Invest.*, **98**, 216-24.
23. Muller, M., Fleischmann, B.K., Selbert, S., Ji, G.J., Endl, E., Middeler, G., Muller, O.J., Schlenke, P., Frese, S., Wobus, A.M., Hescheler, J., Katus, H.A., and Franz, W.M. (2000). *FASEB J.*, **14**, 2540-8.
24. Fijnvandraat, A.C., van Ginneken, A.C., Schumacher, C.A., Boheler, K.R., Lekanne Deprez, R.H., Christoffels, V.M., and Moorman, A.F. (2003). *J. Mol. Cell. Cardiol.*, **35**, 1461-72.
25. Chirgwin, J.M., Przybyla, A.E., MacDonald, R.J., and Rutter, W.J. (1979). *Biochemistry*, **18**, 5294-9.
26. Moorman, A.F.M., de Boer, P.A.J., Vermeulen, J.L.M., and Lamers, W.H. (1993). *Histochem. J.*, **25**, 251-66.
27. Moorman, A.F., Houweling, A.C., de Boer, P.A., and Christoffels, V.M. (2001). *J. Histochem. Cytochem.*, **49**, 1-8.
28. Fijnvandraat, A.C., De Boer, P.A., Deprez, R.H., and Moorman, A.F. (2002). *Microsc. Res., Tech.*, **58**, 387-94.
29. Yang, H.T., Tweedie, D., Wang, S., Guia, A., Vinogradova, T., Bogdanov, K., Allen, P.D., Stern, M.D., Lakatta, E.G., and Boheler, K.R. (2002). *Proc. Natl. Acad. Sci. USA*, **99**, 9225-30.
30. Hasty, P. and Bradley, A. (1994). Gene targeting vectors for mammalian cells. In *Gene Targeting - a Practical Approach* (ed. A.L. Joyner), pp. 1-30, IRL Press, Oxford.
31. Pasumarthi, K.B. and Field, L.J. (2002). *Methods. Mol. Biol.*, **185**, 157-68.
32. Kubalak, S.W., Miller-Hance, W.C., O'Brien, T.X., Dyson, E., and Chien, K.R. (1994). *J. Biol. Chem.*, **269**, 16961-70.
33. Ganim, J.R., Luo, W., Ponniah, S., Grupp, I., Kim, H.W., Ferguson, D.G., Kadambi, V., Neumann, J.C., Doetschman, T., and Kranias, E.G. (1992). *Circ. Res.*, **71**, 1021-30.
34. Anisimov, S.V., Tarasov, K.V., Riordon, D., Wobus, A.M., and Boheler, K.R. (2002). *Mech. Dev.*, **202**, 25-74.
35. Koban, M.U., Brugh, S.A., Riordon, D.R., Dellow, K.A., Yang, H.T., Tweedie, D., and Boheler, K.R. (2001). *Mech. Dev.*, **109**, 267-79.
36. Ratajczak, M.Z., Kucia, M., Reca, R., Majka, M., Janowska-Wieczorek, A., and Ratajczak, J. (2004). *Leukemia*, **18**, 29-40.
37. Anisimov, S.V., Tarasov, K.V., Stern, M.D., Lakatta, E.G., and Boheler, K.R. (2002). *Genomics*, **80**, 213-22.
38. Karbanova, J. and Mokry, J. (2002). *Acta Histochem.*, **104**, 361-5.
39. Fijnvandraat, A.C., van Ginneken, A.C., de Boer, P.A., Ruijter, J.M., Christoffels, V.M., Moorman, A.F., and Lekanne Deprez, R.H. (2003). *Cardiovasc. Res.*, **58**, 399-409.

40. Johkura, K., Cui, L., Suzuki, A., Teng, R., Kamiyoshi, A., Okamura, S., Kubota, S., Zhao, X., Asanuma, K., Okouchi, Y., Ogiwara, N., Tagawa, Y., and Sasaki, K. (2003). *Cardiovasc. Res.*, **58**, 435–43.
41. Ikeyama, S., Wang, X.T., Li, J., Podlutsky, A., Martindale, J.L., Kokkonen, G., van Huizen, R., Gorospe, M., and Holbrook, N.J. (2003). *J. Biol. Chem.*, **278**, 16726–31.

Chapter 7
In vitro differentiation of mouse ES cells into haematopoietic cells

Osamu Ohneda[1,2] and Masayuki Yamamoto[1,3,4]

[1]Graduate School of Comprehensive Human Science, [2]Department of Regenerative Medicine, [3]Center for Tsukuba Advanced Research Alliance, and [4]JST-ERATO Environmental Response Project, University of Tsukuba, 1–1–1 Tennoudai, Tsukuba 305–8575, Japan.

1 Introduction
1.1 Haematopoietic development

In mammals, the haematopoietic system originates from the mesodermal layer in the developing embryo, and is established by a complex hierarchy of cell differentiation pathways associated with so-called 'primitive' and 'definitive' haematopoiesis (1). Primitive haematopoiesis occurs in the yolk sac (YS) and gives rise to one main cell lineage, the primitive erythrocytes (eryP); and definitive erythropoiesis takes place in the YS and diverse sites in the embryo proper, giving rise to multiple lineages constituting all haematopoietic cell types, apart from eryP (1).

1.2 Haematopoietic differentiation in mouse ES cells

Following the establishment of the first mouse ES (mES) cell lines, it was soon recognized (2; and see Chapter 5) that this uncommitted, embryonic cell type displays a propensity for haematopoietic differentiation *in vitro*, as evidenced by the presence of red blood cells in mES cell-derived embryoid bodies (EBs). Consequently, haematopoietic cell development has been studied extensively in the mES cell/EB system, and cells belonging to erythroid, granulocytic, macrophage, and mast-cell lineages all have been shown to differentiate from mES cells by this method (3–7). It was demonstrated also that eryP and other haematopoietic lineages arise from a common multipotent precursor that develops within EBs (8); and so studies of the mES cell/EB system have lead ultimately to the identification of the haemangioblast, a 'primordial' stem cell of endothelial as well as haematopoietic lineages, with the capacity to differentiate into either primitive or definitive haematopoietic derivatives (9).

The subsequent elaboration of EB-mediated differentiation systems has greatly extended opportunities for analysing mechanisms of haematopoiesis at the cellular and molecular levels, especially when coupled with techniques for genetically engineering the mES cells to harbour targeted mutations in one or a few critical genes, such as those encoding transcription factors SCL/tal-1, Lmo2, AML1, GATA-1, GATA-2, and PU.1 (10,11). These *in vitro* studies have complemented *in vivo* studies involving the production of genetically modified mice by the mES-cell methodology; and moreover have facilitated functional studies of particular mutations that are associated with embryonic lethality. Consequently, over the last several years, progress has accelerated in elucidating mechanisms of haematopoiesis, and the mES cell/EB system has been a central means to this end. Details of the mES cell/EB system for haematopoiesis are beyond the scope of this chapter, but are comprehensively reviewed by Kennedy and Keller (12). It should be emphasized here that certain lineages of haematopoietic cells, including megakaryocytes and B lymphocytes, are not produced using the mES cell/EB system; and it has been proposed that those haematopoietic cells so generated represent the diversity of blood cells arising in the YS *in vivo* (4,12). Also, the production of reconstituting haematopoietic stem cells (HSCs) from mES cell-derived EBs has proved problematical to date (see Chapter 5). Ideally, mES cell-based systems are required which allow the generation of the full panoply of haematopoietic cells in an *in vitro* environment.

1.3 The mES/OP9-cell coculture system

Recently, Nakano and colleagues have established an alternative method for achieving haematopoietic differentiation in mES cells, which differs significantly from that mediated by EBs (13). It exploits the capacity of stromal cells, which normally support the proliferation and differentiation of lympho-haematopoietic lineages *in vivo*, to support the expansion of mES cell-derived, multipotent haematopoietic progenitors *in vitro*. A particular stromal cell line, OP9, was established originally from the calvarias of newborn, {C57BL/6 × C3H} F_2, osteopetrotic (op/op) mice by Kodama and colleagues (14). OP9 cells, like the mice from which they were derived, lack functional macrophage colony-stimulating factor (M-CSF), and consequently permit efficient differentiation and expansion of haematopoietic cells of various non-macrophage lineages from cocultured mES cells (13). Nakano's group has subsequently refined the mES/OP9-cell coculture procedure so that two waves of erythroid cell production, primitive and definitive, can be observed separately on day 6 and day 13 of coculture, respectively, when mature erythrocytes are released into the supernatant medium (15). Their results demonstrate that the mES/OP9-cell coculture system (i) recapitulates *in vitro* – albeit partially – the phases of primitive and definitive erythropoiesis occurring *in vivo*, and (ii) constitutes a most useful tool for analysing mechanisms of haematopoietic cell development.

We have adopted the mES/OP9-cell coculture system for the study of haematopoietic differentiation *in vitro*, and have identified therein two progenitor populations that differ in their relative status of differentiation, as assessed by their colony formation under defined conditions: (i) haematopoietic progenitors with the capacity to develop in semisolid methylcellulose culture medium as colony forming unit-erythroid (CFU-E), or colony-forming unit-granulocyte and macrophage (CFU-GM); and (ii) immature haematopoietic progenitors with the capacity for further proliferation as colony-forming units on OP9 cells, to which we refer as CFU-OP9 (16). This chapter describes our protocols for OP9-cell culture, for mES/OP9-cell coculture to obtain primitive and definitive haematopoietic lineages (including erythroid, myeloid, and lymphoid cells), and for the harvesting and expansion of CFU-E, CFU-GM, and CFU-OP9.

2 Routine maintenance of cell lines

2.1 Maintenance of mES cells

To ensure the reproducibility of differentiation experiments it is essential that a high standard of culture of mES cells be maintained throughout; once spontaneous differentiation has initiated due to suboptimal culture conditions, a large expansion of haematopoietic cells can never be achieved using our differentiation procedures.

These protocols for haematopoietic differentiation have been optimized using the mES cell line, E14. The undifferentiated mES cells are maintained by coculture on feeder layers of mouse embryonic fibroblasts (MEFs) in medium supplemented with leukaemia inhibitory factor (LIF), essentially as described in Chapter 2 but with minor modifications, given below. All cell incubations are conducted in a humidified incubator, gassed with 5% CO_2 in air, and set at 37°C.

2.2 MEF culture and feeder-layer preparation

To optimize the routine maintenance of mES cells we particularly recommend that: (i) medium for MEF culture (MEF medium, Chapter 2, *Protocol 1*) be supplemented with 1 mM sodium pyruvate, 0.1 mM non-essential amino acids, 50 U/ml penicillin, and 50 μg/ml streptomycin; (ii) MEFs be used at no higher passage number than five for feeder-layer preparation; and (iii) over-confluent MEF cultures are best discarded, as their capacity to support the undifferentiated state in mES cells may be compromised. Mitotic inactivation of MEFs may be performed either by treatment with mitomycin C or by γ-irradiation. For our purposes, the precoating of culture dishes with gelatin prior to feeder-layer preparation is unnecessary.

2.3 Maintenance of OP9 stromal cells, and preparation for coculture

OP9 cells also require careful maintenance, and should be passaged on attaining between 80% and 90% confluence. Over-confluence of OP9 cells especially should be avoided during routine culture, as their proliferation is contact-inhibited; and when allowed to reach confluence these cells may start spontaneously to

Protocol 1
Maintenance of OP9 cells and preparation for coculture with mES cells

Reagents and equipment

- Frozen vial of OP9 cells, available from Prof. Toru Nakano (Osaka University, Japan), the RIKEN depository (http://www.brc.riken.jp/), or the ATCC (#CRL-2749FL)
- OP9-cell medium: alpha MEM (Cat. No. 12571-063, Invitrogen) supplemented with 20% batch-selected FCS
- Dulbecco's phosphate-buffered saline (DPBS; Cat. No. D8537, Sigma)
- Trypsin/EDTA solution: 0.05% trypsin and 0.53 mM EDTA in HBSS (Cat. No. 25300-054, Invitrogen)
- Freezing medium (1 ×): combine (by volume) 90% FCS and 10% DMSO (Cat. No. 045-24511, Wako Chemicals)

- 10 cm tissue-culture dishes (Cat. No. 353003, BD Biosciences)
- Sterile, Pasteur and 15 ml transfer pipettes
- 15 ml centrifuge tube
- Water bath set at 37 °C
- Benchtop, refrigerated centrifuge
- Cryogenic vials (e.g. Nunc Cryo Tube® No. 375418, Cat. No. 960-210, Fisher Scientific)
- Freezing container (Cat. No. 5100-0001, Nalgene)
- Haemocytometer
- Inverted phase-contrast microscope

Methods

Reviving frozen OP9 cells:

1. Quickly thaw a stock vial of OP9 cells in a water bath set at 37 °C.
2. Aseptically transfer the cells into a 15 ml centrifuge tube containing 10 ml OP9-cell medium, and centrifuge for 5 min at 1000 rpm and 4 °C.
3. Aspirate the supernatant, resuspend the cell pellet in 10 ml OP9-cell medium, dispense the cell suspension into a 10 cm dish evenly, and incubate.
4. Monitor the culture daily by phase-contrast microscopy until ~80% confluence is reached, when the cells are ready to be passaged.

Expansion of OP9 cells:

5. Aspirate culture medium from the dish and wash the cells twice with 10 ml DPBS.
6. Add 1.0–1.5 ml trypsin/EDTA solution to the dish and incubate for 3–5 min, until the cells detach from the substratum.
7. Add 6–8 ml OP9-cell medium to the dish, and disaggregate the cells with pipetting.
8. Transfer the cell suspension into a 15 ml tube, and centrifuge for 5 min at 1000 rpm and 4 °C.
9. Aspirate the supernatant and resuspend the cell pellet in OP9-cell medium.

> **Protocol 1** continued
>
> 10 Count the cells and adjust the density to $2–5 \times 10^4$ cells/ml; dispense 10 ml cell suspension per 10 cm dish, and incubate until 80–90% confluent. At this stage cells may be either passaged further (at a split ratio of 1:4), cryopreserved, or cocultured with mES cells to induce haematopoietic differentiation.
>
> *Cryopreservation of OP9 cells:*
>
> 11 Harvest a subconfluent 10 cm dish of OP9 cells by trypsinization and pellet by centrifugation (as above), resuspend in 4 ml freezing medium, and dispense into cryogenic vials (using 1 ml suspension per vial). Place the vials into a freezing container and store overnight at $-80\,°C$. Next day, transfer the vials into liquid nitrogen for long-term storage.
>
> *Preparation of OP9-cell monolayers for coculture:*
>
> 12 After reaching 80–90% confluence, OP9-cell cultures may be incubated further for up to 1 week prior to use in mES-cell differentiation experiments, with one change of medium being required during this time.

differentiate, and their ability to support haematopoiesis in mES cells is concomitantly impaired. On differentiation, OP9 cells show characteristic features of preadipocytes, such as the formation of lipid-containing vesicles. In contrast to MEF feeder layers, monolayers of OP9 cells are used without prior mitotic inactivation for coculture with mES cells in haematopoietic differentiation experiments. This necessitates the regular passaging of mES/OP9-cell cocultures, to prevent the overgrowth of mES cell-derivates by the OP9 cells (see below).

2.4 Fetal-calf serum (FCS)

Fetal-calf serum (FCS) is a most important reagent serving various functions in the mES/OP9-cell coculture system; and consequently different batches are used in these differentiation experiments. Separate batches are selected for the capacity to: (i) maintain undifferentiated mES cells (as described in detail in Chapter 2, Section 3.4 and *Protocol 4*); and (ii) support differentiation of mES cells into haematopoietic derivatives. The capacity of FCS to support haematopoietic differentiation may be assayed from the lineages produced, and from the yield of desired haematopoietic cells, at specific times during the differentiation protocol, for example by measuring the frequency of haematopoietic lineage-specific markers in a population generated on day 12.

3 mES/OP9-cell coculture system for haematopoietic differentiation

Our mES/OP9-cell coculture regime (*Protocol 2*) is summarized schematically in *Figure 1*. It represents a modification of the original system (13,15), in order to allow the sequential expansion of primitive and definitive haematopoietic cells. The key steps are as follows:

- Undifferentiated mES cells, growing exponentially on MEF feeder layers, are reseeded sparsely onto confluent monolayers of OP9 stromal cells, in medium lacking LIF but supplemented with a higher level of FCS, and with vascular endothelial cell growth factor (VEGF) and bone morphogenetic protein 4 (BMP-4), to promote mesoderm induction.

- Following the preliminary, mesodermal differentiation of mES cells, coculture is continued in the presence of the growth factors, erythropoietin (Epo) and stem cell factor (SCF), and the cytokine, interleukin-3 (IL-3), to induce primitive and definitive haematopoiesis from mesodermal derivatives. Subsequently, mature erythroid cells (i.e. eryP) are released into the supernatant medium, and these may be readily harvested.

- Adherent cells in the cocultures are periodically harvested and reseeded onto fresh monolayers of OP9 cells, to avoid overgrowth of mES-cell derivatives by the proliferating OP9 cells.

- A population of definitive erythroid progenitor cells remains attached to the OP9 monolayer, and may be harvested with relative facility. The efficiency of production of erythroid progenitor cells is evaluated by colony assay (*Protocol 3*).

The sequential stages of haematopoietic cell differentiation in this system are described in the following sections.

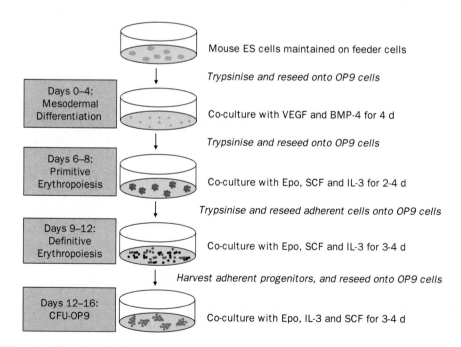

Figure 1 Schematic of mES/OP9-cell coculture system for haematopoietic differentiation.

3.1 Primitive erythroid cell differentiation

After seeding mES cells onto OP9-cell monolayers in the presence of VEGF and BMP-4, mES cells start to differentiate and mesoderm-like colonies are observable around day 4, where the day of initiating coculture is designated as 'day 0'. Cells within these colonies are highly refractile and morphologically distinguishable from those in undifferentiated mES-cell colonies; and fibroblastic cells may be observed to emanate from and surround these differentiated colonies (*Figure 2A*). At this stage, the OP9 cells and mesodermal-differentiated colonies are harvested together by trypsinization, and the single-cell suspension so generated is reseeded onto fresh monolayers of OP9 cells. In order to induce primitive erythroid cell development from these mesodermal derivatives, the cells are cocultured in the presence of Epo, SCF, and IL-3. After a further 2 d of culture (i.e. on day 6 of differentiation), clumps of rounded, haematopoietic cells appear on the OP9 cells (*Figure 2B, D*), and are easily detached from the OP9-cell monolayer. These cells correspond to eryP, the first (primitive) erythrocytes to be formed during embryonic development. They show characteristic morphological features when examined by May–Grünwald/Giemsa staining, and also may be stained using anti-ϵy-globin antibody (17) or benzidine (18). Primitive erythroid cells are distinguishable from definitive erythroid cells by RT-PCR analysis of expression of genes for ϵy-globin (for primitive erythroid cells), and for β-major

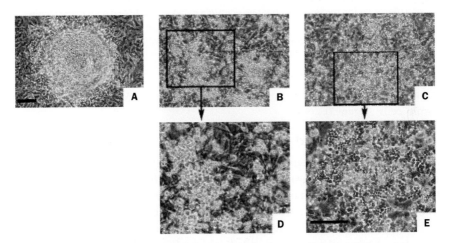

Figure 2 Photomicrographs of differentiating mES cells during coculture with OP9 stromal cells. (A) On day 4 of coculture under mesoderm-inducing conditions, mES cells can be observed to have differentiated and to have formed colonies of mesodermal cells, with outgrowths of fibroblastic cells. (B–E) With continued coculture in the presence of haematopoietic factors, the mesodermal derivatives differentiate further into primitive erythroid cells by day 6 (B, D) and definitive erythroid cells by day 12 (C, E). Note that definitive erythroid cells (E) are relatively smaller than primitive erythroid cells (D), and adhere more tightly to the stromal layer. Micrographs (A–C) and (D–E) are at the same magnification, with the scale bars representing 50 µm.

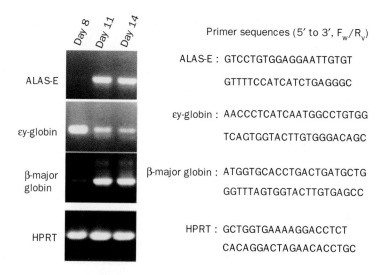

Figure 3 RT-PCR analysis of primitive and definitive erythroid cells. Mature haematopoietic cells were harvested from supernatant medium of mES/OP9-cell cocultures on days 8, 11, and 14. Total RNA was extracted from the cells, and cDNAs reverse transcribed according to (16). PCR was performed on cDNAs for 30 to 36 cycles, using the indicated primers and the following conditions: 94 °C for 20 s, ramp time for 1 min, 55 °C for 1 s, and 72 °C for 1 min. Amplified products were separated electrophoretically on a 1% agarose gel and stained with ethidium bromide. Amplifications for εy-globin transcript revealed strong induction of expression (denoting the presence of primitive erythroid cells) at day 8; and for β-major globin and ALAS-E (denoting the presence of definitive erythroid cells) at day 11, which was sustained at day 14. Amplifications for HPRT transcript represent the internal control.

globin and erythroid-specific 5-aminolevulinate synthase (ALAS-E) (for definitive erythroid cells) (*Figure 3*).

3.2 Definitive erythroid cell differentiation

When the mES cells have been cocultured with OP9 cells for 8 d in total in the presence of the appropriate growth factors and cytokine (firstly VEGF and BMP-4, and subsequently SCF, Epo, and IL-3), OP9 cells and adherent mES-cell derivatives are harvested and reseeded onto fresh monolayers of OP9 cells. The mES-cell derivatives now recommence proliferation and differentiation on the OP9 cells. Around day 12 of coculture, adherent cells start forming ostensible haematopoietic colonies as denoted by their cobblestone appearance (*Figure 2C, E*). Simultaneously, large numbers of cells detach from the stromal cell layer into the supernatant medium. We have ascertained that the majority of these non-adherent cells are definitive haematopoietic cells, which include mature (enucleated) erythrocytes (eryD) and cells of the myeloid lineage (16). Cell-specific features are revealed by May–Grünwald/Giemsa staining (*Figure 4A*); and expression of lineage-specific cell-surface antigens may be analysed by using a fluorescence-activated cell sorter (FACS®) (*Figure 4B–O* and *Figure 5*).

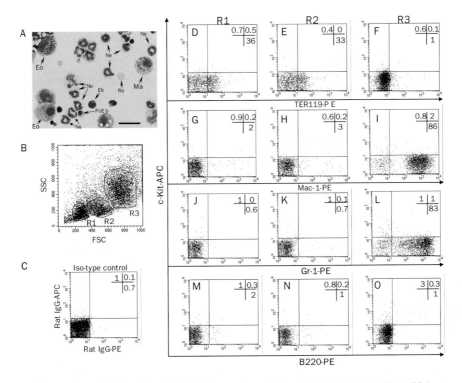

Figure 4 (see Plate 3) Diversity of definitive haematopoietic cells obtained using the mES/OP9-cell coculture system on day 12 of differentiation. For (A) to (O), non-adherent cells in the supernatant medium, representing the mature fraction of definitive haematopoietic derivatives, were harvested on day 12 of differentiation, as described in *Protocol 4*.
(A) May–Grünwald/Giemsa staining of definitive haematopoietic cells. Both erythroid and myeloid lineages are detectable according to the morphological criteria provided in *Protocol 4*. Eb, indicates an erythroblast; Po, late polychromatic normoblast; Re, reticulocyte; Ne, neutrophil granulocytes; Eo, eosinophil granulocytes; and Ma, macrophages. Scale bar: 25 μm. (B)–(O) Flow cytometric analysis of lineage-specific cell-surface antigens expressed by definitive haematopoietic cells. Cells were analysed for forward light scatter (FSC) versus side light scatter (SSC), revealing that three distinct populations (R1, R2, and R3) were present in the culture. (C) For the following FACS® analyses, haematopoietic cells were incubated with isotype-matched fluorochrome-conjugated antibodies, or lineage-specific fluorochrome-conjugated antibodies. Percentages of positive cells in three quadrants are indicated. Haematopoietic cells incubated with test antibodies were then gated with three distinct areas (for R1, R2, and R3 populations), and expression of c-Kit and lineage-specific markers was examined: for population R1, see panels D, G, J, and M; for R2, panels E, H, K, and N; and for R3, panels F, I, L, and O. Cells positive for the erythroid-specific marker, TER119, were detectable in populations R1 and R2 (at 36% and 33%, respectively). In contrast, cells positive for the macrophage marker, Mac-1, and the neutrophil marker, Gr-1, were mainly detectable in population R3, and at high frequency (over 80%). Cells positive for the B-cell marker, B220, also were observed in populations R1 and R2, but at low frequency (below 2%). Cells positive for c-Kit represent either uncommitted progenitors or lineage-committed progenitors.

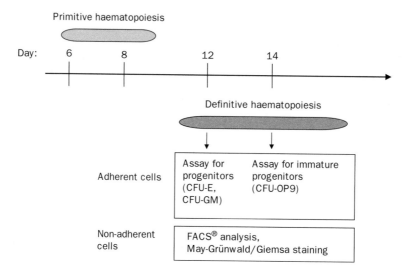

Figure 5 Schematic of mES/OP9-cell coculture system for the analysis of definitive haematopoietic cells. Non-adherent cells (differentiating haematopoietic cells) and adherent cells (both haematopoietic progenitors and immature haematopoietic progenitors) are analysed for their relative status of differentiation by utilizing appropriate methods, as indicated in this figure.

Protocol 2

Induction of primitive and definitive haematopoietic differentiation in mES cells using the mES/OP9-cell coculture system

This adaptation of the original OP9 cell-based system (15) features the use of growth factors during an initial phase of coculture with OP9 cells, specifically to enhance mesodermal differentiation of mES cells and thereby increase the efficiency of haematopoietic differentiation.

Reagents and equipment

- Culture of undifferentiated, exponentially-growing mES cells on a MEF feeder layer in a 35 mm culture dish[a]
- OP9-cell monolayers in 10 cm dishes (Protocol 1)[b]
- Mouse ES-cell medium: Dulbecco's modified Eagle's medium (DMEM; Cat. No. 11960–044, Invitrogen) supplemented with 15% batch-tested FCS (qualified for mES-cell proliferation), 1 mM sodium pyruvate (Cat. No. 11360–070, Invitrogen), 0.1 mM non-essential amino acids (Cat. No. 11140–050, Invitrogen), 2 mM L-glutamine (Cat. No. 25030–081, Invitrogen), 0.1 mM

Protocol 2 continued

- β-mercaptoethanol (Cat. No. M7522, Sigma), 50 U/ml penicillin and 50 μg/ml streptomycin (Cat. No. 15140-122, Invitrogen), and 10^3 U/ml LIF (Cat. No. ESG1107, Chemicon International)
- OP9-cell medium (*Protocol 1*)
- Alpha MEM, serum free (*Protocol 1*)
- Mesodermal Differentiation Medium (MDM): alpha MEM supplemented with 20% FCS (qualified for haematopoietic differentiation), 5 ng/ml mouse vascular endothelial cell growth factor (mVEGF; Cat. No. 450-32, PeproTech), 50 ng/ml human bone morphogenetic protein-4 (hBMP-4; Cat. No. 314-BP, R&D Systems), and 0.1 mM β-mercaptoethanol
- Haematopoietic Differentiation Medium (HDM): alpha MEM supplemented with 20% FCS (qualified for haematopoietic differentiation), 50 ng/ml stem cell factor (SCF; Cat. No. 455-MC, R&D Systems), 10 ng/ml interleukin 3 (IL-3; Cat. No. 403-ML, R&D Systems), and 2 U/ml erythropoietin (Epo; Cat. No. 959-ME, R&D Systems).
- DPBS (*Protocol 1*)
- Trypsin/EDTA solution (*Protocol 1*)
- Sterile, 10 ml transfer pipette
- 15 ml centrifuge tubes
- Benchtop, refrigerated centrifuge
- Haemocytometer
- Inverted phase-contrast microscope
- Cell strainer with mesh size 40 μm (Cat. No. 3523-40, BD Biosciences)

Methods

Mesoderm induction:

1. Aspirate medium from the culture of mES cells, and wash the cells twice with 2 ml DPBS.
2. Add 0.5 ml trypsin/EDTA solution to the dish and incubate for 3-5 min, until the cells detach.
3. Add 2 ml alpha MEM to the dish, and disaggregate the cells by pipetting gently.
4. Transfer the cell suspension into a tube and centrifuge for 5 min at 1500 rpm and 4 °C.
5. Aspirate the supernatant, resuspend the cell pellet in MDM, and count the cells.[c]
6. Seed $5-7 \times 10^3$ mES cells in 10 ml MDM onto each monolayer of OP9 cells in 10 cm dishes, and incubate for 3 d.
7. On day 3 of coculture replace half the volume of conditioned medium with fresh MDM, and continue incubation.
8. On day 4, monitor cultures for the appearance of mesodermal colonies (*Figure 2A*).[d]

Primitive haematopoietic differentiation:

9. On day 4, aspirate medium from the cocultures and wash the cells twice with 10 ml DPBS.
10. Add sufficient trypsin/EDTA solution to cover the cells (1-1.5 ml per 10 cm dish), and incubate dishes for 3-5 min at 37 °C, until the cells detach.

Protocol 2 continued

11. Add alpha MEM to the dishes (using 6–8 ml per 10 cm dish), and pipette gently to disaggregate the cells.

12. Transfer the cell suspension from each dish into a tube, and centrifuge for 5 min at 1500 rpm and 4 °C.

13. Aspirate the supernatant, and resuspend each cell pellets in 20 ml HDM (for a split ratio of 1:2 for further incubation). Reseed the cell suspension (containing both OP9 and mES cells) onto two fresh monolayers of OP9 cells in 10 cm dishes (using 10 ml cell suspension per dish) and incubate.

14. On day 6, non-adherent haematopoietic cells, including eryP, should be observable in the cultures by phase-contrast microscopy. Harvest these cells in the supernatant medium, and replenish the cultures with fresh HDM.

15. By day 8, a further population of non-adherent, primitive haematopoietic cells should appear in the supernatant. Harvest these cells also for analysis (17, 18; and Figure 3).

Definitive haematopoietic differentiation:

16. On day 8 of coculture after removing the supernatant, wash the monolayers twice with DPBS.

17. Add 8 ml alpha MEM to each dish, and detach adherent, haematopoietic cells from the underlying OP9 cells by pipetting vigorously.[e]

18. Collect the supernatant cell suspension, and pass through a cell strainer to remove as many OP9 cells as possible.

19. Centrifuge the filtered cell suspension for 5 min at 1500 rpm and 4 °C, and aspirate the supernatant; resuspend the cell pellet in 40 ml HDM (for a split ratio of 1:4 for further incubation) with gentle pipetting, reseed the cell suspension onto fresh monolayers of OP9 cells in 10 cm dishes (using 10 ml of cell suspension per 10 cm dish) and incubate. Under these conditions, the haematopoietic cells start expanding relatively rapidly on the OP9 cells.

20. On day 12 of differentiation, non-adherent haematopoietic cells should be observable in the culture medium; these can be harvested and analysed further for definitive haematopoietic cell development.

[a] A relatively small number of mES cells are required to initiate cocultures, and therefore it is convenient to cryopreserve at a given passage numerous stock vials containing sufficient mES cells for individual mES/OP9-cell coculture experiments. These are recovered on feeder layers in 35 mm tissue-culture dishes, as described in Chapter 2, *Protocol 3*.

[b] As numerous 10 cm dishes of OP9 cells are required during the course of mES/OP9-cell coculture experiments, it is advisable to prepare beforehand numerous cultures at different states of confluence.

> **Protocol 2** continued
>
> ^{c} The counting of viable mES cells is described in Chapter 6, *Protocol 1*. The contribution of feeder cells to this suspension may be regarded as negligible.
>
> ^{d} The development of mesodermal colonies on OP9-cell monolayers is highly dependent on the following parameters:
>
>   (a) Mouse ES cells: if mES cells are not in an undifferentiated state at the initiation of coculture, mesodermal colonies will hardly be observable on OP9 cells.
>
>   (b) FCS: only batch-selected FCS should be used for mES/OP9-cell coculture.
>
>   (c) OP9 cells: well-maintained OP9 cells are crucial for not only mesoderm induction but also haematopoietic differentiation.
>
> ^{e} Trypsin/EDTA solution should not be used for this procedure, as such treatment appears to inhibit the readherence of haematopoietic cells to OP9 cells, and consequently their further expansion.

3.3 Definitive haematopoietic progenitors and CFU-OP9 (*Figure 5*)

By day 12–14 of coculture, two categories of mES cell-derived, definitive haematopoietic cells are observable in the non-adherent and adherent fractions: (i) fully-differentiated cells in the supernatant medium, which can be analysed by FACS® for the expression of lineage-specific markers, and analysed morphologically by May–Grünwald/Giemsa staining (*Protocol 4*); and (ii) haematopoietic progenitors attached to the OP9-cell monolayer, which may be subjected to colony-forming unit-erythroid (CFU-E), or colony-forming unit-granulocyte and macrophage (CFU-GM) assays in semisolid medium supplemented with appropriate cytokines (16).

For the latter category, colonies showing a cobblestone appearance – which is a characteristic feature of immature haematopoietic progenitors – are also routinely observed (*Figure 2C, E*). The cobblestone area-forming cells (CAFCs) are considered to be immature haematopoietic progenitors, including (reconstituting) HSCs (19, 20). The CFACs whose frequency is determined at early time points (day 7) correlate with murine early stem cells, so-called 'CFU-S day 8'. Late CFACs, as determined on day 28, correlate with stem cells having long-term bone-marrow-repopulating ability.

We have modified the original CAFC assay (19) by employing OP9, instead of primary bone-marrow-derived, stromal cells. When mES cell-derived, immature haematopoietic progenitor cells (typically, harvested at day 14 of differentiation) are subjected to secondary reseeding onto OP9 cells, and following coculture in growth factor- and cytokine-supplemented medium, haematopoietic colonies with cobblestone appearance are regenerated (by day 18–20). These colonies consist of immature erythroid and/or myeloid progenitors (16), and we refer to them collectively as CFU-OP9. The total number of CFU-OP9 in a culture at this stage is a measure of the proliferative activity of mES cell-derived, definitive, immature haematopoietic progenitor cells.

Protocol 3

Isolation of definitive haematopoietic progenitors: the CFU-E, CFU-GM and CFU-OP9 assays

Reagents and equipment

- A culture of adherent, differentiating mES cells at day 12–14 of coculture with OP9 cells (*Protocol 2*), showing the cobblestone morphology characteristic of immature haematopoietic cells (see *Figure 2*): for CFU-E and CFU-GM assays, isolate progenitors on day 12 of coculture; and for CFU-OP9, on day 14
- DPBS (*Protocol 1*)
- Trypsin/EDTA solution (*Protocol 1*)
- FCS-supplemented medium: alpha MEM with 10% FCS (qualified for haematopoietic differentiation); and Iscove's Modified Dulbecco's Medium (IMDM; Cat. No. 12440–053, Invitrogen) with 2% FCS (qualified for haematopoietic differentiation)
- Methylcellulose medium (Methocult 3234, StemCell Technologies)
- Granulocyte-macrophage-colony stimulating factor (GM-CSF; Cat. No. 415-ML, R&D Systems)
- CFU-OP9 medium: alpha MEM supplemented with 10% FCS, 50 ng/ml SCF, 2 U/ml Epo, and 10 ng/ml IL-3
- 35 mm non-treated Petri dishes (Cat. No. 351008, BD Biosciences)
- 10 cm culture dishes
- Haemocytometer
- Inverted microscope
- 15 ml centrifuge tubes
- Benchtop, refrigerated centrifuge

Methods

Separation of definitive haematopoietic progenitor cells from OP9 cells:

This method exploits the greater adhesiveness of OP9 cells than of haematopoietic progenitor cells to achieve effective fractionation of the latter, with removal of >90% of OP9 cells.

1. After harvesting the supernatant with suspended cells, wash the adherent cells on the stromal layer twice with 10 ml DPBS.[a]

2. Add trypsin/EDTA solution (1.5–2 ml per 10 cm dish), and incubate for 3–5 min at 37 °C until the cells detach.[b,c]

3. Collect and centrifuge the cells for 5 min at 1500 rpm and 4 °C.

4. Aspirate the supernatant and resuspend the cells in 10 ml FCS-supplemented alpha MEM, reseed the cell suspension into an empty 10 cm dish, and incubate for 1 h to allow the OP9 cells preferentially to adhere to the dish to deplete them from the suspension.

5. Harvest non-adherent cells in the supernatant (most of which are haematopoietic cells), and centrifuge for 5 min at 1500 rpm and 4 °C.

6. Aspirate the supernatant, and resuspend the cells in FCS-supplemented IMDM for either the CFU-E or CFU-GM colony assays, or in CFU-OP9 medium for the CFU-OP9 assay. Count the cells.

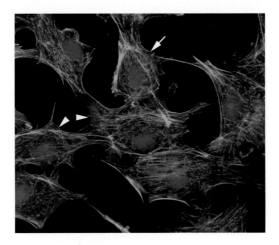

Plate 1 For caption see Chapter 4, Figure 2 (page 76).

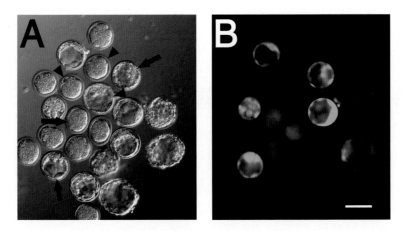

Plate 2 For caption see Chapter 4, Figure 6 (page 102).

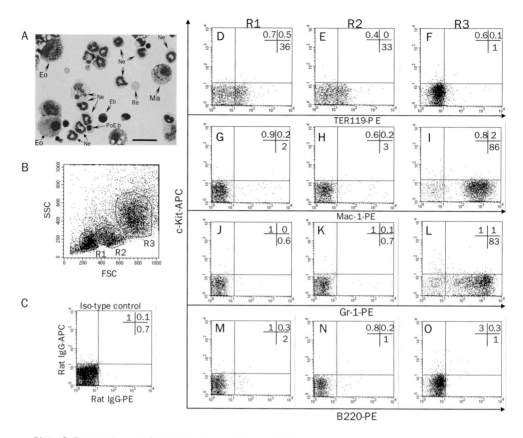

Plate 3 For caption see Chapter 7, Figure 4 (page 177).

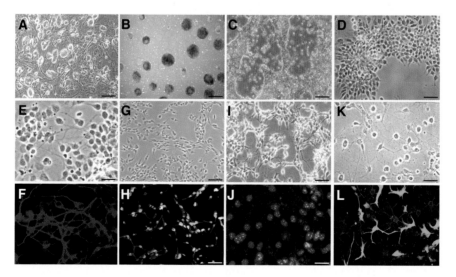

Plate 4 For caption see Chapter 8, Figure 2 (page 192).

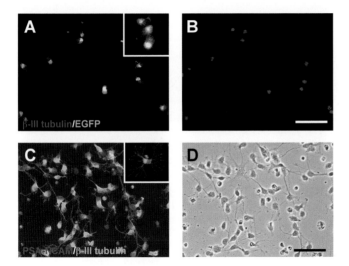

Plate 5 For caption see Chapter 8, Figure 3 (page 200).

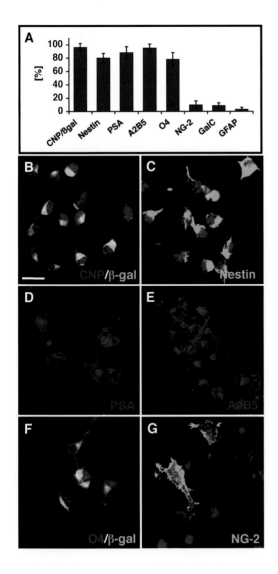

Plate 6 For caption see Chapter 8, Figure 4 (page 204).

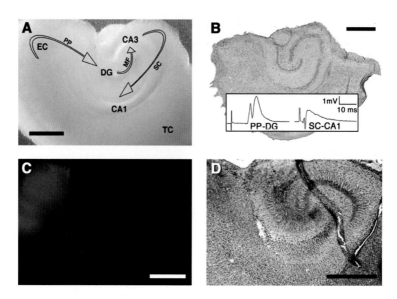

Plate 7 For caption see Chapter 8, Figure 5 (page 205).

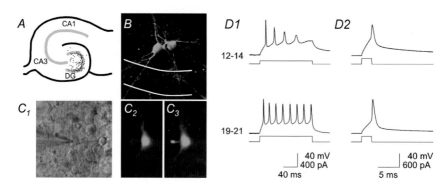

Plate 8 For caption see Chapter 8, Figure 6 (page 206).

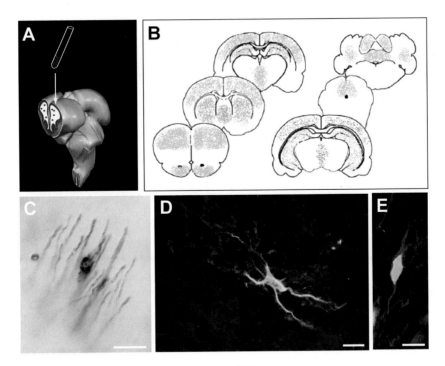

Plate 9 For caption see Chapter 8, Figure 7 (page 209).

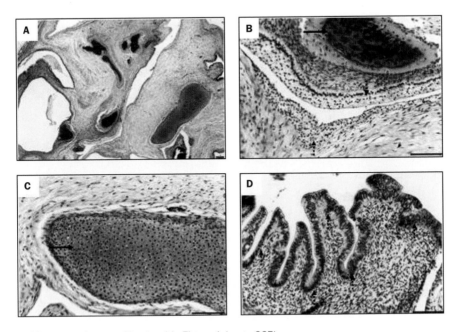

Plate 10 For caption see Chapter 11, Figure 1 (page 265).

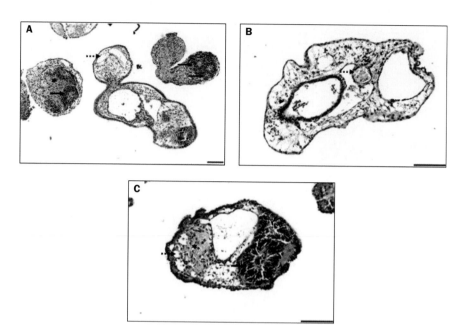

Plate 11 For caption see Chapter 11, Figure 3 (page 269).

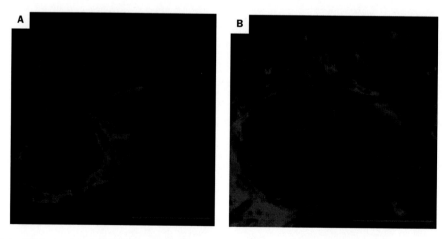

Plate 12 For caption see Chapter 11, Figure 4 (page 275).

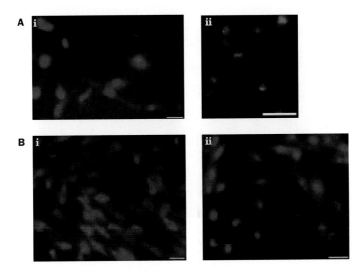

Plate 13 For caption see Chapter 11, Figure 7 (page 280).

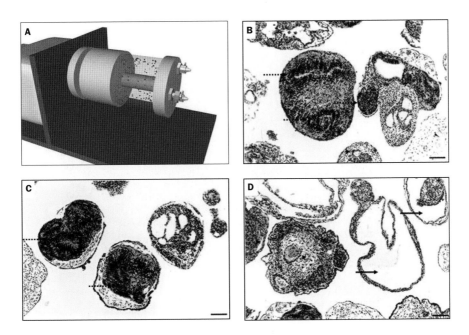

Plate 14 For caption see Chapter 11, Figure 9 (page 286).

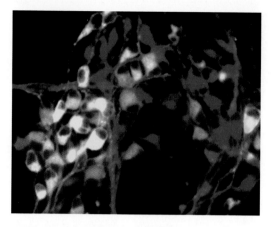

Plate 15 For caption see Chapter 12, Figure 3 (page 309).

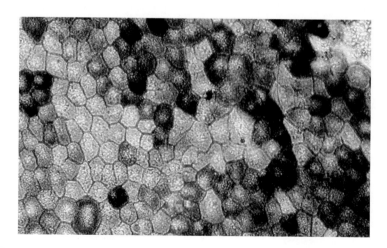

Plate 16 For caption see Chapter 12, Figure 4 (page 310).

Protocol 3 continued

CFU-E/CFU-GM colony assays:

7. Adjust the density of the cell suspension to 1×10^6 cells/ml in FCS-supplemented IMDM; and in a 35 mm Petri dish combine well 0.1 ml suspension with 1 ml methylcellulose medium and Epo (to a final concentration of 2 U/ml) or GM-CSF (100 ng/ml) for CFU-E or CFU-GM, respectively. Incubate cultures for 2 d for CFU-E, or 7 d for CFU-GM.

8. Score the number of developing colonies using inverted microscopy.

CFU-OP9 colony assay:

Note that the isolation of adherent cells at later stages than day 14 of differentiation may allow further enrichment of immature haematopoietic progenitor cells capable of forming CFU-OP9.

9. Seed 1×10^5 cells in 10 ml CFU-OP9 medium per fresh monolayer of OP9 cells in a 10 cm dish, and incubate for 4–6 d.

10. Monitor for the appearance of CFU-OP9 colonies with characteristic cobblestone appearance, and which tightly adhere to the OP9 cell-monolayer; and evaluate the proliferative activity of immature haematopoietic progenitors by scoring the foci of cobblestone-area formation per dish.[d]

[a] Cells in the supernatant are used for analysis of definitive haematopoietic cells (*Protocol 2*).

[b] At this stage the cell suspension contains both mES-cell derivatives, which tend to adhere to OP9 cells, and OP9 cells.

[c] Trypsin/EDTA solution, although detrimental at day 8 (see *Protocol 2*), may safely be used to harvest adherent cells at day 12 or 14.

[d] Some floating cells may be observed in the supernatant medium during this time, but the number is quite low compared with earlier stages of mES/OP9-cell coculture.

4 Analysis of definitive haematopoietic cells produced using the mES/OP9-cell coculture system

Definitive haematopoietic cells appear after the differentiation of primitive haematopoietic cells on day 8 of coculture. On reseeding the adherent, progenitor fraction onto fresh OP9 cells around day 10, definitive haematopoietic cells start rapidly to proliferate and differentiate. These derivatives may be characterized morphologically, and by analysis of the expression of lineage-specific markers. Both erythroid (late polychromatic erythroblasts and reticulocytes) and myeloid (neutrophils, eosinophils, and macrophages) lineages are detectable in the supernatant medium by the May–Grünwald/Giemsa staining method (*Figure 4A*). The large numbers of floating cells produced are amenable to FACS® analysis (*Figure 4*), for detailed investigation of cell-surface markers. We detect three distinct populations (termed R1, R2, and R3) in the pool of definitive haematopoietic cells by analysis of forward scatter (FSC) and side scatter (SSC), as shown in *Figure 4B*.

To characterize cell types within the three populations, we examine the expression of the following markers: c-Kit for the immature, haematopoietic progenitor cells and for lineage-committed progenitors; TER119 for the erythroid lineage (*Figure 4D, E, F*); Mac-1 for macrophages (*Figure 4G, H, I*); Gr-1 for neutrophils (*Figure 4J, K, L*); and B220 for B cells (*Figure 4M, N, O*). CD4- and/or CD8-positive cells are not detectable under our conditions (data not shown), from which we conclude that another key factor(s), for example delta-like 1, may be required for the development of the T-cell lineage in the mES/OP9-cell coculture assay (21).

Protocol 4
Analysis of definitive haematopoietic cells

Reagents and equipment

- Mouse ES/OP9-cell coculture at day 12 of induction of haematopoietic differentiation
- Alpha MEM supplemented with 10% FCS (qualified for haematopoietic differentiation)
- CFU-OP9 medium (*Protocol 3*)
- Washing buffer: DPBS (*Protocol 1*) supplemented with 2% FCS
- Fluorochrome-conjugated monoclonal antibodies (see *Table 1*)
- Mouse BD Fc Block™ (CD16/CD32; Cat. No. 553141, BD Biosciences)
- Isotype control immunoglobulins: R-phycoerythrin (PE)-conjugated rat IgG monoclonal immunoglobulin isotype control (either Cat. No. 553930 or 553989, BD Biosciences) and allophycocyanin (APC)-conjugated rat IgG monoclonal immunoglobulin isotype control (either Cat. No. 553932 or 553991, BD Biosciences)
- Haemocytometer
- Fluorescence-activated cell sorter, e.g. FACSCalibur (BD Biosciences)

- Benchtop, refrigerated centrifuge
- Benchtop, Cytospin® centrifuge (Thermo Shandon)
- May–Grünwald stain solution (Cat. No. 131-1281, Wako Chemicals)
- M/15 phosphate buffer (Cat. No. 1561, Muto Pure Chemicals)
- Giemsa stain solution: Giemsa (Cat. No. 079-04391, Wako Chemicals) diluted 1:10 in M/15 phosphate buffer
- Malinol (Cat. No. 2009-3, Muto Pure Chemicals)
- Round-bottom centrifuge tubes (BD Falcon™ No. 352058, BD Biosciences)
- Tube with cell strainer cap incorporating a 30 μm nylon mesh (Falcon 352235, BD Biosciences)
- Non-coated cytoslides (Cat. No. S-0021, Matsunami Glass Industries)
- Slide chamber (Cat. No. 5991040, Shandon)

Methods

Cell preparation:

1. On day 12 of coculture, transfer the supernatant with suspended, non-adherent cells into a centrifuge tube for characterization. (Replenish the culture dish with fresh CFU-OP9 medium for further experimentation.)

Protocol 4 continued

2. Centrifuge the harvested medium with suspended cells for 5 min at 1500 rpm and 4 °C.

3. Aspirate the supernatant, and resuspend the cell pellet in 1 ml washing buffer. Count the cells. Divide the cell suspension into 0.5 ml aliquots for FACS® analysis, and for May–Grünwald/Giemsa staining.

Preparation of cells for FACS® analysis:

4. Adjust the density of the cell suspension to $5 \times 10^6 - 2 \times 10^7$ cells/ml in washing buffer. Dispense 100 μl aliquots of the cell suspension into round-bottom centrifuge tubes, and adjust the volume to 200 μl with washing buffer (giving a total of $5 \times 10^5 - 2 \times 10^6$ cells/sample).

5. Add Fc Block™ (1:200 dilution) to the samples, and incubate the tubes for 20 min on ice.

6. Add fluorochrome-conjugated antibodies to the samples at the appropriate dilutions (*Table 1*), and incubate for 30 min at 4 °C.

7. Add 4 ml of ice-cold washing buffer per tube, and centrifuge for 3 min at 1800 rpm and 4 °C.

8. Aspirate the supernatant, and loosen the cell pellet gently by tapping the tube.

9. Add 4 ml cold washing buffer to resuspend the cells, and centrifuge the cells again (for 3 min at 1800 rpm and 4 °C).

10. Aspirate the buffer and resuspend the cells in a further 0.5 ml washing buffer, and filter the cells through a cell strainer to remove debris.

11. Perform FACS® analysis, according to the equipment manufacturer's instructions.

May–Grünwald/Giemsa staining:

4. Dispense the cell suspension into Eppendorf tubes, at a density of $1 \times 10^4 - 1 \times 10^5$ cells in 120 μl washing buffer.

5. Set a cytoslide into the holder, according to the equipment manufacturer's instructions.

6. Before applying samples to the holder, apply 120 μl washing buffer supplemented with 10% FCS (no cells) and centrifuge for 3 min at 300 rpm and r.t.[a]

7. Apply samples carefully to the holder and centrifuge for 3 min at 300 rpm and r.t., to cause the haematopoietic cells to adhere to the slide.

8. After centrifugation, air-dry samples for ~15 min.

9. Apply May–Grünwald stain solution to the samples to cover, and incubate for 2 min at r.t.

10. Discard the solution (do not wash the samples), apply Giemsa stain solution, and incubate samples for 8 min at r.t.

11. Wash samples by stirring vigorously in a beaker of water.

12. Air-dry the samples, and mount with Malinol.

> **Protocol 4 continued**

13 Observe the stained cells using light microscopy; and identify cell types according to the following morphological criteria (22).

Erythroid lineages:

Erythroblasts (Eb in Figure 4A): large, rounded nucleus surrounded by a relatively small amount of darkly stained cytoplasm; cytoplasm containing a pale area adjacent to the nucleus.

Late polychromatic normoblasts (PoEb): faintly polychromatic cytoplasm and a small eccentric nucleus.

Reticulocytes (Re): rounded, anucleate cells with faintly polychromatic cytoplasm and containing a basophilic reticulum.

Myeloid lineages:

Neutrophil granulocytes (Ne): nucleus composed of two to five segments; cytoplasm containing many very fine granules staining faint purple.

Eosinophil granulocytes (Eo): nucleus composed of one to four segments; cytoplasm packed with large rounded granules staining reddish-orange.

Macrophages (Ma): nucleus with a pale-staining, lace-like chromatin structure; cytoplasm staining pale blue and containing azurophilic granules and vacuoles.

[a] This step is to prevent the subsequent loss of cells.

5 Evaluation of the mES/OP9-cell coculture system as a model for haematopoietic differentiation

In this *in vitro* differentiation system two successive waves of haematopoiesis occur, with production of eryP on days 6–8 and of definitive haematopoietic cells from day 10 onwards. This timescale is roughly congruent with haematopoiesis occurring during embryogenesis, where primitive erythroid cells arise from committed progenitors in the YS at E7.0 (23); and definitive haematopoietic progenitors (lymphoid–myeloid–erythroid) are detected in the para-aortic splanchnopleura of the embryo at E8.5 (24), and in the fetal liver at E10.0 (25). Thus it is likely that the mES/OP9-cell coculture system recapitulates aspects of both the primitive and definitive phases occurring *in vivo*.

From the consecutive appearance of first primitive and then definitive erythroid lineages in this *in vitro* system, it has previously been postulated that distinct progenitors exist in the mES cell culture that differentiate toward either primitive or definitive haematopoietic lineages. In contrast, common precursor cells for both primitive and definitive haematopoietic cells have been found in the blast colony-forming cells derived from mES cells differentiated via EBs, in secondary methylcellulose culture (8). We therefore envisage that distinct types of progenitors may be detected preferentially, depending on which of the two mES cell-based *in vitro* differentiation systems is used. It is essential to clarify the

Table 1 Fluorochrome-conjugated monoclonal antibodies (all from BD Biosciences Pharmingen) for FACS® analysis of haematopoietic derivatives

Antibody	Clone	Cat. No.	Specificity	Recommended dilution	Fluorochrome
CD16/CD32 Fc Block™	2.4G2	553142	–	1/200	–
TER119	TER119	553673	Erythroid cells	1/200	PE
Mac-1 (CD11b)	M1/70	553311	Macrophages	1/200	PE
Gr-1	RB6–8C5	553128	Neutrophils	1/200	PE
CD45R/B220	RA3–6B2	553090	B cells	1/200	PE
c-Kit	2B8	553356	Immature haematopoietic progenitors	1/400	APC

PE = R-phycoerythrin; APC = allophycocyanin.

mechanism by which mES cells commit to differentiate into either the primitive or definitive haematopoietic lineages; and the present mES/OP9-cell differentiation system may provide additional insights into these processes.

There are other, practical advantages of the OP9 cell-based differentiation system over EB-based approaches, including: the ease of monitoring differentiation throughout, by viewing cell morphology under light microscopy; the facility of fractionating mature (non-adherent) cells without recourse to proteolytic enzymes, which thereby permits flow cytometric analysis; and the substantially lower level of contamination in differentiating cultures by non-haematopoietic derivatives. Moreover, the versatility of the OP9-cell system is demonstrated by other studies where, by changing the microenvironment (via cytokines and growth factors), other lineages may be produced, such as megakaryocytes, osteoclasts, and neutrophils (26); and where primate ES cells are similarly induced to differentiate into primitive and definitive haematopoietic derivatives (see Chapter 12).

In conclusion, the mES/OP9-cell coculture system is, at present, greatly facilitating the study of molecular mechanisms of transcription factor-regulated gene expression in haematopoiesis (16), as well as aspects of this multilineage developmental process that hitherto were refractory to study, such as the early stages of haematopoiesis from mesoderm. In particular, the potential to isolate and identify progenitor populations in that system by means of colony assays should assist in elucidating differentiation in the haematopoietic stem cell hierarchy. Therefore we regard the OP9-based system for the haematopoietic differentiation of mES cell as a valuable adjunct to *in vivo* and other *in vitro* approaches to elucidating molecular mechanisms involved in haematopoietic stem cell proliferation and differentiation, with a view to eventual therapeutic applications.

Acknowledgements
We thank Drs H. Kodama and T. Nakano for providing us with OP9 stromal cells.

References

1. Dzierzak, E. and Medvinsky, A. (1995). *Trends in Genetics*, **11**, 359-66.
2. Doetschman, T.C., Eistetter, H., Katz, M., Schmidt, W., and Kemler, R. (1985). *J. Embryol. Exp. Morphol.*, **87**, 27-45.
3. Schmitt, R.M., Bruyns, E., and Snodgrass, H.R. (1991). *Genes Dev.*, **5**, 728-40.
4. Burkert, U., vonRuden, R.T., and Wagner, E.F. (1991). *New Biol.*, **3**, 698-708.
5. Wiles, M.V. and Keller, G. (1991). *Development*, **111**, 259-67.
6. Chen, U. (1992). *Dev. Immunol.*, **2**, 29-50.
7. Keller, G., Kennedy, M., Papayannopoulou, T., and Wiles, M.V. (1993). *Mol. Cell. Biol.*, **13**, 472-86.
8. Kennedy, M., Firpo, M., Choi, K., Wall, C., Robertson, S., Kabrun, N., and Keller, G. (1997). *Nature*, **386**, 488-93.
9. Huber, T., Kouskoff, V., Fehling, H., Palis, J., and Keller, G. (2004). *Nature*, **432**, 625-30.
10. Shivsadani, R. and Orkin, S. (1996). *Blood*, **87**, 4025-39.
11. Barreda, D. and Belosevic, M. (2001). *Dev. Com. Immunol.*, **25**, 763-89.
12. Kennedy, M. and Keller, G. (2003). *Methods Enzymol.*, **365**, 39-59.
13. Nakano, T., Kodama, H., and Honjo, T. (1994). *Science*, **265**, 1098-101.
14. Kodama, H., Nose, M., Niida, S., Nishikawa, S., and Nishikawa, S. (1994). *Exp. Hematol.*, **22**, 979-84.
15. Nakano, T., Kodama, H., and Honjo, T. (1996). *Science*, **27**, 722-4.
16. Suwabe, N., Takahashi, S., Nakano, T., and Yamamoto, M. (1998). *Blood*, **92**, 4108-18.
17. Onodera, K., Takahashi, S., Nishimura, S., Ohta, J., Motohashi, H., Yomogida, K., Hayashi, N., Engel, J.D., and Yamamoto, M. (1997). *Proc. Natl. Acad. Sci. USA*, **94**, 4487-92.
18. Ohneda, O., Yanai, N., and Obinata, M. (1990). *Development*, **110**, 379-84.
19. Ploemacher, R.E., van der Sluijs, J.P., van Beurden, C.A., Baert, M.R., and Chan, P.L. (1991). *Blood*, **78**, 2527-33.
20. Nebsen, S., Anklesaria, P., Greenberger, J., and Mauch, P. (1993). *Exp. Hematol.*, **21**, 438-43.
21. Schmitt, T.M. and Zuniga-Pflucker, J.C. (2002). *Immunity*, **17**, 749-56.
22. Wickramasinghe, S.N. and McCullough, J., eds (2003). *Blood Bone Marrow Pathology*, Churchill Livingstone, Edinburgh.
23. Palis, J., Robertson, S., Kennedy, M., Wall, C., and Keller, G. (1999). *Development*, **126**, 5073-84.
24. Godin, I., Dieterlen-Lievre, F., and Cumano, A. (1995). *Proc. Natl. Acad. Sci. USA*, **92**, 773-7.
25. Johnson, G. and Moore, M. (1975). *Nature*, **258**, 726-8.
26. Suzuki, A. and Nakano, T. (2001). *Int. J. Hematol.*, **73**, 1-5.

Chapter 8

Lineage selection and transplantation of mouse ES cell-derived neural precursors

Tanja Schmandt, Tamara Glaser and Oliver Brüstle

Institute of Reconstructive Neurobiology, Life and Brain Center, University of Bonn and Hertie Foundation, Sigmund-Freud-Strasse 25, 53105 Bonn, Germany.

1 Introduction

Embryonic stem (ES) cells have become an indispensable experimental resource for studying cellular and molecular processes of cell differentiation under controlled conditions *in vitro*. Furthermore, their capacities to proliferate indefinitely and to differentiate into all somatic cell types make them a precious donor source for cell replacement strategies. The success of ES cell-based applications will critically depend on the availability of: (i) protocols for controlled *in vitro* differentiation; and (ii) assay systems for validating the functional performance of ES cell-derived somatic cells – the two key aspects covered in this chapter. The protocols described here are concerned specifically with mouse ES (mES) cells, which remain a paradigmatic model system for the development of novel differentiation strategies.

1.1 Neural differentiation potential of mES cells *in vitro*

Due to their enormous potential for the treatment of neurological disorders, neural precursors have always been a prime target of controlled ES-cell differentiation studies; and a variety of cell culture conditions, growth factors, and differentiation stimuli have been explored for the generation of enriched neuronal and glial precursors. Conventionally, neural differentiation has been initiated by aggregating mES cells into EBs in the presence of all-*trans* retinoic acid (RA) (1–5). Alternatively, serum-free culture systems without RA treatment have been used to generate neurons and glia (6–9). Recently, the field has experienced a renaissance of RA-based protocols for the enrichment of radial glia and oligodendrocytes (10, 11). Another emerging method for neural differentiation of mouse and monkey ES cells is based on coculture with bone marrow-derived stromal cells (specifically, PA6 cells (12); and see Chapter 12). In this system an unidentified stromal cell-derived activity induces efficient neural differentiation of the cocultured ES cells under serum-free conditions – without EB formation or the use of RA. Coculture with stromal cells, in conjunction with defined growth and differentiation factors, also has been employed to induce neurons with distinct regional phenotypes (13). Efficient neural differentiation of mES cells has further been achieved by culturing mES-cell aggregates in medium conditioned by human hepatocellular carcinoma cells (14). Furthermore, it has been shown that mES cells cultured at low density in serum-free conditions can undergo direct conversion into a neuroectodermal fate (15–16).

Here, we provide protocols for the enrichment and isolation of neural lineages from precursors originating in EBs, the multicellular aggregates which are formed by mES cells upon their withdrawal from feeder-layer support, and in the absence of leukaemia inhibitory factor (LIF) or other differentiation-inhibiting factors (see also Chapters 5, 6, and 9).

EBs recapitulate many aspects of cell differentiation observed during normal mammalian embryogenesis (17–19). Initially, they generate an outer layer of endodermal-like cells, which is followed by the appearance internally of ectodermal cells, and subsequent specification of mesodermal progeny. However, the lack of precise structural organization and positional information within EBs results in a pronounced cellular heterogeneity both within and between individual EBs. Upon plating of EBs onto tissue-culture plates, they typically give rise to an outgrowth of a diversity of cells emanating from the three germ layers at different developmental stages. Attempts to derive purified somatic cells from mES cells reproducibly therefore aim at restricting this broad developmental potential to distinct lineages. This is achieved by using cell culture conditions that promote differentiation and proliferation of defined lineages (i.e. controlled/lineage-restricted differentiation) from the mixed progenitor populations, and/or genetic selection strategies (i.e. lineage selection).

2 Controlled differentiation of mES cells into neural precursors

During normal development, neural precursors undergo a stepwise transition from a multipotent to a more lineage-restricted neuronal or glial progenitor stage. Although neural differentiation has been considered a default pathway of ES cell differentiation (15, 16), the generation of high numbers of neurons and glia requires protocols that promote the survival and proliferation of neural precursors. The experimentally induced differentiation pathway presented here (*Protocol 1*) recapitulates the gradual transition of neurogenic into gliogenic precursors observed during CNS maturation. Based on a differentiation paradigm described by Okabe *et al.* (6,20), EB-derived neural precursors are sequentially propagated with discrete growth factors, and develop from a multipotent into a predominantly glial precursor fate (7,20). *Figure 1* shows a schematic of the temporal sequence of events.

An important prerequisite for the successful application of this protocol is the use of mES cell cultures that are well maintained, that is with little or no signs of spontaneous differentiation. Undifferentiated mES cells are refractile and small, and the cell clusters are rounded and well defined (*Figure 2A*). On withdrawal of LIF from the medium and reseeding of mES cells onto a non-adhesive substratum, EB formation is initiated (*Figure 2B*). After 4 to 5 days, the newly formed EBs are plated onto an adhesive substratum and propagated in a serum-free medium (ITSFn; *Figure 2C*), which favours the survival of neuroepithelial cells and inhibits the growth of other cell types (6). Following this selection step, the resulting

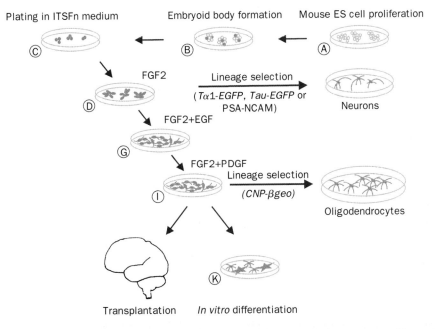

Figure 1 Stepwise differentiation of mES cells into multipotent neural precursors, with subsequent conversion into glial precursors (schematic). Mouse ES cells are proliferated to large numbers, on feeder layers and in the presence of LIF, in preparation for controlled neural differentiation (*Protocol 1*), as follows. Inducing aggregation by suspension culture in the absence of LIF initiates mES cell differentiation via EBs. The EBs are plated in defined medium promoting the survival of neural precursors. These cells are proliferated in the presence of FGF2; and sequential culture in defined media with FGF2 and EGF, followed by FGF2 and PDGF-AA, gradually converts the cells into glial precursors. Terminal differentiation is induced by growth factor withdrawal. By combining controlled differentiation with lineage selection, highly purified neurons and glia can be isolated from populations at individual stages (for details see *Protocols 3 and 4*). Letters A–K refer to corresponding phase-contrast micrographs shown in *Figure 2*.

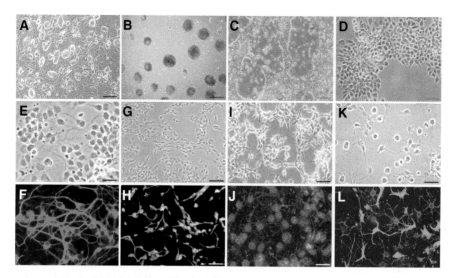

Figure 2 (see Plate 4) Representative photomicrographs of differentiating mES cells at various stages of neural development. (A) Undifferentiated mES cells grown on a feeder monolayer of mitotically inactivated MEFs prior to differentiation according to *Protocol 1*. (B) EBs generated by aggregation of undifferentiated mES cells. (C) Selection of neural precursors after plating EBs in ITSFn medium. (D) Proliferation of multipotent neural precursors (i.e. ESNPs) in FGF2-supplemented medium. After withdrawal of FGF2, cells differentiate into post-mitotic neurons and glia: (E) Phase-contrast micrograph of cells 4 d following FGF2 withdrawal; the corresponding immunofluorescence micrograph (F) depicts β-III-tubulin in mES cell-derived neurons (red). Further proliferation of ESNPs with FGF2 and EGF (G, H), and subsequently with FGF2 and PDGF-AA (I, J), enables the generation of highly purified glial precursors (ESGPs). (K, L) After growth factor withdrawal, ESGPs differentiate into GFAP-positive astrocytes (green) and O4-positive oligodendrocytes (red). (G–L) Phase-contrast micrographs (G, I, K) and corresponding immunfluorescence analyses with antibodies to nestin (H), A2B5 (J), GFAP (L, green), and O4 (L, red). Nuclei are counterstained with Hoechst (blue). Scale bars = 80 μm in (A–C) and (G–H), 40 μm in (D) and (I), 20 μm in (E–F) and (J–L). Adapted from Wernig *et al.* (65), with kind permission of Springer Science + Business Media.

outgrowths are dissociated into a single-cell suspension, replated, and further proliferated in the presence of fibroblast growth factor-2 (FGF-2). At this stage the cells represent multipotent, mES cell-derived neural precursors (ESNPs (6), and see *Figure 2D*) which can be induced to differentiate into neurons and glia by growth factor withdrawal (*Figure 2E, F*). Alternatively, proliferating ESNPs can be shifted towards a predominantly glial precursor stage by further propagation with FGF-2 and epidermal growth factor (EGF). These mES cell-derived glial precursors (ESGPs; *Figure 2G, H*) are stable for several passages, and also can be differentiated further by growth factor withdrawal. An additional culture step in the presence of FGF-2 and platelet-derived growth factor-AA (PDGF-AA) can be used to obtain oligoastrocytic precursors, which efficiently give rise to myelinating oligodendrocytes ((7) and see *Figure 2I–L*).

Protocol 1
Controlled differentiation of mES cells into neural precursors

These methods were developed using mES cell lines CJ7 (21) and J1 (22). Routine maintenance of mES cells, preparation of MEF feeder layers, and cryopreservation of cells are described in detail in Chapter 2; and alternative methods for EB formation from mES cells are provided in Chapters 5, 6, and 9. Cell culture throughout is in a humidified incubator gassed with 5% CO_2 in air, and at 37 °C.

Reagents and equipment

Reagents are from Sigma, unless otherwise indicated.

- Vial of cryopreserved mES cells
- Basal medium, for mES cells: DMEM (Cat. No. 11966, Invitrogen) supplemented with 1 × minimal essential medium (MEM)-non-essential amino acids (Cat. No. 11140, Invitrogen), 4.5 g/L glucose (Cat. No. G-7021), 0.12 g/L sodium pyruvate (Cat. No. P-5280), 8 mg/L adenosine (Cat. No. A-4036), 8.5 mg/L guanosine (Cat. No. G-6264), 7.3 mg/L cytosine (Cat. No. C-4654), 7.3 mg/L uridine (Cat. No. U-3003), 2.4 mg/L thymidine (Cat. No. T-1895), 0.1 mM β-mercaptoethanol (Cat. No. M-7522), and 26 mM Hepes (Cat. No. H-3784)
- Complete medium: basal medium supplemented with 20% mES cell-qualified fetal-calf serum (FCS; Cat. No. 10270106, Invitrogen) and 1000 U/ml LIF (Cat. No. ESG1107, Chemicon International)
- Differentiation medium: basal medium supplemented with 10% mES cell-qualified FCS and 1 × penicillin/streptomycin (Cat. No. 15140, Invitrogen)
- FCS-supplemented medium: DMEM with 10% FCS
- Flame-polished Pasteur pipettes
- Haemocytometer
- Trypan blue (Cat. No. 15250, Invitrogen)
- DMSO-based freezing medium (2×): combine 22.8 ml basal medium, 5 ml mES cell-qualified FCS, and 3.6 ml DMSO (Cat. No. D-2650)
- Two 60 mm culture dishes precoated for 20 min with a sterile solution of 0.1% gelatin (Cat. No. G-1890) in ddH_2O (see Chapter 6, Protocol 1)
- Hanks' balanced salt solution (HBSS; Cat. No. H-9394)
- Trypsin/EDTA solution: 0.5 g trypsin (1:250) and 0.2 g EDTA*4Na/L final concentrations, from 10 × stock (Cat. No. 15400, Invitrogen) diluted in HBSS
- Hepes/HBSS: 15.95 mM Hepes (Cat. No. H-3375) in HBSS
- EDTA/HBSS: 0.04% EDTA (Cat. No. E6511) in HBSS
- Eight 6 cm bacteriological-grade dishes (Cat. No. 240043, Nunc)
- ITSFn, or neural selection, medium: DMEM/F12 medium (Cat. No. 11320, Invitrogen) with 5 µg/ml insulin (Cat. No. S-6634), 50 µg/ml human apo-transferrin (Cat. No. T-2036), 30 nM sodium selenite (Cat. No. S-5261), 2.5 µg/ml fibronectin (Cat. No. 33010, Invitrogen), and 1 × penicillin/streptomycin (Cat. No. 15140, Invitrogen)
- 10 cm culture dishes precoated with poly-ornithine (15 µg/ml solution in ddH_2O; Cat. No. P-3655) for 2 h
- DNase solution: 0.1% DNase I (Cat. No. LS02141, CellSystems) in HBSS
- 40 µm nylon mesh (Cat. No. 352340, BD Biosciences)

Protocol 1 continued

- N2 medium: DMEM/F12 medium with 72 mg/L glutamine (Cat. No. G-5763), 1.54 g/l glucose, 25 μg/ml insulin, 100 μg/ml human apo-transferrin, 20 nM progesterone (Cat. No. P-8783), 100 nM putrescine (Cat. No. P-5780), 30 nM sodium selenite, and 1 × penicillin/streptomycin
- N3 medium: DMEM/F12 medium with 25 μg/ml insulin, 50 μg/ml human apo-transferrin, 20 nM progesterone, 100 nM putrescine, 30 nM sodium selenite, and 1 × penicillin/streptomycin
- Cell scraper (Cat. No. 3008, Corning)
- Supplements and growth factors: laminin (Cat. No. L2020, Sigma), FGF2 (Cat. No. 233-FB, R&D Systems), EGF (Cat. No. 236-EG, R&D Systems), PDGF-AA (Cat. No. 221-AA, R&D Systems)
- 6 and 10 cm culture dishes (Cat. No. 3004 and 3003, BD Biosciences)
- Sterile, 15 ml and 50 ml centrifuge tubes (all BD Biosciences)
- Benchtop refrigerated centrifuge (e.g. Megafuge 1.0R, Heraeus)
- Inverted phase-contrast microscope (e.g. Axioscop 2, Carl Zeiss)

Methods

Preparation of mES cells for EB formation:

1. Thaw and culture mES cells for 2–3 d in complete medium (*Figure 2A*). Change medium daily.
2. When mES cells reach 80% confluence, aspirate medium, wash cells with EDTA/HBSS, and add trypsin/EDTA solution to cover the monolayer. Incubate the dish until the cells detach.
3. Triturate using a flame-polished Pasteur pipette to obtain a single-cell suspension.
4. Stop trypsinization by adding 5 ml FCS-supplemented medium, transfer the suspension into a tube, and centrifuge cells for 5 min at 1000 rpm and 4 °C.
5. Resuspend cells in complete medium, and determine number of vital cells by trypan blue exclusion using a haemocytometer.
6. Seed 2.5×10^6 mES cells in 3 ml complete medium onto a gelatin-coated, 6 cm culture dish. Freeze the remaining cells for future use at 1.25×10^6 cells/1.5 ml cryovial in DMSO-freezing medium/complete medium (1:2 ratio; see Chapter 2).
7. Incubate cultures for 20 min to allow feeder cells to settle and adhere to the substratum.[a]
8. Take the supernatant with the non-attached mES cells, dispense into an additional gelatin-coated, 6 cm culture dish, and incubate.
9. After 2 d, induce aggregation and differentiation of mES cells in EBs as follows: harvest cells from a 6 cm dish by trypsinization, resuspend in 32 ml differentiation medium, and seed into eight 6 cm bacteriological-grade dish (*Figure 2B*). Incubate the cultures. The day of initiating EB formation is designated 'day 0'.

Protocol 1 continued

Plating and disaggregation of EB-derivatives to induce neural precursors:

10 Plate and evenly distribute day 4–5 EBs onto a 10 cm culture dish. There should be sufficient space between plated EBs for spreading and outgrowth (see also Chapter 9, *Protocol 3*).

11 Next day, aspirate medium, wash cells three times with 1 × HBSS, and change to ITSFn medium to promote selection of neural progenitors.

12 Change medium every second day (for cell morphology at this stage, see *Figure 2C*).

13 After 4–6 d (i.e. on day 8–11 of EB differentiation), harvest cells by trypsinization as in Step 2, stop trypsination as in Step 4, and centrifuge the cells for 5 min at 1000 rpm and 4 °C.

14 Use flame-polished Pasteur pipettes with decreasing pore size to disaggregate cell clusters into a single-cell suspension in DNase solution.[b]

15 Filter the cells through a nylon mesh, wash the mesh with HBSS to dislodge retained single cells, and centrifuge filtered cells for 5 min at 1000 rpm and 4 °C. Determine the cell number using an haemocytometer.

16 Plate cells onto polyornithine-coated 10 cm culture dishes at a density of 4×10^6 cells/dish, in N3 medium supplemented with laminin (1 µg/ml) and FGF2 (10 ng/ml), and incubate for 4 d to promote proliferation of mES cell-derived neural precursors (ESNPs).

17 Replace medium every second day, and add factors daily (for cell morphology, see *Figure 2D*). Further differentiation of ESNPs into specific cell types may then be achieved as follows.

Differentiation of ESNPs into neurons and glia:

18a Withdraw FGF2 from the medium, and continue incubation for 4–8 d (*Figure 2E, F*).

Differentiation of ESNPs into glial precursors:

18b Harvest proliferating ESNPs by scraping cells from the dish into DNase solution. Gently dissociate cells using a flame-polished Pasteur pipette, and filter cells through a nylon mesh to remove any residual cell clusters.

19 Replate ESNPs at a 1:3 split ratio, and culture cells to subconfluence in N3 medium supplemented with FGF2 (10 ng/ml) and EGF (20 ng/ml).

20 Replace medium every second day, and add factors daily (for cell morphology, see *Figure 2G, H*).

21a After five to seven passages, induce terminal differentiation by withdrawing growth factors from the medium.

or

21b For further enrichment of oligodendroglial progenitors, change to N2 medium supplemented with FGF2 (10 ng/ml) and PDGF-AA (10 ng/ml).

> **Protocol 1 continued**
>
> 22 Culture cells under these conditions for at least another 4 d. Change medium every second day, and add factors daily (for cell morphology, see *Figure 2I, J*).
>
> 23 Induce differentiation into astrocytes and oligodendrocytes by growth factor withdrawal (*Figure 2K, L*).
>
> [a] This procedure effectively eliminates feeder cells from the suspension.
> [b] Material becomes intractable without the use of DNase solution.

3 Selection of lineage-restricted cell populations from differentiated mES cells

Mouse ES cell-derived neural precursors generated solely by epigenetic modulation of cell culture conditions have been shown to survive transplantation and contribute to neural repair in a variety of CNS disease models (7, 10, 13, 23–25); yet the donor cells used in these experiments typically represent a mixture of different cell types. Further purification of the desired phenotype can be achieved by additional lineage-selection steps. For example immunopanning, fluorescence-activated cell sorting (FACS®) or magnetic-activated cell sorting (MACS®) can be used to purify mES cell-derived neural subpopulations expressing specific cell-surface markers. Alternatively, mES cells can be genetically engineered to express antibiotic-resistance or fluorescent marker genes under the control of lineage-specific promoters. Thus, upon induction of differentiation, the desired cell type can be isolated by antibiotic treatment or FACS®. Typically, such lineage-selection strategies are combined with optimized epigenetic differentiation protocols to provide a high degree of purification of neural cells ((26–29); and see *Protocols 2–5*).

3.1 Transfection of mES cells with selectable marker genes, for neuronal and oligodendroglial lineage selection

Lineage selection involves the isolation of a desired phenotype from a heterogeneous cell population. This can be facilitated by transfecting undifferentiated mES cells initially with selectable marker genes that are expressed under control of a cell type-specific promoter. Following cell differentiation, and if the marker is an antibiotic-resistance gene, selection is performed by adding the respective antibiotic to the culture medium. Alternatively, reporter genes such as *EGFP* permit enrichment of the desired, differentiated phenotype by FACS®.

Protocols 1–5 encompass controlled differentiation and lineage selection: mES cells are first transfected with lineage-selection constructs (*Protocol 2*), differentiated according to *Protocol 1*, and subsequently subjected to lineage-selection strategies such as FACS®, immunopanning or antibiotic selection (see *Protocols 3, 4* and *5*). To enable the isolation of transfected clones, mES cells should be cotransfected with constructs carrying an additional, distinct antibiotic resistance gene under the control of a constitutively active promoter.

Protocol 2
Transfection of mES cells by electroporation

Reagents and equipment

- A minimum of 1×10^6 mES cells (~80% confluent culture)
- Two 10 cm feeder dishes, prepared with γ-irradiated multiple-drug-resistant (γmdr) MEFs (30)[a]
- Feeder (γmdrMEFs) layers in 96-well and 24-well plates, and in 3 cm dishes [a,b]
- Plasmids with cassettes for lineage selection (e.g. Tα1-EGFP, CNP-βgeo) and antibiotic resistance (e.g. pPGK-neo or -hyg)[c]
- PBS, Ca^{2+}- and Mg^{2+}-free (Cat. No. H15-002, PAA Laboratories)
- Trypsin/EDTA solution (Protocol 1)
- Complete medium (Protocol 1)
- Selection medium: complete medium with antibiotics, e.g. 400 µg/ml G418 (Cat. No. 11811, Invitrogen) for the neomycin-resistance gene, neo
- DMSO-based freezing medium (Protocol 1)
- Trypan blue (Protocol 1)
- Haemocytometer
- Flame-polished Pasteur pipettes
- Sterile, 15 ml Falcon tubes
- Empty 96-well plate (Cat. No.03599, Corning)
- Gene Pulser Xcell eukaryotic system (Cat. No. 165-2661, Bio-Rad Laboratories)
- Electroporation cuvette with 0.4 cm electrode gap (Cat. No. 165-2081, Bio-Rad Laboratories)
- Benchtop refrigerated centrifuge
- Inverted phase-contrast microscope

Methods

Electroporation:

1. Change medium in mES-cell culture 1 h before electroporation.
2. Wash mES cells with PBS, and remove cells from the dish with trypsin/EDTA solution (Protocol 1). Triturate into a single cell suspension with a flame-polished Pasteur pipette, and stop trypsinization by addition of complete medium.
3. Centrifuge cells for 5 min at 1000 rpm and 4 °C, and determine the cell number by trypan blue exclusion.
4. Resuspend the cell pellet in cold PBS at a concentration of 1×10^7 mES cells/ml.
5. Transfer 800 µl of the cell suspension into an electroporation cuvette.
6. Add plasmid DNA (100 µg of selectable marker construct, plus 10–20 µg of construct with a constitutively active promoter; either circular or linearized).[c]
7. Incubate cells for 5 min at r.t.
8. Perform electroporation at 230 V and 490 mF using a Gene Pulser II®.
9. Incubate cells in the cuvette for 5 min on ice, and transfer them into fresh complete medium.
10. Plate cells from one electroporation onto two 10 cm (γmdrMEFs) feeder dishes, and incubate.
11. Next day, replace medium.

Protocol 2 continued

12 Two days after electroporation, commence selection by adding antibiotics to the medium (selection medium), and change selection medium daily.

Isolation of genetically modified clones/colonies:

Isolation of antibiotic-resistant mES-cell colonies is started 9–15 d after the onset of selection, or when large and undifferentiated colonies are visible.

13 Dispense 10 µl PBS into each well of an empty 96-well plate.

14 Aspirate medium from cultures (in 10 cm dishes) having undergone selection, wash cells twice with PBS, and add 3 ml PBS per dish.

15 Using an inverted microscope placed under a laminar flow hood, identify undifferentiated mES cell colonies; scrape them off the dish using the tip of a micropipette, and transfer each colony into a separate well of the 96-well plate containing PBS.

16 Add 25 µl of trypsin/EDTA solution to each well, and incubate the plate for 5 min.

17 Add 25 µl selection medium to each well, and triturate cells to a single cell suspension using a flame-polished Pasteur pipette or a micropipette with a 100 µl plastic tip.

18 Transfer each cell suspension into a separate well of a 96-well (γmdrMEF) feeder plate, add 150 µl selection medium per well, and incubate. Change medium daily.

19 When cultures reach 60–80% confluence, expand mES cells from a 96-well into a 24-well feeder plate, and then into a 3 cm feeder dish, whilst maintaining selection conditions for several weeks. For each clonal subline, if possible, freeze an aliquot at every passage in DMSO-based freezing medium/selection medium (see *Protocol 1*).

[a] Required only for feeder-dependent mES cells.

[b] These feeder layers are required for expanding the selected antibiotic-resistant mES-cell clones.

[c] Alternatively, a single plasmid carrying both selectable genes may be used (see Chapter 6, *Protocol 7*)

The concentration of the lineage-selection construct should exceed that of the constitutively active construct five to ten fold, to ensure cotransfection of the former (28, 29).

3.2 Neuronal lineage selection

Here we describe two protocols that permit the isolation of mES cell-derived neurons by subjecting predifferentiated ESNPs, generated according to *Protocol 1*, to FACS® or immunopanning. The selection markers used are standard, and the protocols may be extended to other selection markers with similar expression profiles.

3.2.1 FACS®-based neuronal lineage selection

This technique requires mES cells carrying an *EGFP* reporter gene under the transcriptional control of a neuron-specific promoter, to enable FACS® of

Protocol 3
FACS®- based neuronal lineage selection

Reagents and equipment

- Culture of predifferentiated ESNPs, generated from mES cells carrying an *EGFP* reporter gene controlled by a neuron-specific promoter, and following withdrawal of FGF2 from the medium for 2–7 d[a] (*Protocol 1*, step 18a): a minimum of 1×10^7 ESNPs is required
- HBSS (*Protocol 1*)
- Trypsin/EDTA solution (*Protocol 1*)
- Trypsin inhibitor, 1 × solution (Cat. No. T6414, Sigma)
- 40 µm nylon mesh (*Protocol 1*)
- DNase solution (*Protocol 1*)
- Trypan blue (*Protocol 1*)
- Polyornithine-coated coverslips (Cat. No. L-4097-1, Plano) in a 24-well plate (*Protocol 1*)
- N3 medium (*Protocol 1*)
- Laminin (*Protocol 1*)
- Flame-polished Pasteur pipette
- Sterile, 15 ml Falcon tubes
- Preparative fluorescence-activated cell sorter (e.g. FACSAria®; BD Biosciences)
- Haemocytometer
- 4% paraformaldehyde in PBS (Chapter 9, *Protocol 6*)
- 0.1% sodium azide in PBS

Methods

1. Aspirate medium and wash ESNPs three times in HBSS.
2. Add trypsin/EDTA solution to cover, incubate until the cells detach, and add trypsin inhibitor to stop trypsinization.
3. Transfer detached cells into a 15 ml centrifuge tube; wash the dish twice with HBSS to remove residual cells, and add to the contents of the tube.
4. Centrifuge cells for 5 min at 1000 rpm and 4 °C, and resuspend the cell pellet in 3 ml DNase solution.
5. Triturate the cells into a single-cell suspension with a flame-polished Pasteur pipette.
6. Filter the suspension through a nylon mesh, wash the mesh with HBSS, and centrifuge the filtered cells for 7 min at 1000 rpm and 4 °C.
7. Determine the number of viable cells by trypan blue exclusion.
8. Resuspend cells in DNase solution at a concentration of $0.25-1 \times 10^6$ cells/ml, and maintain them on ice until flow cytometry is completed.
9. Analyse cells by forward-angle and side-angle light scatter with an argon-ion laser operating at 488 nm. Sorting procedures should only be based on fluorescence intensity, and are performed with flow rates of up to 10 000 events/second.
10. Replate the sorted neuronal cells onto polyornithine-coated coverslips (or culture dishes) in N3 medium supplemented with laminin (1 µg/ml) for further culture.

Protocol 3 continued

For quality control, fix an aliquot of the replated cells in 4% paraformaldehyde for immunofluorescence analysis. Store fixed cells at 4 °C in 0.1% sodium azide solution.

[a] For the selection of *EGFP*-expressing cells, withdraw FGF2 from the medium and incubate cells for 2 d (for *Tα1-EGFP* selection construct) or 7 d (for *Tau-EGFP* selection construct).

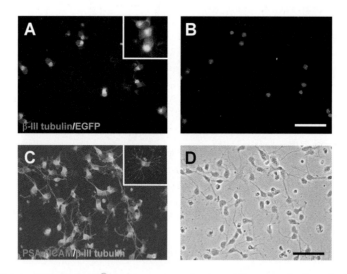

Figure 3 (see Plate 5) FACS®- and immunopanning-based lineage selection of mES cell-derived neurons. (A, B) FACSorting of *Tα1-EGFP*-positive cells results in efficient enrichment of neuronal cells, yielding purities of >98% β-III-tubulin-positive cells. (A) *EGFP* expression (green) and β-III-tubulin immunofluorescence (red) in the sorted cell population 1 d after plating. The inset in (A) shows a high-power view. (B) Corresponding nuclear Hoechst stains. (C, D) Following a 2 d period of growth factor withdrawal, immunopanning of ESNPs with an antibody to PSA-NCAM yields purities of >95% β-III-tubulin-positive neurons. (C) PSA-NCAM-positive cells (red) double labelled with an antibody to β-III-tubulin (green) 2 d after immunopanning (Inset: detail). Nuclei are counterstained with Hoechst (blue). (D) Corresponding phase-contrast micrograph. Scale bars = 50 μm in C and F. Adapted from Schmandt et al. (28).

EGFP-positive neurons generated from ESNPs. Transgenes that have been used successfully to this end include *Tα1-EGFP* ((28); and see *Figure 3A, B*) and *Tau-EGFP* (27). The time point of FACS® has to be adapted to the kinetics of the individual reporter gene activity. Ideally, FACS® of *Tα1-EGFP*- and *Tau-EGFP*-positive neurons is performed 2 d and 7 d after initiation of differentiation, respectively. Both methodologies typically yield neuronal populations with purities exceeding 90% (27, 28).

Protocol 4
Neuronal lineage selection using immunopanning

Reagents and equipment

- Culture of ESNPs following withdrawal of FGF2 from the medium for 2 d (*Protocol 1*, step 18): a minimum of 4×10^6 ESNPs is required
- 10 cm OPTILUX™ Petri dish (Cat. No. 1001, BD Biosciences)
- Biotinylated goat anti-mouse IgM antibody (Cat. No. ab 9167-1, Abcam)
- 50 mM Tris-HCl (pH 9.5; Cat. No. T1503, Sigma)
- Anti-PSA-NCAM mouse IgM antibody (Cat. No. MAB 5324, Chemicon International)
- BSA/PBS solution: 2% bovine serum albumin (BSA; Cat. No. A8806, Sigma) in PBS
- PBS (*Protocol 2*)
- HBSS (*Protocol 1*)
- DNase solution (*Protocol 1*)
- Flame-polished Pasteur pipette
- Cell scraper (*Protocol 1*)
- 40 μm nylon mesh (*Protocol 1*)
- Trypsin/EDTA solution (*Protocol 1*)
- Trypsin inhibitor (*Protocol 3*)
- Haemocytometer
- Trypan blue (*Protocol 1*)
- Polyornithine-coated 3 cm culture dish (*Protocol 1*)
- N3 medium (*Protocol 1*)
- Laminin (*Protocol 1*)
- 4% paraformaldehyde (*Protocol 3*)
- 0.1% sodium azide solution (*Protocol 3*)

Methods

Preparation of the immunopanning dish:

1. One day prior to immunopanning, cover the base of a 10 cm OPTILUX™ Petri dish with a solution of biotinylated anti-mouse IgM antibody (diluted 1:100 in 50 mM Tris-HCl, pH 9.5), and store the dish overnight at 4 °C.

2. Next day, aspirate the antibody solution, wash the dish three times with PBS, cover the base of the dish with the anti-PSA-NCAM antibody (diluted 1:1000 in 2% BSA/PBS), and incubate for 1 h at r.t.

Preparation of cells:

3. Aspirate medium and wash ESNPs three times with HBSS, add 3 ml DNase solution, and scrape cells off the dish.

4. Triturate cells into a single cell suspension with a flame-polished Pasteur pipette, and filter them through a nylon mesh.

5. Centrifuge filtered cells, and resuspend cells in N3 medium. Count viable cells.

6. Plate $5-6 \times 10^6$ cells in 5 ml N3 medium onto the immunopanning dish, and incubate for 30 min.

7. Remove the medium with non-panned (i.e. non-bound) cells, and wash the surface of the dish three times with PBS to ensure removal of non-panned cells. Pool these fractions and collect the suspended cells for use as a control.

Protocol 4 continued

8. Remove panned (i.e. bound) cells from the dish by trypsinization, and stop trypsinization with trypsin inhibitor.

9. Centrifuge the panned cells for 7 min at 1000 rpm and 4 °C.

10. Resuspend the panned (neuronal) cells in N3 medium supplemented with laminin (1 µg/ml), plate 3×10^5 cells per polyornithine-coated 3 cm dish, and incubate.

Protocol 5
Oligodendroglial lineage selection using selectable marker genes

Reagents and equipment

- Culture of ESGPs, derived from mES cells expressing an antibiotic resistance gene (e.g. *neo*) under control of an oligodendrocyte-specific (e.g. *CNP*) promoter: the cells are expanded in the presence of FGF2 and PDGF-AA as described in *Protocol 1*, steps 21–22, and a minimum of 4×10^6 cells are required

- 3,3,5-triiodothyronine hormone (T3; Cat No. T-5518, Sigma)
- N2 medium (*Protocol 1*)
- FCS[a]
- FGF2 and PDGF-AA (*Protocol 1*)
- G418 (*Protocol 2*)

Method

1. Initiate differentiation in ESGPs by withdrawal of growth factors FGF2 and PDGF-AA, but addition of T3 (30 nM) to the N2 medium, and culture for 4 d.

2. Aspirate medium, and replace with N2 medium supplemented with 0.5% FCS, 30 nM T3, 10 ng/ml FGF2, and 10 ng/ml PDGF-AA; and culture cells for 2 d.

3. Subject cells to antibiotic selection (e.g. 200 µg/ml G418) for 2–3 d in the same medium.

4. For terminal differentiation, withdraw growth factors from the medium, and culture cells for 4 d in N2 medium supplemented with T3.

[a] The efficiency of oligodendroglial lineage selection may vary with different serum batches.

3.2.2 Neuronal lineage selection using immunopanning

Immunopanning is an antibody-based strategy for the isolation of cells expressing specific cell-surface antigens, which does not require cells to carry a selectable marker construct. For neuronal lineage selection, one suitable target antigen is polysialylated neural cell adhesion molecule (PSA-NCAM). Although, in the developing CNS, PSA-NCAM is expressed in both neurons and glia (31–33), it is virtually restricted to neuronal cells during the early stages of mES cell differentiation. Thus, immunopanning of predifferentiated mES cell-derived neural precursors with an antibody to PSA-NCAM permits the enrichment of neurons at purities of more than 95% ((28); and see *Figure 3C,D*).

3.3 Oligodendroglial lineage selection

A suitable marker gene for the oligodendroglial lineage is *CNP*, encoding 2'3'-cyclic nucleotide 3'-phosphodiesterase, one of the earliest known myelin-associated proteins synthesized in developing oligodendrocytes and Schwann cells (34–36). Constructs encoding selection markers under transcriptional control of the *CNP* promoter have been successfully used to enrich oligodendrocytes from both brain tissue (36) and mES cells (29). *Protocol 5* describes how ESGPs generated according to *Protocol 1* are differentiated to a preoligodendrocyte stage by a specific sequence of culture conditions, and subsequently subjected to *CNP*-based lineage selection ((29); and see *Figure 4*). The protocol may be adapted to other oligodendroglial marker genes with similar expression profiles.

4 Transplantation models for studying functional integration of mES cell-derived neural precursors

The potential therapeutic value of ES cell-derived donor cells generated by a particular *in vitro* differentiation protocol depends critically on their potential to integrate into the recipient tissue, respond to environmental signals, undergo further regional differentiation, and take over specific functions of lost or injured cells. Suitable model systems are required to validate the performance of ES cell-derived donor cells in these complex and interrelated processes under controlled conditions.

For the transplantation protocols described below, all animal procedures must be sanctioned by the appropriate licensing authorities, and conducted humanely.

4.1 Neural precursor integration in organotypic slice cultures

Organotypic slice cultures provide an opportunity to explore migration and incorporation of donor cells under controlled conditions in a real-time setting (*Figures 5* and *6*). In our '*in vitro* transplantation' system, the donor cells are deposited at the surface of cultured brain slices from where they migrate and incorporate into the host tissue (*Protocol 6*). When combined with *EGFP* genetic-labelling of the engrafted cells, this system permits direct experimental access to migration, differentiation, and electrophysiological activity at the single-cell level. Indeed, hippocampal slice cultures have been successfully

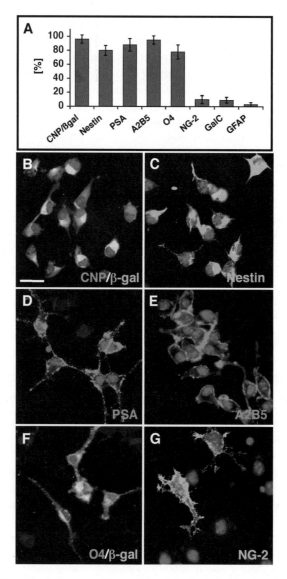

Figure 4 (see Plate 6) Antigenic profile of mES cell-derived oligodendrocytes generated by *CNP-βgeo*-based lineage selection. *CNP-βgeo* transfected mES cells were first subjected to controlled differentiation into glial precursors. Subsequent induction and stabilization of a CNP-positive fate permitted the isolation of mES cell-derived oligodendrocyte progenitors. Quantitative marker expression and representative immunofluorescence images of the selected cells are shown in (A) and (B–G), respectively. More than 96% of the cells coexpress CNP and β-gal (A-B). In addition, the majority of cells express nestin (C), PSA (D), A2B5 (E), and O4 (F). The selected population also includes GalC- and NG2-positive cells (G). Nuclei of cells are stained with DAPI (blue). Scale bar = 50 μm. For details see Glaser *et al.* (29).

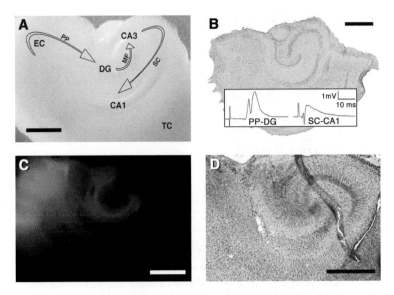

Figure 5 (see Plate 7) Morphological and functional integrity of hippocampal slice cultures used as recipient tissue for 'in vitro transplantation' of mES cell-derived neural precursors. (A) A 400 μm-thick slice 1 d after explantation. Dentate gyrus (DG), pyramidal-cell layer (CA3–CA1), entorhinal cortex (EC), and adjacent regions of the temporal cortex (TC) are well delineated. SC, Schaffer collaterals; MF, mossy fibres; PP, perforant path. (B) Cryostat section (10 μm) of a slice preparation maintained in culture for 31 d. Note the morphological preservation of the key anatomic structures (haematoxylin and eosin stain). The inset shows typical field potentials following orthodromic stimulation of the perforant path (PP-DG) and the Schaffer collaterals (SC-CA1) at the end of the culture period. (C) Anterograde axonal tracing with rhodamine-conjugated dextran (Micro-Ruby®) confirms the integrity of the perforant path at 33 d in culture. (D) TIMM stain demonstrating the histological preservation of the mossy fibre system of a slice culture maintained for 33 d (20 μm cryostat section). Scale bars = 1 mm. For details see Scheffler et al. (37).

used to demonstrate network integration of mES cell-derived neurons and glia (37, 38).

Candidate regions for the preparation of slice cultures include the cerebellum and the hippocampus; and the latter offers the possibility to study donor cell integration in the context of ongoing neurogenesis in the dental granule cell layer (37). Typically, organotypic slices are propagated as interface cultures on polyester membranes.

According to our method (Protocol 6), hippocampal slice cultures are prepared from 9-day-old Wistar rats. Using a vibroslicer, 400 μm-thick slices encompassing the dentate gyrus, hippocampus, and entorhinal/temporal cortex (Figure 5A) are propagated on a porous (0.4 μm pore size) polyester membrane in a humidified incubator gassed with 5% CO_2 in atmospheric air, and set at 35 °C (39). Cultures are initiated in a horse serum-containing medium, which is gradually replaced over 5 d in culture with a serum-free, defined medium based on DMEM/F12 and including the N2 and B27 supplements. Under these conditions, cultured slices can be maintained for a period of up to 35 d ((37); and see Figure 5).

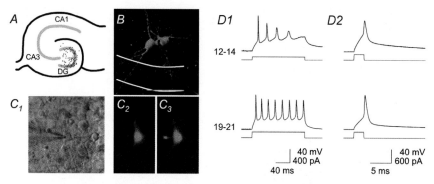

Figure 6 (see Plate 8) Functional integration of mES cell-derived neurons into hippocampal slice cultures. (A) Schematic representation of the implantation site: CA1, CA1 subfield; CA3, CA3 subfield of the hippocampus; DG, dentate gyrus. (B) Confocal image of two EGFP-positive neurons 3 weeks after engraftment (three-dimensional reconstruction of 16 individual planes taken from a fixed slice). Scale bar = 10 μm. (C_1–C_3) Infrared DIC image (C_1) and fluorescence image (C_2) of an EGFP-positive donor neuron after formation of a gigaseal. Diffusion of EGFP into the pipette serves as confirmation of the donor cell identity (C_3). (D) Functional maturation of engrafted mES cell-derived neurons. Current-clamp recordings during prolonged (D1) and brief (D2) current injections at different time points after transplantation (12–14 d, and 19–21 d, as indicated at the left). Top traces represent voltage recordings, whereas bottom traces indicate current injections. Note the progressive development of repetitive discharge properties and action potential morphology with time in culture. For details see Benninger et al. (38). Copyright, the Society for Neuroscience.

4.1.1 Functional validation of hippocampal slice cultures

The integrity of slice cultures may be assessed using the following techniques:

(a) *Recording of field excitatory postsynaptic potentials.* Field potential recordings can be used to validate the synaptic connectivity between perforant path and dentate gyrus, as well as between Schaffer collaterals and CA1 pyramidal neurons ((40); and see *Figure 5B*, inset).

(b) *Anterograde axonal tracing.* Anterograde tracing of perforant path axons is performed by depositing rhodamine-conjugated dextran (e.g. Micro-Ruby®, Molecular Probes) on top of the entorhinal cortex, and confirming anterograde labelling of the perforant path ((41); and see *Figure 5C*).

(c) *Specialized histological staining.* TIMM staining can be used to delineate the mossy fibre system ((42); and see *Figure 5D*).

For a more detailed description of these methods see Scheffler et al. (37).

4.2 *In utero* transplantation of mES cell-derived neural progenitors

An important issue in stem-cell research, with a view to therapeutic applications, is whether stem cell-derived precursors can participate in normal development

Protocol 6

Establishment of hippocampal slice cultures and transplantation procedures

Reagents and equipment

- Culture of ESNPs or ESGPs generated according to *Protocol 1*: a minimum of 1×10^6 cells are required[a]
- Nine-day-old, male P9 Wistar rats (Charles River)
- 70% ethanol
- Horizontal flow hood (e.g. HeraGuard, Heraeus)
- Forene® (B506, Abbott), or other inhalable anaesthetic
- Stereotactic frame (Stoelting)
- Decapitator (Stoelting)
- Sterile surgical instruments (e.g. FST)
- Vibroslicer (Cat. No. VSLM1, Campden Instruments)
- Glass capillary
- Transwell® permeable supports: these are clear, polyester membranes, 24 mm diameter and with 0.4 µm pore size, supplied in 6-well multidishes (Cat. No. 3450, Corning)
- Slice culture medium: chemically defined, serum-free culture medium based on DMEM/F12 (Cat. No. 11320, Invitrogen), $1 \times$ N2 supplement (Cat. No. 17502-048, Invitrogen) and $1 \times$ B27 supplement (Cat. No. 17504-044, Invitrogen), $1 \times$ penicillin/streptomycin (Cat. No. 15140, Invitrogen); and if necessary use amphotericin B as Fungizone® ($1 \times$ antibiotic-antimycotic; Cat. No. 15240-062, Invitrogen)
- 'Initial' slice culture medium; this is slice culture medium with 20% normal horse serum (Cat. No. 26050-088, Invitrogen) on day 0, which is reduced to 14% (day 1), 7% (day 3), and 0% (day 5) over 5 days
- HBSS, HBSS/Hepes, and Trypsin/EDTA solution (all *Protocol 1*)
- Trypsin inhibitor (*Protocol 3*)
- DMEM/F12 (*Protocol 1*)
- Flame-polished Pasteur pipette
- Haemocytometer
- Trypan blue (*Protocol 1*)
- 15 ml Falcon tubes
- 500 µl Eppendorf tube
- 4% paraformaldehyde and 0.1% sodium azide solutions (*Protocol 3*)
- 30% saccharose solution in PBS

Methods

Setting up brain slice cultures:
Establish brain slice cultures 8–12 d prior to transplantation.

1 Anaesthetize a rat with Forene®.

2 Swiftly remove the head using a decapitator, and carefully remove intact brains.

3 Use the vibroslicer, placed under the horizontal flow hood, to prepare 400 µm-thick brain slices, according to the manufacturer's instructions (39).

> **Protocol 6** continued
>
> 4. Incubate slices as interface cultures (39) on the polyester membranes with slice culture medium containing 20% normal horse serum, at 35 °C and with 5% CO_2 in air. The day of explantation is designated 'day 0'.
> 5. Change medium on day 1 of culture, and then on every other day.
> 6. Gradually replace initial slice culture medium with serum-free medium by day 5.
>
> *Preparation of donor cells:*
>
> 7. On the day of transplantation, aspirate medium from ESNP or ESGP cultures, and wash cells three times with HBSS.
> 8. Harvest cells from the dish with trypsin/EDTA solution, and stop trypsinization with trypsin inhibitor (*Protocol 3*, Step 2).
> 9. Triturate cells to a single-cell suspension with a flame-polished Pasteur pipette, and count viable cells.
> 10. Centrifuge cells for 5 min at 1000 rpm and 4 °C, and resuspend cells in HBSS/Hepes or DMEM/F12 at a concentration of 7.5×10^5 cells/ml, for transplantation into hippocampal slice cultures; *or* $2-4 \times 10^5$ cells in 4–8 µl for *in utero* transplantation (see *Protocol 7*).
> 11. Transfer cells into a 500 µl Eppendorf tube, and keep on ice until transplantation.
>
> *Donor cell transplantation:*
> Perform transplantation of neural cells at day 10±2.
>
> 12. Use a glass capillary connected to the manipulator of a stereotactic frame to deposit donor cells on the surface of the slice preparation at the desired location. Resume culture of the transplanted slices.
> 13. Two days after deposition, wash the slices thoroughly with medium to remove non-adherent donor cells, replenish medium and continue incubation. During culture, the migration and differentiation of *EGFP*-labelled donor cells may be monitored directly by epifluorescence microscopy. To avoid phototoxicity, keep observation times as short as possible.
> 14. Fix slices 10–20 d after transplantation with 4% paraformaldehyde, and use whole or cryosectioned slices for further immunfluorescence analysis. For cryoprotection, fixed slices are incubated in 30% saccharose until they sink (43). Store fixed slices at 4 °C in 0.1% sodium azide solution.
>
> [a] For real-time tracing of the donor cells, use ESNPs or ESGPs expressing *EGFP*. (For details see Scheffler *et al.* (37) and Benninger *et al.* (38).)

in vivo. To address this question with respect to the nervous system, the derivative cells can be introduced into the ventricular zone of the fetal brain via intrauterine transplantation ((27, 44–46); and see *Figure 7*). This procedure involves direct injection of donor cells through the uterine wall into the lateral ventricle

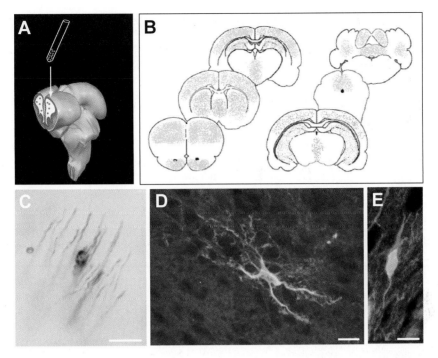

Figure 7 (see Plate 9) Transplantation as a tool to study the developmental potential of ES cell-derived neural precursors. (A) Schematic representation of the transplant approach. Following injection into the cerebral ventricles of rat embryos, the donor cells distribute throughout the ventricular system. (B) Frequent integration sites observed after transplantation of neural precursors into E15-E17 rat embryos include cortex, hippocampus, corpus callosum, septum, striatum, thalamus, hypothalamus, and tectum. (C) Xenografted mES cell-derived neural precursors can be reliably detected by *in situ* hybridization with a mouse-specific DNA probe. The example depicts a hybridized donor-derived oligodendrocyte double-labelled with an antibody to proteolipid protein. (D) Transplanted mES cell-derived astrocytes can be detected with an antibody to the mouse-specific antigen, M2 (red). The astrocytic identity is confirmed by double labelling with an antibody to glial fibrillary acidic protein (green). (E) Human donor cells transplanted into a neonatal mouse brain can be visualized with an antibody to human-specific nuclear antigen (HuNu, green). Double labelling with an antibody to β-III-tubulin (red) confirms the neuronal identity of this hES cell-derived donor cell. Scale bar = 20 μm in C, 10 μm in D, E. See refs (45, 65, 66). Copyright, National Academy of Sciences, U.S.A., and kind permission of Springer Science + Business Media.

of embryonic rats or mice (*Figure 7A, B*). From there, the grafted cells have been shown to distribute widely throughout the host ventricular zone, and to integrate into a variety of host brain regions (7, 44–47). Intrauterine transplantation is particularly advantageous for xenografts, which, due to the early time point of transplantation, can be conducted without the need for immunosuppression. The following protocol summarizes the key steps of this procedure; a detailed description has been reported elsewhere (48). Either ESNPs proliferating in the presence of FGF-2, or ESGPs proliferating in the presence of FGF-2 and EGF, (see *Figure 1* and *Protocol 1*) are appropriate donor cell populations.

Protocol 7

Intrauterine transplantation of mES cell-derived neural progenitors

Reagents and equipment

- Culture of ESNPs or ESGPs: a minimum of 4×10^6 cells are required, triturated to a single-cell suspension (see *Protocol 6*, Steps 7–11)
- Timed-pregnant Sprague Dawley® rats (Charles River) at embryonic day (E) 15–17
- 70% ethanol
- Injectable anaesthetic: ketamine-HCl (80 mg/kg body weight; Ketanest®) and xylazine (10 mg/kg; Rompun®)
- Sterile surgical gown, gloves, and a mask
- Sterile PBS (*Protocol 2*)
- Glass micropipette with a 50–75 μm flame-polished orifice, connected via a polyethylene tube to an injection pump, or to a Hamilton syringe
- Heating pads, at 37 °C
- Tape
- Sterile surgical instruments (scalpel, fine scissors, fine tissue forceps, haemostats, needle holder, sterile gauze)
- Operating microscope
- Fibre-optic light source
- Resorbable and non-resorbable sutures
- Betadine solution
- Isopentane
- 15% saccharose in PBS
- Vibratome (e.g. VT10005, Leica)
- Cryostat (e.g. HM560, MICROM)
- 0.1% sodium azide solution (*Protocol 3*)
- Glass slides (e.g. SuperFrost®Plus, Menzel-Gläser)

Methods

1. Anaesthetize pregnant rats by intraperitoneal injection of ketamine/xylazine.
2. Shave abdomen, and disinfect skin with 70% ethanol.
3. Rest the animal on its back, on the heating pad; and secure it to the pad with tape, in a slightly stretched position.
4. Make a skin incision above the lower abdomen.
5. Open the abdominal cavity and retract the muscles.
6. Very carefully extract one uterine horn, and place it on wet gauze. Make sure that the exposed uterine horn and intestines remain wet, and moisten with PBS if necessary.
7. Load the capillary with $2-4 \times 10^5$ cells in 4–8 μl.
8. Using fibre optic trans-illumination, locate a fetus' head; and quickly penetrate uterine wall, skull, and the lateral ventricle with the loaded capillary. Inject the cells within 2 s, and retract the capillary immediately.
9. After injection of all fetuses in one uterine horn, gently return the horn to the abdominal cavity and repeat the procedure on the other side.

> **Protocol 7 continued**
>
> 10 Suture the muscle and skin.
> 11 Apply betadine solution to the skin, and recover the animal on a heating pad until consciousness is regained.
> 12 After the desired postnatal follow-up period, anaesthetize animals and subject them to transcardial perfusion with 4% paraformaldehyde. Postfix brains overnight with 4% paraformaldehyde (43).
> 13a Remove brains and cut on a vibratome into 50 μm coronal sections. Store fixed sections in 0.1% sodium azide solution at 4 °C.
>
> *or*
>
> 13b For cryosectioning, cryoprotect brains by submersion in 30% saccharose solution until they sink, freeze brains in isopentane, section them at the required thickness (20–50 μm), and mount sections on glass slides (43).

4.3 Identification and characterization of transplanted cells using donor-specific markers

One of the most critical issues in transplant analysis is the unequivocal identification and characterization of the engrafted donor cells. While genetic labelling systems (e.g. using *EGFP* reporter genes) are state of the art, xenotransplants offer an additional set of indisputable methods for donor cell detection. *Protocol 8* provides methods for identifying engrafted donor cells using species-specific antibodies and DNA probes. These methods can be used to tackle critical issues such as transgene downregulation and cell fusion. Combined with multilabelling immunofluorescence, they have been successfully used for the phenotypic characterization of engrafted cells in numerous neural transplantation studies (7, 44–47, 49).

4.4 Immunohistochemical detection of transplanted cells with donor-specific antibodies

For the analysis of xenografts, antibodies to species-specific antigens represent an attractive alternative for the rapid detection of integrated cells. These antigens can be visualized with established immunohistochemical protocols (e.g. see Chapter 9 or (51)). *Table 2* contains an overview of antibodies frequently used for detecting mouse and human cells in a xenogeneic context.

5 Translation to human ES cell-derived neural precursors

The availability of human ES (hES) cells (52) provides exciting prospects for the development of lineage-selection and transplantation strategies for biomedical applications. Although conditions for the maintenance and neural differentiation

Protocol 8
Species-specific *in situ* hybridization

Here we describe a protocol to detect xenografted mouse cells based on species-specific *in situ* hybridization with a digoxigenin (DIG) end-labelled oligonucleotide probe to the mouse-specific major satellite DNA (44). The probe recognizes a highly repetitive tandem sequence located in the centromeric heterochromatin ((50); and see *Figure 7C*). The protocol can be easily adapted to detect human donor cells with a probe to the human *alu* repeat sequence (47).

Reagents and equipment

- Cryoprotected tissue sections mounted on glass slides
- Pronase solution in ddH$_2$O, final concentration 50 µg/ml (Cat. No. P-6911, Sigma)
- 50 ml of 70%, 80%, 90%, and 100% EtOH at r.t.
- 50 ml of 70%, 80%, 90%, and 100% EtOH at −20 °C
- 20 × SSC: dissolve 175.3 g NaCl (Cat. No. 1.06404.5000, VWR international) and 88.2 g sodium citrate (Cat. No. 1.06448.1000, VWR international) in 1 L of ddH$_2$O; adjust to pH 7.0, and sterilize by autoclaving
- 2 × SSC: combine 5 ml 20 × SSC and 45 ml ddH$_2$O
- 2 × SSC/EDTA: combine 5 ml 20 × SSC, 0.5 ml 0.5 M EDTA (Cat. No. E-6511, Sigma), and 44.5 ml ddH$_2$O; and incubate separate solutions at r.t. and 37 °C
- 0.5 × SSC: combine 1.25 ml of 20 × SSC and 48.75 ml ddH$_2$O, and incubate at 37 °C
- Formamide solution I: combine 5 ml of 20 × SSC, 10 ml ddH$_2$O, and 35 ml deionized formamide (Cat. No. 9342, Ambion); and adjust to pH 7.0
- Formamide solution II: combine 5 ml of 20 × SSC, 20 ml ddH$_2$O, and 25 ml formamide (Cat. No. 1.09684.1000, VWR international); and adjust to pH 7.0 (prepare this solution twice)
- Hybridization buffer: combine 1 ml salmon sperm DNA (250 µg/ml; Cat. No. D-7656, Sigma), 3.8 ml of 20 × SSC, and 24.7 ml formamide (Cat. No. 1.09684.1000, VWR international), make up to 38 ml with ddH$_2$O; and adjust to pH 7.0
- Hybridization solution: combine 1.2 µl DIG-labelled mouse satellite DNA probe and 60 µl hybridization buffer
- Buffer 1: combine 100 ml of 0.5 M Tris (pH 7.5; Cat. No. T-1503, Sigma) and 15 ml of 5 M NaCl, make up to 500 ml with ddH$_2$O, and sterilize
- Blocking solution: 2% normal goat serum (NGS; Cat. No. G-6767, Sigma) and 0.2% Triton-X100 (Cat. No. T-8787, Sigma) in Buffer 1
- Antibody solution: 1% NGS, 3% normal rat serum (NRS; Cat. No. R-9759, Sigma), and 0.1% Triton-X100 in Buffer 1
- Rhodamine-conjugated anti-DIG antibody (1:500 dilution; Cat. No. 1207750, Roche)
- Heating block at 90 °C
- Heating plate at 37 °C
- Water bath at 37 °C, placed under a fume hood
- Hybridization oven at 37 °C (e.g. OV3, Biometra)
- Heat-resistant 50 ml cuvettes (for holding slides)
- Rubber cement (e.g. Fixogum; Cat. No. 290117000, Marabu)
- Coverslips (e.g. 24 × 46 mm, Engelbrecht)

Protocol 8 continued

- Humidified chamber
- Vectashield Mounting Medium (Cat. No. H-1000, Vector Laboratories)
- Confocal microscope (e.g. LSM-510; Carl Zeiss)

Methods

Before starting *in situ* hybridization, preheat the water bath, heating block, heating plate, and hybridization oven. Prepare a row of ice-cold ethanol solutions (70%, 80%, 90%, and 100%).

1. Thaw and dry the sections carefully.
2. Place one cuvette filled with 50 ml $2 \times$ SSC/EDTA into the water bath at 37 °C.
3. Add 50 µl pronase solution to the contents of the cuvette shortly before starting the procedure.
4. Transfer the slides into a separate cuvette, add 50 ml $2 \times$ SSC/EDTA to cover the slides, and incubate for 5 min at r.t.
5. Transfer the slides into the cuvette with $2 \times$ SSC/EDTA + pronase at 37 °C, and incubate for 10–30 min for digestion of sections.
6. Pour off the solution, add 50 ml $2 \times$ SSC, and incubate for 3 min at r.t.
7. Place a cuvette with formamide solution I into the water bath and heat to 90 °C.
8. Dehydrate sections in ascending concentrations of ethanol (70%, 80%, 90%, and 100%; 4 min each) at r.t.
9. Air-dry slices for 20–30 min.
10. Denature sections by incubating slides for 12 min in the cuvette with formamide solution I at 90 °C.
11. Re-dehydrate sections in ascending concentrations of ice-cold ethanol (70%, 80%, 90%, and 100%; 4 min each).
12. Meanwhile, incubate hybridization solution for 3 min in the heating block at 90 °C, and quickly transfer onto ice.
13. Transfer and air-dry sections on the 37 °C heating plate.
14. Apply 20–60 µl hybridization solution to each slide, mount sections with a coverslip and seal with rubber cement. Maintain the slides at 37 °C throughout the procedure.
15. Denature mounted specimens for 3 min on the 90 °C heating block.
16. Transfer sections into a humidified chamber, and incubate overnight in the hybridization oven at 37 °C.
17. Next day, preheat formamide solution II to 37 °C in the water bath.
18. Transfer hybridized sections to the 37 °C heating plate, and gently remove the coverslips.
19. Incubate sections in a cuvette as follows: 2×10 min with formamide solution II at 37 °C; 1×15 min with $0.5 \times$ SSC at 37 °C; and 1×5 min with Buffer 1 at r.t.

Protocol 8 continued

20 Incubate sections in a cuvette with blocking solution for 30 min in at r.t.
21 Apply rhodamine-conjugated antibody to DIG (1:500 dilution in antibody solution) to the sections, and incubate overnight in a humidified chamber at r.t.
22 Next day, transfer sections into a cuvette with Buffer 1 for 10 min at r.t.
23 Optional: use conventional immunofluorescence for double-labelling the hybridized cells with cell type-specific antibodies (*Table 1*).
24 Mount sections with Vectashield and analyse on a confocal microscope.

Table 1 Cell type-specific antibodies for characterization of mES cell-derived neural populations

Antibody name (or Cat. No.)	Source	Antigen	Antibody reactivity (according to data sheets)	Dilution
(MAB312R)	Chemicon International	A2B5	Antibody recognizes an epitope common to sialogangliosides and sulfatides associated with the plasma membrane of glial precursors and certain neuronal and neuroendocrine cell types	1:1000
TUJ1 (MMS-435P)	BAbCo	β-III tubulin	Antibody recognizes neuron-specific β-III-tubulin	1:500
(ab6319)	Abcam	CNPase	Antibody reacts with both isoforms of CNPase, CNP1 (46 kDa) and CNP2 (48 kDa), expressed by oligodendrocytes and Schwann cells	1:1000
(69110)	ICN	GFAP	Antibody detects glial fibrillary acidic protein, which is typically expressed in astrocytes but has also been found in neural precursors (58)	1:1000
RAT-401	Hybidoma Bank	Nestin	Antibody recognizes the intermediate filament protein, nestin, which is typically expressed in neural precursor cells (59)	1:50
(MAB377)	Chemicon International	NeuN	Antibody detects neuron-specific nuclear protein, expressed by most neuronal cell types throughout the nervous system	1:200
(AB5320)	Chemicon International	NG2	Antibody reacts with a high-molecular-weight, integral membrane chondroitin sulfate proteoglycan found on the surface of neuroglial precursor cells	1:150–600
(MAB5324)	Chemicon International	PSA-NCAM	Antibody reacts with the polymer polysialic acid (PSA), which is linked to neural cell adhesion molecule (NCAM); in the CNS, PSA-NCAM is expressed in both neuronal and glial cell types and their respective precursors; PSA-NCAM is also found in a variety of non-neural cells and tumours	1:1000

Table 1 Continued

Antibody name (or Cat. No.)	Source	Antigen	Antibody reactivity (according to data sheets)	Dilution
(MAB1326)	R&D Systems	O4	Antibody reacts with an unidentified antigen that appears on the surface of oligodendrocyte progenitors and is commonly used as an early marker for the oligodendroglial lineage	1:100

Table 2 Species-specific antibodies for characterizing mouse and human ES cell-derived donor cells in xenotransplantation studies

Antibody	Reactivity/antigenicity	Source	Dilution	References
Mouse-specific:				
M2	Antigen expressed in astrocytes (*Figure 7D*)	Hybridoma Bank	1:10	(44, 45, 60, 61)
M6	Antigen found in both neurons and glia. Outside the CNS, M6 is found in kidney, olfactory epithelium and choroid plexus	Hybridoma Bank	1:10	(44, 45, 62, 63)
Human-specific:				
HuNu (Cat. No. MAB1281)	Antibody stains nuclei of all human cell types (*Figure 7E*)	Chemicon International	1:5000	(49)
HO-14	Antibody recognizes phosphorylated medium-molecular-weight human neurofilament	Non-commercial	1:50	(49, 64)

of hES cells differ from those required by their murine counterparts, the basic principles of embryoid body formation and FGF-2-mediated expansion of ES cell-derived neural precursors can be applied also to human cells (49, 53). Similarly, transfection methods including lipofection, electroporation, and nucleofection®, as well as viral vectors, have been successfully used to introduce selectable markers into hES cells (54–56).

Moreover, transplantation into the immature and adult rodent brain has proven a valuable tool to study the *in vivo* differentiation of hES cell-derived neural precursors (49, 53, 57). As with mouse-to-rat xenografts, these human-to-rodent systems benefit from the availability of an array of suitable (i.e. human-specific) antibodies and DNA probes, which permit reliable identification and phenotypic characterization of engrafted donor cells using the methodology described in Section 4.3 ((47, 49); see also *Table 2* and *Figure 7E*).

References

1. Bain, G., Kitchens, D., Yao, M., Huettner, J.E., and Gottlieb, D.I. (1995). *Dev. Biol.*, **168**, 342–57.
2. Bain, G., Ray, W.J., Yao, M., and Gottlieb, D.I. (1996). *Biochem. Biophys. Res. Commun.*, **223**, 691–4.

3. Fraichard, A., Chassande, O., Bilbaut, G., Dehay, C., Savatier, P., and Samarut, J. (1995). *J. Cell Sci.*, **108**, 3181–8.
4. Strübing, C., Ahnert-Hilger, G., Shan, J., Wiedenmann, B., Hescheler, J., and Wobus, A.M. (1995). *Mech. Dev.*, **53**, 275–87.
5. Finley, M.F.A., Kulkarni, N., and Huettner, J.E. (1996). *J. Neurosci.*, **16**, 1056–65.
6. Okabe, S., Forsberg-Nilsson, K., Spiro, A.C., Segal, M., and McKay, R.D.G. (1996). *Mech. Dev.*, **59**, 89–102.
7. Brüstle, O., Jones, K.N., Learish, R.D., et al. (1999). *Science*, **285**, 754–6.
8. Lee, S.H., Lumelsky, N., Studer, L., Auerbach, J.M., and McKay, R.D. (2000). *Nat. Biotechnol.*, **18**, 675–9.
9. Rolletschek, A., Chang, H., Guan, K., Czyz, J., Meyer, M., and Wobus, A.M. (2001). *Mech. Dev.*, **105**, 93–104.
10. Liu, S., Qu, Y., Stewart, T.J., et al. (2000). *Proc. Natl. Acad. Sci. USA*, **97**, 6126–31.
11. Bibel, M., Richter, J., Schrenk, K., et al. (2004). *Nat. Neurosci.*, **7**, 1003–9.
12. Kawasaki, H., Mizuseki, K., Nishikawa, S., et al. (2000). *Neuron*, **28**, 31–40.
13. Barberi, T., Klivenyi, P., Calingasan, N.Y., et al. (2003). *Nat Biotechnol.*, **21**, 1200–7.
14. Rathjen, J., Haines, B.P., Hudson, K.M., Nesci, A., Dunn, S., and Rathjen, P.D. (2002). *Development*, **129**, 2649–61.
15. Tropepe, V., Hitoshi, S., Sirard, C., Mak, T.W., Rossant, J., and van der Kooy, D. (2001). *Neuron*, **30**, 65–78.
16. Ying, Q.L., Stavridis, M., Griffiths, D., Li, M., and Smith, A. (2003). *Nat. Biotechnol.*, **21**, 183–6.
17. Doetschman, T.C., Eistetter, H., Katz, M., Schmidt, W., and Kemler, R. (1985). *J. Embryol. Exp. Morph.*, **87**, 27–45.
18. Keller, G. (1995). *Curr. Opinion Cell Biol.*, **7**, 862–9.
19. Lake, J., Rathjen, J., Remiszewski, J., and Rathjen, P.D. (2000). *J. Cell Sci.*, **113**, 555–66.
20. Okabe, S. (1997). Differentiation of embryonic stem cells. In *Protocols in Neuroscience* (eds J. Crawley, C. Gerfen, R. McKay, M. Rogawski, D. Sibley, and P. Skolnick), 3.6.1–3.6.13, John Wiley and Sons, New York.
21. Swiatek, P.J. and Gridley, T. (1993). *Genes Dev.*, **7**, 2071–84.
22. Li, E., Bestor, T.H., and Jaenisch, R. (1992). *Cell*, **69**, 915–26.
23. McDonald, J.W., Liu, X.Z., Qu, Y., et al. (1999). *Nat Med.*, **5**, 1410–2.
24. Kim, J.H., Auerbach, J.M., Rodriguez-Gomez, J.A., et al. (2002). *Nature*, **418**, 50–6.
25. Chiba, S., Iwasaki, Y., Sekino, H., and Suzuki, N. (2003). *Cell Transplant.*, **12**, 457–68.
26. Li, M., Pevny, L., Lovell-Badge, R., and Smith, A. (1998). *Curr. Biol.*, **8**, 971–4.
27. Wernig, M., Tucker, K.L., Gornik, V., et al. (2002). *J. Neurosci. Res.*, **69**, 918–24.
28. Schmandt, T., Meents, E., Gossrau, G., Gornik, V., Okabe, S., and Brustle, O. (2005). *Stem Cells Dev.*, **14**, 55–64.
29. Glaser, T., Perez-Bouza, A., Klein, K., and Brustle, O. (2005). *Faseb J.*, **19**, 112–4.
30. Tucker, K.L., Wang, Y., Dausman, J., and Jaenisch, R. (1997). *Nucleic Acids Res.*, **25**, 3745–6.
31. Kiss, J.Z. and Rougon, G. (1997). *Curr. Opin. Neurobiol.*, **7**, 640–6.
32. Seki, T. and Arai, Y. (1993). *Neurosci Res.*, **17**, 265–90.
33. Minana, R., Sancho-Tello, M., Climent, E., Segui, J.M., Renau-Piqueras, J., and Guerri, C. (1998). *Glia*, **24**, 415–27.
34. Reynolds, R. and Wilkin, G.P. (1988). *Development*, **102**, 409–25.
35. Vogel, U.S. and Thompson, R.J. (1988). *J. Neurochem.*, **50**, 1667–77.
36. Chandross, K.J., Cohen, R.I., Paras, P., Jr, Gravel, M., Braun, P.E., and Hudson, L.D. (1999). *J. Neurosci.*, **19**, 759–74.
37. Scheffler, B., Schmandt, T., Schroder, W., et al. (2003). *Development*, **130**, 5533–41.

38. Benninger, F., Beck, H., Wernig, M., Tucker, K.L., Brustle, O., and Scheffler, B. (2003). *J. Neurosci.*, **23**, 7075–83.
39. Stoppini, L., Buchs, P.A., and Muller, D. (1991). *J. Neurosci. Methods*, **37**, 173–82.
40. Newman, G.C., Hospod, F.E., Qi, H., and Patel, H. (1995). *J. Neurosci. Methods*, **61**, 33–46.
41. Kluge, A., Hailer, N.P., Horvath, T.L., Bechmann, I., and Nitsch, R. (1998). *Hippocampus*, **8**, 57–68.
42. Zimmer, J. and Gahwiler, B.H. (1984). *J. Comp Neurol.*, **228**, 432–46.
43. Gerfen, C. (1997). Basic neuroanatomical methods. In *Protocols in Neuroscience* (eds J. Crawley, C. Gerfen, R. McKay, M. Rogawski, D. Sibley, and P. Skolnick), 1.1.1–1.1.11. John Wiley and Sons, New York.
44. Brüstle, O., Maskos, U., McKay, R.D.G. (1995). *Neuron*, **15**, 1275–85.
45. Brüstle, O., Spiro, C.A., Karram, K., Choudhary, K., Okabe, S., and McKay, R.D.G. (1997). *Proc. Natl. Acad. Sci. USA*, **94**, 14809–14.
46. Wernig, M., Benninger, F., Schmandt, T., et al. (2004). *J. Neurosci.*, **24**, 5258–68.
47. Brüstle, O., Choudhary, K., Karram, K., et al. (1998). *Nature Biotechnol.*, **16**, 1040–4.
48. Brüstle, O., Cunningham, M., Tabar, V., and Studer, L.T.E. (1997). Experimental transplantation in the embryonic, neonatal, and adult mammalian brain. In *Current Protocols in Neuroscience*, 3.10.11–13.10.28, John Wiley and Sons, New York.
49. Zhang, S.C., Wernig, M., Duncan, I.D., Brüstle, O., and Thomson, J.A. (2001). *Nat. Biotechnol.*, **19**, 1129–33.
50. Hörz, W. and Altenburger, W. (1981). *Nucl. Acids Res.*, **9**, 683–96.
51. Volpicelli-Dayley, L.A. and Levey, A. (1997). Immunhistochemical localization of proteins in the nervous system. In *Protocols in Neuroscience* (eds J. Crawley, C. Gerfen, R. McKay, M. Rogawski, D. Sibley, P. Skolnick), 1.2.1–1.2.17, John Wiley and Sons, New York.
52. Thomson, J.A., Itskovitz-Eldor, J., Shapiro, S.S., et al. (1998). *Science*, **282**, 1145–7.
53. Reubinoff, B.E., Itsykson, P., Turetsky, T., et al. (2001). *Nat. Biotechnol.*, **19**, 1134–40.
54. Siemen, H., Nix, M., Endl, E., Koch, P., Itskovitz-Eldor, J., and Brüstle, O. (2005). *Stem Cells Dev.*, **14**, 378–83.
55. Zwaka, T.P. and Thomson, J.A. (2003). *Nat. Biotechnol.*, **21**, 319–21.
56. Gropp, M., Itsykson, P., Singer, O., et al. (2003). *Mol. Ther.*, **7**, 281–7.
57. Tabar, V., Panagiotakos, G., Greenberg, E.D., et al. (2005). *Nat Biotechnol.*, **23**, 601–6.
58. Doetsch, F., Caille, I., Lim, D.A., Garcia-Verdugo, J.M., and Alvarez-Buylla, A. (1999). *Cell*, **97**, 703–16.
59. Hockfield, S., and McKay, R. (1995). *J. Neurosci.*, **5**, 3310–28.
60. Lagenaur, C. and Schachner, M. (1981). *J. Supramol. Struct. Cell. Biochem.*, **15**, 335–46.
61. Zhou, H.F., Lee, L.H.C., and Lund, R.D. (1990). *J. Comp. Neurol.*, **292**, 320–30.
62. Wictorin, K., Lagenaur, C.F., Lund, R.D., and Bjorklund, A. (1991). *Eur. J. Neurosci.*, **3**, 86–101.
63. Lagenaur, C., Kunemund, V., Fischer, G., Fushiki, S., and Schachner, M. (1992). *J. Neurobiol.*, **23**, 71–88.
64. Trojanowski, J.Q., Mantione, J.R., Lee, J.H., et al. (1993). *Exp. Neurol.*, **122**, 283–94.
65. Wernig, M., Scheffler, B., and Brüstle, O. (2003). Medizinische Perspektiven der Stammzellforschung. In *Grundlagen der Molekularen Medizin*, 2nd edn (eds D. Ganten and K. Ruckpaul), pp. 680–710, Springer-Verlag, Berlin and Heidelberg.
66. Brüstle, O. (1999). *Brain Pathology*, **9**, 527–45.

Chapter 9
In vitro differentiation of mouse ES cells into pancreatic and hepatic cells

Przemyslaw Blyszczuk, Gabriela Kania and
Anna M. Wobus
In Vitro Differentiation Group, Institute of Plant Genetics and Crop Plant Research (IPK), D-06466 Gatersleben, Germany.

1 Introduction

In mammalian development, both the pancreas and liver originate from the definitive endoderm. The pancreas develops from dorsal and ventral regions of the foregut, and the liver from the foregut adjacent to the ventral pancreas compartment (1, 2). At present, the lack of available methods for enriching either pancreatic or hepatic progenitor cells from their respective organ *in vitro* is a major obstacle for the efficient generation of selected, pure populations of derivative cells. Moreover, the progenitor cells of the liver and pancreas are not yet clearly defined. Nestin-positive cells deriving from pancreatic islets, and generated after *in vitro* culture, are suggested to represent pancreatic progenitor cells (3); and oval cells are regarded as hepatic stem/progenitor cells (4). However, cells residing in the pancreatic (5) and hepatic (6) ductal epithelium and expressing cytokeratin 19 also have been proposed as pancreatic and hepatic stem or progenitor cells *in vivo*.

Pluripotent embryonic stem (ES) cells of the mouse (or mES cells) have, by definition, the capacity for self-renewal together with the potential to differentiate into any cell type of the somatic and germ-cell lineages, including those endodermal cells forming the pancreas and liver (for reviews see (7, 8)). During *in vitro* differentiation, mES cells express hepatic- (9–11) and pancreatic-restricted (12–14) transcripts and proteins, and therefore offer valuable model systems in which to elucidate, at the molecular level, the mechanisms of differentiation of these lineages.

In vitro differentiation of mES cells without the application of lineage-specific differentiation-inducing factors results in a heterogeneous mixture of derivatives and a low yield of endocrine pancreatic (15) and hepatic (16) cells. However, by using specific growth and extrcellular matrix (ECM) factors (11, 13), and/or transgene expression (12, 14, 17), the yield of the desired, differentiated cell types is enhanced significantly and the generation of functional pancreatic or hepatic cells may be achieved. In this chapter we describe our optimized procedures for the efficient induction of differentiation of mES cells into pancreatic and hepatic lineages. Our overall strategy involves: (i) the generation of a pool of progenitor cells – including endodermal precursors – via the differentiation of mES cells in embryoid bodies (EBs; see Chapter 5); and (ii) further, directed differentiation of those endodermal progenitor cells into pancreatic and hepatic cell types using lineage-specific, differentiation-inducing and ECM-associated factors. In addition to these procedures, constitutive overexpression of the pancreatic developmental control gene, *Pax4*, may be induced in mES cells by transgene integration, to increase the number of functional islet-like cells generated (14).

2 Maintenance of undifferentiated mES cells

Reproducibility in mES-cell differentiation experiments is strongly dependent on a high standard of mES cell culture. In order for mES cells to maintain their undifferentiated state they must be cultured at relatively high density, in the presence of mitotically inactivated, mouse embryonic fibroblast (MEF) feeder cells and/or leukaemia inhibitory factor (LIF) (see Chapter 2). We do not recommend the use of the STO line for feeder cells, as in our experience the supportive capacity for mES cells is dependent on specific sublines, which may not be commonly available. Good-quality, batch-tested fetal-calf serum (FCS) is critical for long-term culture of mES cells, and for subsequent, successful differentiation. As mES cells divide every 12–15 h, the culture medium should be replenished daily and the cells passaged every 24–48 h onto fresh feeder layers ((18), and see Chapter 2); and during passaging, mES cells must be dissociated by trypsinization with particular care. If one or more of these requirements are not complied with, mES cells may differentiate spontaneously during culture and become unsuitable for differentiation studies.

All cell cultures are maintained in a humidified incubator, gassed with 5% CO_2 in air, at 37 °C.

3 *In vitro* differentiation procedures

3.1 Generation of mES cell-derived, heterogeneous lineage progenitor cells

On withdrawal of feeder cells and LIF, mES cells start spontaneously to differentiate, which basic propensity is exploited in all mES-cell differentiation procedures. Methods for the controlled production of lineage progenitor cells from mES cells entail two consecutive steps: (i) mES cells are cultured in suspension to form three-dimensional aggregates, or 'embryoid bodies' (EBs), to promote differentiation into early progenitors of all lineages (see Chapter 5); and (ii) EBs are plated onto adhesive substrata to allow the expansion and further differentiation of the progenitor cells.

Protocol 1
Preparation of mES cells for *in vitro* differentiation

Mouse ES cells are maintained on feeder layers as described in detail in Chapters 2 and 6, in gelatinized tissue-culture dishes. For passaging, and prior to differentiation experiments, cultures of mES cells are harvested and suspensions prepared as follows.

Reagents and equipment

- Culture of mES cells, e.g. line R1 (A. Nagy, Toronto), in a 60 mm feeder dish
- MEF feeder layers (see Chapter 2, *Protocols 2* and *3*) in 60 mm tissue-culture dishes pre-coated with gelatin[a]
- Phosphate-buffered saline (PBS), without Ca^{2+} or Mg^{2+} ions: dissolve 10 g NaCl, 0.25 g KCl, 1.44 g Na_2HPO_4, and 0.25 g $KH_2PO_4.2H_2O$ in 1 L distilled water, and sterilize through a 0.22 μm pore-size filter
- Trypsin/EDTA solution: prepare separate, stock solutions of 0.2% (w/v) trypsin (Cat. No. 37290, Serva) and 0.02% (w/v) EDTA (Cat. No. E-6758, Sigma) in PBS, filter sterilize the solutions, and combine them in equal volumes[b]
- Mouse ES-cell medium: Dulbecco's modified Eagle's medium (DMEM), 1 × liquid formulation with 4.5 g/L glucose (Cat. No. 52100-047, Invitrogen), supplemented with 2 mM L-glutamine (Cat. No. 25030-024, Invitrogen), 1 × non-essential amino acids (Cat. No. 11140-035, Invitrogen), 100 μM β-mercaptoethanol (Cat. No. 28625, Serva), 0.05 mg/ml streptomycin and 0.03 mg/ml penicillin (Cat. No. 15070-063, Invitrogen), 15% FCS (Cat. No. 10207-106, Invitrogen), and 1000 U/ml leukaemia inhibitory factor (LIF; Cat. No. ESG1106, Chemicon International)[c]
- Differentiation Medium I: Iscove's modification of DMEM (IMDM; Cat. No. 42200-030, Invitrogen) supplemented with 2 mM glutamine, 1 × non-essential amino acids, 450 μM monothioglycerol (or 3-mercapto-1, 2-propanediol; Cat. No. M-6145, Sigma), 0.05 mg/ml streptomycin, 0.03 mg/ml penicillin, and 20% FCS
- Sterile, 2 ml glass transfer pipette
- Haemocytometer, e.g. THOMA Chamber (Cat. No. 9.161 080, Schütt Labortechnik)

Method

1. Change medium 1–2 h before passaging mES cells.
2. Aspirate medium from the culture, and quickly wash cells with 2 ml trypsin/EDTA solution to remove residual serum-containing medium.
3. Add a further 2 ml trypsin/EDTA solution to the culture and quickly aspirate most of this, leaving sufficient to cover the cell monolayer (~200 μl). Allow trypsinization to occur for 30–60 s at r.t.
4. Halt trypsinization by adding 2 ml medium to the cell monolayer: for routine passaging use mES-cell medium; and for EB production, Differentiation Medium I.

Protocol 1 continued

5 Draw cells repeatedly up and down a transfer pipette in order to obtain a single-cell suspension, and either passage at a split ratio of 1:3 onto fresh feeder layers, or use immediately for the production of EBs.

[a] To precoat culture dishes with gelatin, see Protocol 3, Step 1.
[b] Trypsin/EDTA solution should be freshly prepared every week.
[c] An alternative formulation for mES-cell culture medium is provided in Chapter 2, Protocol 1.

Protocol 2
Generation of EBs by the 'hanging-drop' method

Reagents and equipment

- Single-cell suspension of mES cells in Differentiation Medium I (Protocol 1)
- PBS (Protocol 1)
- 100 mm and 60 mm bacteriological-grade plastic Petri dishes, e.g. Falcon® dishes (Cat. Nos. 351029 and 351007, respectively, Bacto Laboratories Pty Ltd)
- Haemocytometer
- Inverted phase-contrast microscope
- Sterile, 2 ml glass transfer pipette

Method

1 Adjust the density of the mES cell suspension in Differentiation Medium I to give 30 000 cells/ml, i.e. 600 cells per 20 μl.[a]

2 Dispense single drops (n = 50 to 60) of 20 μl of the mES cell suspension onto the undersides of lids of 100 mm bacteriological-grade Petri dishes, each containing 10 ml PBS, and quickly replace the lids onto the plates. The drops of cell suspension are now suspended from the lids, over the PBS.

3 Incubate the mES cells in the hanging drops for 2 d, during which time they aggregate and form one EB per drop. Monitor the development of EBs by light microscopy.

4 Carefully take up the EBs from the lids into 2 ml Differentiation Medium I in a pipette, transfer them into 60 mm bacteriological-grade Petri dishes containing an additional 5 ml Differentiation Medium I, and continue their incubation in suspension for a further 3 d.[b]

5 After a total of 5 d in suspension culture, plate EBs onto an adhesive substratum (Protocol 3).

[a] For accurate assessment of mES cell density and viability, see Chapter 6, Protocol 1.
[b] EBs fail to adhere to the bacteriological-grade plastic and remain in suspension.

Protocol 3

Differentiation of EBs on an adhesive substratum

To promote further differentiation of lineage progenitor cells, EBs should be plated onto an adhesive surface at day 5 of suspension culture, at a proper density and with homogeneous distribution. The distance between individual EBs should be sufficient to allow the differentiating cells to proliferate and to migrate for several days after plating – if EBs are plated at too high a density, differentiation may be inhibited. Continued culture of EB outgrowths results in multilayered cell clusters.

Reagents and equipment

- Day 5 EBs, generated as described in Protocol 2
- Gelatin solution: prepare a suspension of 0.1% (w/v) gelatin (Cat. No. 48720, Fluka) in distilled water, and dissolve and sterilize by autoclaving
- Differentiation Medium I (Protocol 1)
- 60 mm or 35 mm tissue-culture dishes, or 24-well plates[a]

Method

1. To precoat the culture dishes with gelatin, add sufficient gelatin solution to cover their bases, incubate for 1 h to overnight at 4 °C, and aspirate the residual solution.[b]

2. Into each gelatin-coated dish dispense an appropriate volume of Differentiation Medium I (i.e. 5 ml per 60 mm dish, 2 ml per 35 mm dish, and 1 ml per well of a 24-well plate), followed by EBs at the following densities: plate 20–30 EBs per 60 mm dish; 5–10 EBs per 35 mm dish; and one EB per well of a 24-well plate.

3. Incubate EB cultures, changing medium every second or third day, until 9 d after EB plating (i.e. 5 + 9 d in total from the start of aggregation); and proceed to lineage-specific differentiation (Protocols 4 and 5).

[a] Multi-well plates are convenient to use for the morphological analysis of EB outgrowths.
[b] For immunofluorescence analysis, spread sterile, glass coverslips over the bases of the dishes before gelatinization.

For EB formation, mES cells may be induced to aggregate either by the 'hanging-drop' method (18) or by 'mass culture' in bacteriological-grade dishes ((19); and see Chapter 6, Protocol 3). The advantages of the 'hanging-drop' method, employed here, include the low variation in size of EBs produced due to a defined number of mES cells in the starting aggregates, as well as greater reproducibility of differentiation. Incubating EBs for 5 d prior to plating ensures the development of progenitor cells deriving from all three primary germ layers.

3.2 Induction of lineage-specific differentiation

Although EB-mediated differentiation of mES cells generates cellular derivatives of all three primary germ layers, the proliferation and differentiation of heterogeneous lineage-specific progenitor cells thus generated are considerably enhanced by their culture in the presence of defined growth- and differentiation-inducing factors. In addition, the following parameters are known to affect the efficiency of differentiation of mES cell-derived progenitors into pancreatic and hepatic cells in such systems, and should therefore be taken into consideration when devising and optimizing *in vitro* differentiation experiments:

(a) *Dissociation of the compact structure of EB outgrowths.* Cell-to-cell interactions within the complex and heterogeneous structure of EB outgrowths may influence the fate of progenitor cells.

(b) *A suitable, adhesive substratum.* ECM factors determine adhesion, proliferation, and migration of specific progenitor cells after replating of dissociated EBs.

(c) *Cell density after replating.* To support the proliferation and migration of a specific progenitor population, the concentration of cells should be optimal to prevent overgrowth resulting in metabolic starvation, necrosis, and cell death. However, if the initial density of cells is too low, the level of autocrine factors may not reach a required, threshold level; and consequently the low growth-factor activity and reduced cell-to-cell contacts may result in poor efficiency of differentiation of the desired cell types.

3.2.1 Induction of pancreatic differentiation

The medium used for the differentiation of pancreatic cells from EB-derived progenitor cells contains factors required for pancreatic-cell survival, such as progesterone, putrescine, insulin, sodium selenite, and transferrin, in addition to factors promoting pancreatic differentiation, such as nicotinamide (20) and laminin (21). This pancreatic differentiation medium (PDM) is serum-free. The addition to PDM of other substances promoting pancreatic differentiation, such as glucagon-like peptide-1 (GLP-1, (22)) and growth inhibitors (23), also may prove effective. The precoating of tissue-culture dishes with poly-L-ornithine and laminin is necessary for pancreatic differentiation to occur. After replating and culture of EBs under these conditions for 7 d, pancreatic progenitor cells with a characteristic morphology (*Figure 1A*) become apparent. Continued incubation and concomitant differentiation results in the formation of typical islet-like cell clusters (*Figure 1B*; and see (24)).

3.2.2 Induction of hepatic differentiation

Mouse ES cells are induced to differentiate into hepatic cells when incubated in Hepatocyte Culture Medium (HCM). HCM contains factors required for hepatic-cell survival, such as insulin and transferrin, as well as hepatic differentiation factors, including hydrocortisone and epidermal growth factor (EGF). Supplementation of HCM with FCS also is required for cell survival; and the culture dishes are precoated with collagen Type I (or collagen I), which further promotes

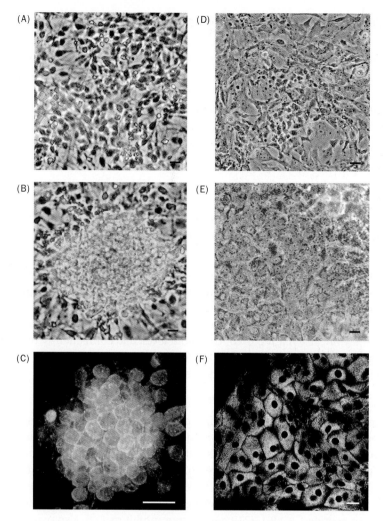

Figure 1 Morphology of pancreatic (A–C) and hepatic (D–F) cells differentiating from mES cells *in vitro*, and viewed by light (A, B, D, E) and confocal (C, F) microscopy. (A) Pancreatic progenitor cells 7 d after induction of pancreatic differentiation in plated EBs, that is at stage 5 + 16 d (according to *Protocol 4*). (B) Morphology of an islet-like cluster, at stage 5 + 28 d; and (C) confocal image of immunolabelled, C-peptide-positive cells constituting an islet-like cluster. (D) Hepatic progenitor cells 10 d after induction of hepatic differentiation in plated EBs, that is at stage 5 + 19 d (according to *Protocol 5*). Continued culture in HCM leads to the generation of further differentiated and specialized cell types: (E) Morphology of cuboidal, hepatocyte-like (and frequently binucleated) cells, at stage 5 + 39 d; and (F) confocal image of immunolabelled, α1-antitrypsin-positive cells. Scale bars = 20 μm.

hepatic differentiation (9). After replating and culture of EBs under these conditions, hepatic progenitor cells develop with characteristic morphology (*Figure 1D*). With continued incubation and further differentiation, large cuboidal and epithelioid, hepatocyte-like cells become apparent, a proportion of which are binucleated (*Figure 1E*; and see (25)).

Protocol 4

Induction of pancreatic differentiation

To increase the yield of mES cell-derived pancreatic cells by this procedure, we recommend the use of transgenic mES cells that constitutively express pancreatic developmental-control genes. For example we have shown that over-expression of *Pax4* in the R1 line of mES cells significantly up-regulates pancreatic β cell-specific mRNA and protein levels, resulting in an increased number of insulin-expressing cells (14).

Reagents and equipment

- 5 + 9 d, adherent cultures of EBs: these consist of EBs plated at day 5 into 60 mm tissue-culture dishes, and incubated for 9 d further (*Protocol 3*)
- PBS (*Protocol 1*)
- Trypsin/EDTA solution (*Protocol 1*)
- Pancreatic Differentiation Medium (PDM): 1 × liquid formulation DMEM/F12 with 4.5 g/L glucose (Cat. No. 32500–043, Invitrogen), supplemented with 20 nM progesterone (Cat. No. P-7556, Sigma), 100 μM putrescine (Cat. No. P-5780, Sigma), 1 μg/ml laminin (Cat. No. L-2020, Sigma), 10 mM nicotinamide (Cat. No. N-3376, Sigma), 25 μg/ml insulin (Cat. No. I-1882, Sigma), 30 nM sodium selenite (Cat. No. S-5261, Sigma), 50 μg/ml transferrin (Cat. No. T-1147, Sigma), B27 media supplement (Cat. No. 17504–044, Invitrogen), 0.05 mg/ml streptomycin and 0.03 mg/ml penicillin (Cat. No. 15070–063, Invitrogen)[a]
- FCS (*Protocol 1*)
- Sodium borate buffer: prepare a solution of 10 mM H_3BO_3 in distilled water, and adjust to pH 8.4 with NaOH
- Poly-L-ornithine solution: dissolve 0.1 mg/ml poly-L-ornithine (Cat. No. P-2533, Sigma) in sodium borate buffer, and filter sterilize
- Laminin solution: dissolve 0.001 mg/ml laminin (Cat. No. L-2020, Sigma) in sterile PBS
- Sterile, distilled water
- 60 mm tissue-culture dishes
- Drying oven or non-gassed incubator, set at 40 °C
- Cell scraper (Cat. No. 541070, Greiner)
- Sterile, 5 ml glass transfer pipette

Methods

Preparation of poly-L-ornithine/laminin-coated dishes:[b]

1. Add poly-L-ornithine solution to culture dishes (sufficient to cover the bases), and incubate for 3 h at 37 °C.[c]

2. Aspirate the poly-L-ornithine solution, wash dishes three times with distilled water, and incubate with 5 ml distilled water for 12 h at r.t.

3. Wash the dishes three times with distilled water, and dry at 40 °C.

4. Add laminin solution to the poly-L-ornithine-coated dishes, sufficient to cover the bases, and incubate at 37 °C for 3 h.

5. Aspirate the laminin solution, and wash the dishes twice with PBS.

Protocol 4 continued

Initiation of pancreatic differentiation:

6. Aspirate medium from cultures of differentiating EB outgrowths, and quickly wash the cells with 2 ml PBS to remove residual medium.
7. Add 2 ml trypsin/EDTA solution, and incubate for 1 min at r.t.
8. Aspirate trypsin/EDTA solution completely, and gently detach the cells using a cell scraper.
9. Add to each dish 4 ml PDM supplemented with 10% FCS.
10. Resuspend the cells using a transfer pipette to obtain a suspension of single cells and small clusters, and plate 1 ml of suspension onto each of four poly-L-ornithine/laminin-coated, 60 mm dishes; add to each dish 3 ml PDM supplemented with 10% FCS, and incubate overnight.[d]
11. The next day check that all cells are attached, aspirate medium, and add PDM *without* FCS.
12. Change medium every second or third day, and also 1 day prior to analysis.

[a] This medium should be freshly prepared

[b] Poly-L-ornithine/laminin-coated dishes should be freshly prepared, therefore commence precoating of dishes 1 day in advance of pancreatic differentiation.

[c] For immunofluorescence analysis, spread sterile, glass coverslips over the bases of the dishes before adding poly-L-ornithine solution.

[d] For mouse insulin ELISA, plate 0.5 ml cell suspension onto poly-L-ornithine/laminin-coated, 35 mm tissue-culture dishes, adding 1.5 ml FCS-supplemented PDM.

Protocol 5
Induction of hepatic differentiation

Reagents and equipment

- 5+9 d, adherent cultures of EBs: these consist of EBs plated at day 5 into 60 mm tissue-culture dishes, and incubated for 9 d further (*Protocol 3*)
- PBS (*Protocol 1*)
- Trypsin/EDTA solution (*Protocol 1*)
- Hepatocyte culture medium (HCM): Hepatocyte basal, modified Williams 'E' medium (HBM medium; Cat. No. CC-3199, Cambrex) supplemented per 500 ml with the following quantities of sterile additives (from Cambrex, unless otherwise indicated): 0.5 ml ascorbic acid (Cat. No. CC-4316); 10 ml bovine serum albumin (BSA), fatty-acid-free (Cat. No. CC-4362); 0.5 ml hydrocortisone (Cat. No. CC-4335); 0.5 ml transferrin (Cat. No. CC-4313); 0.5 ml insulin (Cat. no. CC-4321); 0.5 ml human epidermal growth factor (Cat. No. CC-4317); 0.5 ml gentamycin-amphotericin (Cat. No. CC-4381); 0.05 mg/ml streptomycin and 0.03 mg/ml penicillin (*Protocol 1*)[a]
- FCS (*Protocol 1*)
- 0.02 N acetic acid in distilled water, filter sterilized

Protocol 5 continued

- Collagen I solution: 0.05 mg/ml collagen Type I (Cat. No. 354236, BD Biosciences) in 0.02 N acetic acid
- PBS (*Protocol 1*)
- 60 mm tissue-culture dishes
- Cell scraper (*Protocol 4*)
- Sterile, 5 ml glass transfer pipette

Methods

Preparation of collagen I-coated dishes:[b]

1. Add sufficient collagen I solution to culture dishes (sufficient to cover the bases), and incubate for 1 h at r.t.[c]
2. Aspirate the collagen I solution, and wash dishes twice with PBS.

Induction of hepatic differentiation:

3. Aspirate medium from cultures of differentiating EB outgrowths, and quickly wash the cells with 2 ml PBS to remove residual medium.
4. Add 2 ml of trypsin/EDTA solution, and incubate for 1 min at r.t.
5. Aspirate trypsin/EDTA solution completely, and gently detach the cells using a cell scraper.
6. Add to each dish 4 ml HCM supplemented with 10% FCS.
7. Resuspend the cells using a transfer pipette to obtain a suspension of single cells and small clusters, and plate 1 ml of suspension onto each of four collagen I-coated, 60 mm dishes; add to each dish 3 ml HCM supplemented with 10% FCS, and incubate.[d]
8. Change medium every second to third day, and additionally 1 day prior to analysis.

[a] This medium should be freshly prepared.

[b] Collagen I-coated dishes should be freshly prepared.

[c] For immunofluorescence analysis, spread coverslips over the bases of dishes before adding collagen I solution.

[d] For mouse albumin ELISA, plate 0.5 ml cell suspension onto collagen I-coated, 35 mm tissue-culture dishes, adding 1.5 ml FCS-supplemented HCM.

4 Characterization of differentiated, pancreatic and hepatic cell phenotypes

The differentiation status of mES cell-derived pancreatic or hepatic cells may be determined by: (i) reverse transcriptase-polymerase chain reaction (RT-PCR), (ii) immunofluorescence analysis, and (iii) enzyme-linked immunosorbent assay (ELISA). Collectively, these methods allow the characterization and quantitation of pancreatic or hepatic differentiation at the transcriptional and protein levels, as well as the detection of differentiation-specific markers.

4.1 RT-PCR

A criterion for successful pancreatic or hepatic differentiation of mES cells *in vitro* is the expression of genes that are involved in pancreas or liver development *in vivo*. We recommend expression analysis of the following genes as specific markers to denote pancreatic differentiation: insulin, islet amyloid pancreatic polypeptide (*IAPP*), *Pax4*, and *Pdx1*; and for hepatic differentiation: albumin, *Cyp2b9*, *Cyp2b13*, and tyrosine aminotransferase (*TAT*).

Detailed protocols for RNA isolation, reverse transcription, cDNA amplification, and separation of PCR products from mES cell-differentiated derivatives are described elsewhere ((18) and see Chapter 6); and a discussion of RT-PCR analysis in the context of mES cell-based differentiation systems is given in Chapter 6. Primer sequences for detecting gene transcripts that are specifically involved in the development and function of pancreatic β cells, and of hepatocytes, are presented in *Table 1*. Some of these gene transcripts may be detected in mES cells and/or mES-cell derivatives after spontaneous differentiation ((14); and Kania *et al.*, unpublished data).

4.2 Immunofluorescence staining

Immunofluorescence analysis allows the characterization of cells in terms of proteins expressed at both intracellular levels (e.g. intermediate-filament

Table 1 Primers for RT-PCR amplification of pancreatic and hepatic differentiation-specific gene transcripts

Gene transcript	Primer sequences (5′ to 3′, F_w/R_v)	Annealing temp. (°C)	Product size (bp)
For pancreatic and hepatic progenitor cells:			
Cytokeratin 19	CTGCAGATGACTTCAGAACC GGCCATGATCTCATACTGAC	62	299
Nestin	CTACCAGGAGCGCGTGGC TCCACAGCCAGCTGGAACTT	60	220
For pancreatic cells:			
Glut-2	TTCGGCTATGACATCGGTGTG AGCTGAGGCCAGCAATCTGAC	60	556
IAPP	TGATATTGCTGCCTCGGACC GGAGGACTGGACCAAGGTTG	65	233
Insulin	GTGGATGCGCTTCCTGCCCCTG ATGCTGGTGCAGCACTGA	64	288
Isl-1	GTTTGTACGGGATCAAATGC ATGCTGCGTTTCTTGTCCTT	60	514
Ngn3	TGGCGCCTCATCCCTTGGATG AGTCACCCACTTCTGCTTCG	60	157
Pax4	ACCAGAGCTTGCACTGGACT CCCATTTCAGCTTCTCTTGC	60	300
Pax6	TCACAGCGGAGTGAATCAG CCCAAGCAAAGATGGAAG	58	332
Pdx1	CTTTCCCGTGGATGAAATCC GTCAAGTTCAACATCACTGCC	60	230
For hepatic cells:			
Albumin	GTCTTAGTGAGGTGGAGCAT ACTACAGCACTTGGTAACAT	58	569
α-1-antitrypsin	CAATGGCTCTTTGCTCAACA AGTGGACCTGGGCTAACCTT	63	518
α-fetoprotein	CACTGCTGCAACTCTTCGTA CTTTGGACCCTCTTCTGTGA	58	301
Cyp2b9	GATGATGTTGGCTGTGATGC CTGGCCACCATGAAAGAGTT	53	153
Cyp2b13	CTGCATCAGTGTATGGCATTTT TTTGCTGGAACTGAGACTACCA	65	166
HNF3 β	GCGGGTGCGGCCAGTAG GCTGTGGTGATGTTGCTGCTCG	63	378
Transthyretin	CTCACCACAGATGAGAAG GGCTGAGTCTCTCAATTC	55	225
Tyrosine aminotransferase	ACCTTCAATCCCATCCGATCCCGACTGGATAGGTAG	50	206
Control, housekeeping gene:			
β-tubulin	TCACTGTGCCTGAACTTACC GGAACATAGCCGTAAACTGC	60	317

Table 2 Primary antibodies and fixation methods used for the detection of pancreatic and hepatic cells by immunofluorescence

Antigen	Antibody isotype	Working dilution	Antibody supplier	Fixation method Met/Ac	PFA
Pancreatic and hepatic cells					
Cytokeratin 18	mouse IgG	1:100	Sigma	+	−
Cytokeratin 19	mouse IgM	1:100	Chemicon	+[a]	+[b]
Nestin	mouse IgG	1:3	Hybridoma Bank	+	+
Pancreatic cells					
Carbonic anhydrase II	rabbit IgG	1:200	Abcam	−	+
C-peptide	guinea pig IgG	1:100	Linco	−	+
Insulin	mouse IgG	1:40	Sigma	−	+
Isl-1	rabbit IgG	1:200	Abcam	+	−
Hepatic cells					
Albumin	sheep IgG	1:100	Serotec	+	−
α-1-antitrypsin	rabbit IgG	1:100	Sigma	+	−
α-fetoprotein	goat IgG	1:100	Santa Cruz	+	−
Amylase	goat IgG	1:100	Santa Cruz	+	−
Cytokeratin 14	mouse IgG	1:100	Sigma	+	−
Dipeptidyl peptidase IV	mouse IgG	1:100	Santa Cruz	+	−

Met/Ac = methanol/acetone solution; PFA = paraformaldehyde solution
[a] Fixation method results in filamentous structures; [b] dot-like structures

proteins) and extracellular levels (e.g. cell-surface antigens), with sufficient sensitivity to detect even a small minority of antigen-positive cells against a largely negative background. Therefore the technique is especially valuable for mES-cell differentiation studies, where cell populations are generated that commonly are heterogeneous with respect to lineage and developmental stage.

Here, we describe immunostaining procedures suitable for detecting pancreatic- and hepatic-cell marker proteins in differentiated mES-cell derivatives, including two sample fixation methods using (i) methanol/acetone (suitable for cytoskeletal proteins, such as intermediate filaments), and (ii) paraformaldehyde (suitable for intracellular proteins). Primary antibodies, dilutions and appropriate fixation methods are given in *Table 2*. (For additional information on fluorescence techniques and fluorescence probes, see Chapter 6.)

It should be emphasized that, when assaying for the presence of cell-specific marker proteins, the endogenous level of a protein may be over-estimated if it is present also in the culture medium. As a case in point, mES cell-derived pancreatic cells were found to be overly immunopositive for insulin as a consequence of uptake of that hormone from the medium (26). Therefore for pancreatic differentiation, the by-product of insulin synthesis, C-peptide, serves as a reliable marker for *de novo* insulin production. Similarly for albumin synthesis following hepatic differentiation, it is notable that albumin is a constituent of hepatocyte differentiation medium, HCM (see *Protocol 5*); and therefore use of elevated immunoreactivity for that protein as a criterion for differentiation must be confirmed by coexpression with other hepatic cell markers. For validation purposes, we also advise that the demonstration of pancreatic or hepatic differentiation by immunofluorescence requires parallel gene expression studies by RT-PCR, including a comparison of diverse marker genes.

Protocol 6
Fixation of cells for immunostaining

Reagents and equipment

- EBs and differentiated derivatives, cultured on coverslips (*Protocols 3, 4,* and *5*)
- PBS (*Protocol 1*)
- Methanol/acetone solution: combine 7 ml methanol and 3 ml acetone, and store at −20 °C; *or* paraformaldehyde solution: add 0.4 g paraformaldehyde to 10 ml PBS, heat to 60 °C with stirring until clarified, and cool to r.t.[a]
- Blunt forceps
- Tissue paper
- Humidified chamber
- 200 ml beaker

Methods

1. Remove coverslips with adherent mES-cell derivatives (handling coverslips by their edge, using blunt forceps) from culture dishes, and carefully place the edges in contact with tissue paper to drain residual medium.
2. Wash coverslips twice by dipping and gently stirring in a beaker of PBS for 30 s, and drain residual PBS as above.

Fixation with methanol/acetone:

3. Fix cells on coverslips by overlaying with 200 μl methanol/acetone solution (pre-cooled at −20 °C), and incubate for 10 min at −20 °C.
4. Drain fixative from coverslips, as above.
5. Wash coverslips three times by dipping and gently stirring in a beaker containing PBS for 30 s.
6. Rehydrate the samples by overlaying with 200 μl PBS and incubating coverslips in a humidified chamber for 5 min at r.t. The cells are now ready for immunostaining (*Protocol 7*).

Or fixation with paraformaldehyde:

3. Fix cells on coverslips by overlaying with 200 μl paraformaldehyde solution, and incubate in a humidified chamber for 15 min at r.t.
4. Drain fixative from coverslips, as above.
5. Wash coverslips three times in PBS, by dipping and gently stirring in a beaker containing PBS for 30 s. The cells are ready now for immunostaining (*Protocol 7*).

[a] Paraformaldehyde solution should be freshly prepared.

Protocol 7
Immunostaining

Reagents and equipment

- EBs and differentiated derivatives, cultured on coverslips (*Protocols 3, 4,* and *5*)
- 10% serum solution: combine 1 ml heat-inactivated serum with 9 ml PBS, and store at 4 °C[a]
- PBS (*Protocol 1*)
- Distilled water
- Humidified chamber
- Primary antibodies (*Table 2*)
- 200 ml beaker
- Fluorescence-conjugated secondary antibodies (commercial preparations)
- Hoechst 33342 solution: dissolve 50 µg Hoechst 33342 in 10 ml PBS (giving 5 µg/ml), protect from light and store at 4 °C
- Mounting medium: Vectashield (Cat. No. L-010, Vector Labs)
- Fluorescence or confocal microscope, e.g. confocal microscope model LSM 510 META (Carl Zeiss)

Method

1. To prevent non-specific binding, incubate fixed cells on coverslips with 10% serum solution (applied as an overlay of ~100 µl per coverslip) in a humidified chamber for 30 min at r.t.
2. Prepare aliquots of 60–100 µl per sample of the primary antibody by dilution in PBS (the optimal dilution must be determined separately).
3. Incubate coverslips with primary antibody in a humidified chamber for 60–90 min at 37 °C.
4. Wash coverslips three times by dipping and gently stirring in a beaker of PBS for 30 s.
5. Prepare 100 µl per sample of the secondary, fluorescence-conjugated antibody by dilution in PBS (the optimal dilution must be determined separately).
6. Incubate coverslips with the secondary antibody in a humidified chamber for 45 min at 37 °C.
7. Wash coverslips three times with PBS at r.t., as in Step 4.
8. To counter-stain nuclei, incubate cells on coverslips with 200 µl per sample Hoechst 33342 solution, in a humidified chamber for 5 min at 37 °C.
9. Wash coverslips three times with PBS at r.t., as in Step 4, and once quickly with distilled water.
10. Embed cells on coverslips in mounting medium, and analyse samples using a fluorescence or confocal microscope.

[a] Do not use serum from the same species as was used for the preparation of the primary antibody!

4.3 ELISA

The Enzyme-Linked Immunosorbent Assay (ELISA) for substance detection and quantification is a valuable tool in ES-cell technology in general due to its high sensitivity and specificity, whilst allowing rapid and simultaneous processing of a large number of samples. Here we highlight ELISA-based methods that are particularly suitable for analysis of pancreatic and hepatic differentiation in mES cell-based culture systems: namely, the immunoassay of secreted and intracellular insulin and albumin by pancreatic and hepatic cells, respectively.

Protocol 8
Sample preparation for insulin ELISA

The mouse insulin ELISA allows determination of glucose-induced insulin secretion by mES cell-derived pancreatic cells. Both released and intracellular insulin levels are determined, relative to total protein content.

Reagents and equipment

- Mouse ES cell-derived pancreatic cells, cultured in 35 mm dishes (*Protocol 4*)
- Krebs Ringer Bicarbonate Hepes (KRBH) buffer: 118 mM NaCl, 4.7 mM KCl, 1.1 mM KH_2PO_4, 25 mM $NaHCO_3$, 3.4 mM $CaCl_2$.$2H_2O$, 2.5 mM $MgSO_4$.$7H_2O$, 10 mM Hepes and 2 mg/ml bovine serum albumin in distilled water
- Glucose/KRBH-buffer solutions: KRBH buffer supplemented with 2.5 mM, 5.5 mM, and 27.7 mM glucose[a]
- FCS-supplemented DMEM: DMEM (*Protocol 1*) supplemented with 10% FCS (*Protocol 1*)
- Trypsin/EDTA solution (*Protocol 1*)
- Distilled water

- Acid ethanol: combine 1 ml hydrochloric acid and 9 ml absolute ethanol
- 15 ml PP-test tubes (Cat. No. 188271, Greiner Bio-One)
- 1.5 ml tubes (Cat. No. 0300 121.848, Eppendorf AG)
- Ultrasonic homogenizer (Sonoplus HD70, Bandelin)
- Sterile, 2 ml glass transfer pipette
- Mouse Insulin ELISA (Cat. No. 10–1149–01, Mercodia AB)
- Bio-Rad Protein Assay, Dye Reagent Concentrate, 450 ml (Cat. No. 500–0006, Bio-Rad Laboratories)

Method

1. Aspirate medium from cultures of mES cell-derived pancreatic cells, and wash the cells five times with 3 ml PBS per dish.
2. Add 3 ml 2.5 mM glucose/ KRBH buffer per dish, and incubate for 90 min at 37 °C.
3. Aspirate 2.5 mM glucose/KRBH buffer, and replace with 5.5 mM glucose/KRBH buffer (for control samples) and 27.7 mM glucose/KRBH buffer (for test samples) in separate tissue-culture dishes, and incubate for 15 min at 37 °C.
4. Collect supernatants and store at −20 °C, for determination of insulin release.
5. Add sufficient trypsin/EDTA solution to cover the whole surface of the dishes, and incubate for 2–3 min at r.t.

Protocol 8 continued

6. Carefully remove the trypsin/EDTA solution from each dish, and add 3 ml FCS-supplemented DMEM.
7. Resuspend the cells using a 2 ml glass pipette and transfer into 15 ml tubes.
8. Centrifuge samples for 5 min at 1000 g and r.t.
9. Aspirate supernatant, resuspend cells in 50 µl acid ethanol, transfer to 1.5 ml tubes, and incubate overnight at 4 °C.
10. Disintegrate cells using an ultrasonic homogenizer, according to the manufacturer's instructions.
11. Add 500 µl distilled water to samples, mix, and centrifuge for 1 min at 13 000 g and r.t.
12. Transfer supernatants to fresh 1.5 ml tubes and store at −20 °C, for determination of total intracellular protein and insulin levels.
13. Determine insulin levels by ELISA using the Mercodia Mouse Insulin kit, according to the manufacturer's instructions.
14. Determine total protein levels by the Bradford method using the Bio-Rad Protein Assay, according to the manufacturer's instructions.

[a] Buffer solutions should be freshly prepared.

Protocol 9
Sample preparation for albumin ELISA

The albumin ELISA allows the determination of both secreted and intracellular albumin in cultures of mES cell-derived hepatic cells. Cultures are preincubated in BSA- and FCS-free medium, to prevent over-estimation of endogenous albumin levels.

Reagents and equipment

- Mouse ES cell-derived hepatic cells, cultured in 35 mm dishes (*Protocol 5*)
- PBS (*Protocol 1*)
- HCM, without BSA or FCS (*Protocol 5*)
- Trypsin/EDTA solution (*Protocol 1*)
- FCS-supplemented DMEM: (*Protocol 8*)
- 1.5 ml Eppendorf tubes (*Protocol 8*)
- Sterile, 5 ml glass transfer pipette
- Haemocytometer
- Mouse Albumin ELISA Quantitation Kit (Cat. No. E90–134, Bethyl Laboratories)

Method

1. One day prior to analysis, aspirate medium from a culture of differentiated hepatic cells; wash the cells five times with 3 ml PBS, add BSA- and FCS-free HCM, and incubate for 24 h at 37 °C.

> **Protocol 9 continued**
>
> 2. Collect and transfer the supernatant into 1.5 ml tubes and store at −20 °C, for determination of albumin secretion.
> 3. Add a sufficient trypsin/EDTA solution to cover the base of the dish, and incubate for 2–3 min at r.t.
> 4. Carefully remove the trypsin/EDTA solution, and add 3 ml FCS-supplemented DMEM; resuspend the cells using a glass pipette to obtain single-cell suspension, and count the cells using a haemocytometer.

4.4 Microscopic analysis and imaging of samples labelled by immunofluorescence

Although immunolabelled preparations of mES cell-derived pancreatic and hepatic cells may be examined by conventional fluorescence microscopy, we recommend the use where possible of a confocal laser-scanning microscope as confocal images are superior. Furthermore, confocal microscopy allows the colocalization of proteins, which is especially relevant to analysis of the three-dimensional structures that are produced by mES cell-derived pancreatic (*Figure 1C*) and hepatic cells (*Figure 1F*).

5 Animal transplantation models

For the validation of *in vitro* generated, mES cell-derivatives as being phenotypically functional and having regenerative capacity, animal disease and transplantation models are essential. These disease models, ideally, should demonstrate similar pathological mechanisms and properties to human disease syndromes; and may be either induced experimentally or naturally occurring. Several significant murine (i.e. mouse and rat) models of pancreatic and liver diseases are now established, and have been adopted successfully for transplantation studies using either mES-cell derivatives or adult, progenitor stem cells. Several particular systems that are likely to play key roles in the development of model, mES cell-based therapeutic strategies in the near future are described below. For all animal procedures, appropriate licenses must be obtained and prescribed guidelines for experimentation adhered to.

5.1 Pancreatic islet regeneration models

The non-obese diabetic (NOD) mouse consists of an inbred strain that was generated in Japan by selection from outbred (Swiss) mice (27, 28). In this and a similar rat model, the Biological Breeding (or 'BB') rats, the insulin-producing β cells are destroyed by an autoimmune process that is similar to that occurring in patients with type 1-diabetes (insulin-dependent diabetes mellitus (IDDM) or juvenile diabetes, see (28)). Although, for NOD mice, understanding of pathogenetic mechanisms underlying diabetes is quite advanced, extrapolation to the human condition is limited by genus-specific differences (28).

The destruction of pancreatic β cells also may be induced experimentally in laboratory animals by multiple, low-dose injections of streptozotocin (STZ) (29). This model has been used successfully to develop treatment for hyperglycaemia (which is a consequence of β-cell destruction) by syngeneic islet transplantation (30). The dose of STZ required is species- and strain-dependent, but generally falls within the range of 150–200 mg/kg body weight (12–14, 23, 31); and suspensions of $1 \times 10^6 - 1 \times 10^7$ stem cell-derived, β-like cells are transplanted either subcutaneously (13, 31) into the renal sub-capsular space (14, 23, 32, 33) or into the spleen (12, 14), rather than the pancreas which is seriously damaged by STZ. This STZ/ diabetes model has the advantage of allowing the use of syngeneic mouse strains to minimize the risk of immune rejections, but the disadvantage that STZ is toxic also to other organs, such as the kidney and liver.

5.2 Liver reconstitution models

In contrast to most other parenchymal organs, such as the kidney and pancreas, the liver is characterized by a very high regenerative capacity. Different cell types and mechanisms are involved in liver organ reconstitution, depending on the type of liver injury, and may involve progenitor cell-dependent regeneration as well as repopulation by transplanted cells (34). Transgenic mice expressing the urokinase plasminogen activator gene under the control of the albumin promoter (*Alb-uPA*) display liver inflammation and necrosis with a paucity of mature hepatocytes (35); and transplanted hepatocytes successfully reconstitute the livers of these recipient mice (36). In an extension of this model, the *Alb-uPA* transgene was incorporated into immunotolerant (RAG-2) mice, which then supported the growth of xenotransplanted hepatocytes from different species, including rat and human (37). However, it should be cautioned that for this and similar transgenic mouse models, regenerative liver nodules can develop spontaneously by DNA rearrangement to give transgene-deficient hepatocytes (37).

The enzyme fumarylacetoacetate hydrolase (FAH) catalyses the final reaction in the tyrosine catabolic pathway. FAH deficiency results in accumulation of a hepatotoxic metabolite and is the cause of tyrosinema type I disease in humans (HTI, (38)). In the FAH knockout (FAH-/-) mouse model, transplanted FAH-positive donor hepatocytes are used to repopulate the FAH-mutant liver (39). Because homozygous FAH deficiency is lethal in the neonatal period, FAH-/- mice are treated with an inhibitor of tyrosine catabolism acting upstream of FAH, thus alleviating neonatal mortality and prolonging the lifespan of mutant animals, which display a phenotype resembling hereditary tyrosinaemia type I (40, 41).

Another liver reconstitution model, based in the rat, involves induction of chronic liver damage by treatment with toxic substances followed by partial hepatectomy (42). Here, the regenerative capacity of the host's liver cells is chemically blocked by treatment with lasiocarpine or retrorsine; these pyrrolizidine alkaloids are selectively metabolized to their active form in hepatocytes, and consequently alkylate hepatocyte DNA and cause G_2-phase cell-cycle arrest (43). A disadvantage of this system, with regard to transplantation studies, is that after partial hepatectomy of retrorsine-treated animals, the proliferation of endogenous hepatocyte-like progenitor cells may be induced (44). Generally, and due to the high regeneration capacity of the liver, in all hepatic disease models the use of genetically labelled donor cells for transplantation is highly recommended.

Acknowledgements

We are grateful to Oda Weiss, Sabine Sommerfeld, and Karla Meier for their expert assistance in the development of the mES cell-differentiation protocols for generating pancreatic and hepatic cells reproducibly and at high efficiency. Financial support provided by the Deutsche Forschungsgemeinschaft (DFG), Ministerium für Bildung und Forschung (BMBF), Fonds der Chemischen Industrie (FCI), and DeveloGen AG Göttingen, Germany, is gratefully acknowledged.

References

1. Slack, J.M. (1995). *Development*, **121**, 1569–80.
2. Zaret, K.S. (1996). *Annu. Rev. Physiol*, **58**, 231–51.
3. Zulewski, H., Abraham, E.J., Gerlach, M.J., Daniel, P.B., Moritz, W., Muller, B., Vallejo, M., Thomas, M.K., and Habener, J.F. (2001). *Diabetes,* **50**, 521–33.
4. Petersen, B.E., Grossbard, B., Hatch, H., Pi, L., Deng, J., and Scott, E.W. (2003). *Hepatology*, **37**, 632–40.
5. Bonner-Weir, S., Taneja, M., Weir, G.C., Tatarkiewicz, K., Song, K.H., Sharma, A., and O'Neil, J.J. (2000). *Proc. Natl. Acad. Sci. USA*, **97**, 7999–8004.
6. Dabeva, M.D., Petkov, P.M., Sandhu, J., Oren, R., Laconi, E., Hurston, E., and Shafritz, D.A. (2000). *Am. J. Pathol.*, **156**, 2017–8004.
7. Czyz, J., Wiese, C., Rolletschek, A., Blyszczuk, P., Cross, M., and Wobus, A.M. (2003). *Biol. Chemistry*, **384**, 1391–409.
8. Wobus, A.M. and Boheler, K. (2005). *Phys. Rev.*, **85**, 635–78.
9. Hamazaki, T., Iiboshi, Y., Oka, M., Papst, P.J., Meacham, A.M., Zon, L.I., and Terada, N. (2001). *FEBS Lett.*, **497**, 15–9.
10. Jones, E.A., Tosh, D., Wilson, D.I., Lindsay, S., and Forrester, L.M. (2002). *Exp.Cell Res.*, **272**, 15–22.
11. Kania, G., Blyszczuk, P., Czyz, J., Navarrete-Santos, A., and Wobus, A.M. (2003). *Methods in Enzymology*, **365**, 287–303.
12. Soria, B., Roche, E., Berna, G., Leon-Quinto, T., Reig, J.A., and Martin, F. (2000). *Diabetes*, **49**, 157–62.
13. Lumelsky, N., Blondel, O., Laeng, P., Velasco, I., Ravin, R., and McKay, R. (2001). *Science*, **292**, 1389–94.
14. Blyszczuk, P., Czyz, J., Kania, G., Wagner, M., Roll, U., St Onge, L., and Wobus, A.M. (2003). *Proc. Natl. Acad. Sci. USA*, **100**, 998–1003.
15. Kahan, B.W., Jacobson, L.M., Hullett, D.A., Ochoada, J.M., Oberley, T.D., Lang, K.M., and Odorico, J.S. (2003). *Diabetes*, **52**, 2016–24.
16. Miyashita, H., Suzuki, A., Fukao, K., Nakauchi, H., and Taniguchi, H. (2002). *Cell Transplant.*, **11**, 429–34.
17. Ishizaka, S., Shiroi, A., Kanda, S., Yoshikawa, M., Tsujinoue, H., Kuriyama, S., Hasuma, T., Nakatani, K., and Takahashi, K. (2002). *FASEB J.*, **16**, 1444–6.

18. Wobus, A.M., Guan, K., Yang, H.-T., and Boheler, K. (2002). *Methods Mol. Biol.*, **185**, 127-56.
19. Doetschman, T.C., Eistetter, H., Katz, M., Schmidt, W., and Kemler, R. (1985). *J. Embryol., Exp. Morphol.* **87**, 27-45.
20. Otonkoski, T., Beattie, G.M., Mally, M.I., Ricordi, C., and Hayek, A. (1993). *J. Clin. Invest.*, **92**, 1459-66.
21. Jiang, F.-X., Cram, D.S., DeAizpurua, H.J., and Harrison, L.C. (1999). *Diabetes*, **48**, 722-30.
22. Abraham, E.J., Leech, C.A., Lin, J.C., Zulewski, H., and Habener, J.F. (2002). *Endocrinology*, **143**, 3152-61.
23. Hori, Y., Rulifson, I.C., Tsai, B.C., Heit, J.J., Cahoy, J.D., and Kim, S.K. (2002). *Proc. Natl. Acad. Sci. USA*, **99**, 16105-10.
24. Blyszczuk, P., Asbrand, C., Rozzo, A., Kania, G., St-Onge, L., Rupnik, M., and Wobus, A.M. (2004). *Int. J. Dev. Biol.*, **48**, 1095-1104.
25. Kania, G., Blyszczuk, P., Jochheim, A., Ott, M., and Wobus, A.M. (2004). *Biol. Chem.*, **385**, 943-53.
26. Rajagopal, J., Anderson, W.J., Kume, S., Martinez, O.I., and Melton, D.A. (2003). *Science*, **299**, 363.
27. Makino, S., Kunimoto, K., Muraoka, Y., Mizushima, Y., Katagiri, K., and Tochino, Y. (1980). *Jikken Dobutsu*, **29**, 1-13.
28. Atkinson, M.A. and Leiter, E.H. (1999). *Nat. Med.*, **5**, 601-4.
29. Like, A.A. and Rossini, A.A. (1976). *Science*, **193**, 415-7.
30. Sandler, S. and Andersson, A. (1982). *Diabetes*, **31** (Suppl. 4), 78-83.
31. Suzuki, A., Nakauchi, H., and Taniguchi, H. (2003). *Proc. Natl. Acad. Sci. U.S.A*, **100**, 5034-9.
32. Ramiya, V.K., Maraist, M., Arfors, K.E., Schatz, D.A., Peck, A.B., and Cornelius, J.G. (2000). *Nat Med.*, **6**, 278-82.
33. Yang, L., Li, S., Hatch, H., Ahrens, K., Cornelius, J.G., Petersen, B.E., and Peck, A.B. (2002). *Proc. Natl. Acad. Sci. USA*, **99**, 8078-83.
34. Grompe, M. and Finegold, M.J. (2001). Liver stem cells. In *Stem Cell Biology* (eds D.R. Marshak, R.L. Gardner, D. Gottlieb), pp. 455-97, Cold Spring Harbor Laboratory Press, New York.
35. Sandgren, E.P., Palmiter, R.D., Heckel, J.L., Daugherty, C.C., Brinster, R.L., and Degen, J.L. (1991). *Cell*, **66**, 245-56.
36. Jamal, H.Z., Weglarz, T.C., and Sandgren, E.P. (2000). *Gastroenterology*, **118**, 390-4.
37. Rhim, J.A., Sandgren, E.P., Palmiter, R.D., and Brinster, R.L. (1995). *Proc. Natl. Acad. Sci. USA*, **92**, 4942-6.
38. Lindbald, B., Lindstedt, S., and Steen, G. (1977). *Proc. Natl. Acad. Sci. USA*, **74**, 4641-5.
39. Overturf, K., Al Dhalimy, M., Tanguay, R., Brantly, M., Ou, C.N., Finegold, M., and Grompe, M. (1996). *Nat. Genet.*, **12**, 266-73.
40. Lindstedt, S., Holme, E., Lock, E.A., Hjalmarson, O., and Strandvik, B. (1992). *Lancet*, **340**, 813-7.
41. Grompe, M., Lindstedt, S., Al Dhalimy, M., Kennaway, N.G., Papaconstantinou, J., Torres-Ramos, C.A., Ou, C.N., and Finegold, M. (1995). *Nat. Genet.*, **10**, 453-60.
42. Laconi, E., Sarma, D.S., and Pani, P. (1995). *Carcinogenesis*, **16**, 139-42.
43. Samuel, A. and Jago, M.V. (1975). *Chem. Biol. Interact.*, **10**, 185-97.
44. Gordon, G.J., Coleman, W.B., Hixson, D.C., and Grisham, J.W. (2000). *Am. J. Pathol.*, **156**, 607-19.

Chapter 10
Isolation and characterization of human ES cells

Martin F. Pera, Andrew Laslett, Susan M. Hawes, Irene Tellis, Karen Koh and Lihn Nguyen

Monash Institute of Medical Research, Monash University, and the Australian Stem Cell Centre, Building 75 STRIP, Wellington Road, Clayton Victoria 3800, Australia.

1 Introduction

This chapter will describe methods for the isolation and characterization of human embryonic stem (hES) cell lines from preimplantation blastocysts. The first derivation of hES cells was reported in 1998 (1), and since that time a number of groups have successfully established other, similar cell lines with the key features of primate pluripotent stem cells (2). There have been some significant improvements to the original culture system for hES cell propagation, and some refinements in the methodology for characterization of the cells. However, the mechanisms that regulate hES cell proliferation and differentiation are still poorly understood; there are practical limitations to existing culture methodology; and detailed, multicentre comparisons to assess the properties of hES cell isolates derived under different conditions and from different laboratories have not yet been carried out. Thus, while the current methodology is certainly adequate to support derivation and expansion of hES cells, the techniques and approaches described herein are likely to undergo substantial modification in the next several years.

Nonetheless, we recommend that laboratories follow these well-established protocols in embarking on work with hES cells. Modifications to cell culture methodology should be adopted only after extensive documentation of their ability to support growth of several hES cell lines through extended passage with retention of pluripotentiality and a normal karyotype. This chapter outlines standard methodology for hES cell isolation, culture, and characterization in widespread use at the present time.

Note: Suppliers and catalogue numbers are indicated only in those instances where the source is unique or critical.

2 Isolation of hES cells
2.1 Human embryos

Human ES cells have been isolated from *in vitro* fertilized embryos, in some cases following cryopreservation. Most jurisdictions that allow the use of human embryos for the production of hES cell lines now have in place a regulatory framework to ensure ethical conduct of the research. These regulations vary considerably internationally, and researchers must ensure that all proposed studies are in compliance with local and national standards and statutory requirements.

The comparative properties of human and mouse ES cells have been reviewed elsewhere (3, 4; and see Chapter 12). While there are certainly similarities in the biology of these pluripotent stem cells in the two species, differences in morphology, growth characteristics, and methods for propagation are considerable. Therefore, researchers new to the field should gain experience culturing extant lines of monkey or human pluripotent stem cells (i.e. ES or EC cell lines) prior to attempting to isolate new hES cell lines.

Protocol 1 outlines our procedure for isolating the inner cell mass (ICM) from human embryos for establishment of hES cell lines (*Figure 1*). All cell culture incubation is carried out in a humidified incubator set at 37 °C with 5% CO_2.

Most, though not all, reports of hES cell derivation have used a culture system incorporating serum-supplemented medium, mouse embryonic fibroblast (MEF)

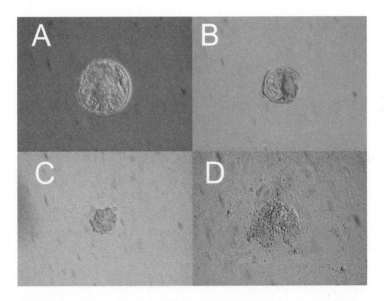

Figure 1 Isolation and primary culture of ICM from human blastocyst. (A) Expanded blastocyst. (B) Blastocyst following removal of zona pellucida. (C) ICM following isolation by immunosurgery. (D) Primary culture of ICM at 3 days. Here the ICM has attached to the MEF feeder layer, and commenced forming an outgrowth. Magnification: (A) 40×; (B–D), 20×. Micrographs provided by Dr Carmel Obrien, Stem Cell Sceiences Pty., and Ms Nicole Merry, Melbourne IVF, Melbourne Australia.

Protocol 1
Isolation of the ICM from human blastocysts

Reagents and equipment

- Blastocyst-stage human embryos 5–7 days postfertilization, in Gardner's G1/G2 or Scandanavian S1/S2 embryo culture medium
- Pronase solution: 10 U/ml pronase in serum-free embryo-culture medium such as Gardner's G2 or S2 medium
- Serum-supplemented hES cell medium (SSM): for 600 ml combine 480 ml DMEM, 120 ml FCS (to give 20%), 6 ml of 10 mM non-essential amino acids (0.1 mM; Cat. No. 11140–050 Invitrogen), 6-ml of 200-mM L-glutamine solution (final concentration 2 mM), 3 ml penicillin/streptomycin 200 × solution (final concentration 50 U/ml penicillin and 50 mg/ml streptomycin), 1.1 ml of 55 mM β-mercaptoethanol solution (final concentration 90 μM), and 6 ml insulin-transferrin-selenium solution (1%, for final concentrations 10 μg/ml insulin, 5.5 μg/ml transferrin and 6.7 ng/ml sodium selenite; Cat. No. 51300–044, Invitrogen)
- Rabbit anti-human serum antiserum[a]
- Baby-rabbit or guinea-pig complement diluted in serum-free embryo culture medium
- Sterile Pasteur pipette, drawn out to a diameter slightly larger than the inner cell mass

Method

(See also Chapter 12, *Protocol 1* for a detailed method for conducting immunosurgery on primate blastocysts.)

1. Set up microdrops of pronase solution in culture dishes. Remove zonae pellucidae from blastocysts by incubation in microdrops of pronase solution for 2 min at 37 °C.
2. Wash an embryo in SSM.
3. Incubate zona-free embryo with rabbit anti-human serum antiserum at a suitable dilution for 30 min at 37 °C.[b]
4. Wash the embryo with SSM.
5. Add embryo to microdrops of baby rabbit or guinea pig complement for 30 min, or the minimum time required to complete lysis of trophectoderm cells at 37 °C.
6. Remove the dead trophectoderm cells by pipetting the embryo gently up and down through a drawn-out Pasteur pipette.
7. Wash the isolated ICM thoroughly by passing through drops of SSM, and proceed immediately to coculture with MEF feeder layers (*Protocol 2*).

[a] Any antiserum that is broadly reactive with human cells will work in this protocol; we have successfully used that supplied by Sigma (Cat. No. H3383).

> **Protocol 1** continued
>
> ᵇ The correct titre of antiserum and complement may be determined empirically using spare embryos or cultured human cells: the combination of antiserum and complement should be used at titres sufficient to lyse >95% of the target cell population, but not in excess. Neither the antiserum nor the complement should display non-specific lytic activity at the titres used (i.e. no cell lysis should be induced by either agent on its own). If the antiserum alone causes cell lysis, heating at 56 °C for 30 min may eliminate this activity.

feeder layers, and mechanical dissection of colonies for subculture of the hES cells. More recently, several laboratories have employed serum substitutes in the establishment of hES cell lines (5), and at least one group have used enzymatic dissociation to passage the cells from an early stage (6). MEF feeder layers have been replaced in some studies by human feeder cells (7, 8).

2.2 Initiation of primary cultures

Following immunosurgery, the ICM is washed thoroughly in hES-cell medium, and plated onto a feeder layer of mitotically inactivated MEFs (below). The ICM cells will not resemble established hES cells during the initial phases of culture, and may grow only slowly at first. However, by 10–14 d, the initial colony should contain discernible hES cells and should be ready to passage. Any colony showing progressive expansion and growth should be passaged since morphology can be variable at this stage.

We recommend that subculture be carried out by mechanical dissection of the colony under a stereo dissecting microscope; subculture in this manner has reliably maintained pluripotent, euploid hES-cell cultures through many generations in our laboratory. The dissection process has two purposes: first, to cut the hES-cell colony into pieces that are large enough to survive subculture, but not so large as to induce formation of embryoid body (EB)-like structures that will differentiate; and secondly, to avoid the transfer of spontaneously differentiated areas of the colony.

Detailed methods for the preparation of MEFs for use in the production of feeder layers are described in Chapter 2, *Protocol 2*. There is no preferred strain of mice for MEF derivation, for hES-cell culture. STO cells or other permanent cell lines derived from MEF cultures can support hES cell maintenance, but in our experience more spontaneous differentiation occurs in cultures grown on STO-cell feeder layers than on MEF feeder layers. *Protocol 2*, here, describes the preparation of MEF feeder layers. Each isolate of MEFs should be tested for its ability to support the growth of established human hES cell lines prior to use in cell-line derivation. It is best to use freshly prepared fibroblasts for hES cell culture, but backup stocks of mitomycin C-treated cells may be cryopreserved using standard methodology (i.e. slow freezing in serum and 10% dimethylsulphoxide, as described for non-inactivated MEFs in Chapter 2, *Protocol 2*) and stored for emergency use.

For primary cultures of hES cells, or other cultures that will be passaged by mechanical dissection of colonies rather than enzymatic dissociation, it is very

Protocol 2

Preparation of MEF feeder layers

Reagents and equipment

- MEFs isolated from fetuses at day 13.5 p.c. (Chapter 2, *Protocol 2*), and cryopreserved between passages 0–3
- MEF medium: for 500 ml combine 450 ml DMEM (high-glucose formulation), 50 ml fetal calf serum, 2.5 ml penicillin/streptomycin 200 × solution (final concentration 50 U/ml penicillin and 50 µg/ml streptomycin), and 5 ml of 200 mM L-glutamine solution (final concentration 1 mM), filter sterilize and store in disposable bottles at 4 °C
- PBS^-: PBS without Ca^{2+} or Mg^{2+} (Chapter 2, *Protocol 1*)
- Trypsin/EDTA solution: 0.05% Trypsin and 200 µM EDTA in PBS^-
- Mitomycin C-supplemented medium: 10 µg/ml mitomycin C in MEF medium[a]
- Gelatin solution: 1% swine-skin gelatin in distilled water[b]
- Organ-culture dishes

Methods

Maintenance of MEFs:

1 Culture fibroblasts revived from frozen stocks in $25 cm^2$ flasks, and passage when they reach 80–90% confluence. To passage, aspirate medium and wash flask twice with 10 ml warm PBS^- and add 1 ml trypsin/EDTA solution. Incubate the flask at 37 °C for 3 min.

2 Rap flask to dislodge cells, then wash cells from flask by flushing with 10 ml warm MEF medium. Wash flask with a further 10 ml medium, and divide evenly across the new flasks. MEF at an early passage (such as P0, P1, or P2) and 80–90% confluence with a healthy, cobblestone appearance can be split 1:5 or 1:6. At a later passage, flasks at 80–90% confluence with the same appearance can be split 1:3 or 1:4.

Mitomycin C-inactivation of MEFs:

Use MEFs between passage 2 and 4 for hES cell culture.

3 Once flask is 95–100% confluent, aspirate medium and add prewarmed mitomycin C-supplemented medium. Incubate at 37 °C for 2.5–3 h.

4 Harvest cells as above (steps 1 and 2). Plate the treated cells 24 h to 5 d prior to hES cell addition, at a density of 6.0×10^4 MEF per cm^2, into the central well of organ-culture dishes, or other appropriate vessels, pretreated with gelatin.[b]

[a] Mitomycin C is cytotoxic and carcinogenic, and appropriate safety precautions (including the use of gloves, gowns, and eye protection, and disposal procedures for cytotoxic agents) should be followed when handling this reagent.

[b] Feeder cells may be mitotically inactivated also by treatment with 75 Gy ionizing radiation.

[c] For the gelatinization of dishes see Chapter 6, *Protocol 1* or Chapter 12, *Protocol 1*.

Protocol 3
Subculture of hES cells using mechanical dissection of colonies under microscopic guidance

Reagents and equipment

- Feeder layers in gelatinized, 60 mm organ-culture dishes
- Dispase solution: dissolve powdered dispase II (Cat. No. 165859, Roche Diagnostics) at 10 mg/ml in SSM, filter sterilize, store at 4 °C, and warm to 37 °C prior to use
- PBS containing calcium and magnesium (PBS$^+$)
- Glass capillary, drawn out over a spirit flame and broken at the tip to have a sharpened end; or a 30 Gauge hypodermic needle
- Stereo dissecting microscope

Method

1. Using a phase-contrast microscope, select best colonies for transfer. Good colonies are of uniform stem-cell morphology throughout, are free of overtly differentiated areas, and will have expanded about five fold in diameter during 1 week of culture. They are separated by a sharp border from surrounding MEFs, and are not surrounded by clouds of apoptotic cells floating in the medium above them.

2. Wash cells with PBS$^+$, and replace the PBS$^+$ to cover the colony.

3. Under a dissecting microscope, slice the colony into small (0.5 mm^2) segments using the tip of a drawn-out capillary or a hypodermic needle. Do not tear the colony or the attached feeder cells, but slice cleanly.

4. Aspirate the PBS$^+$ and add 400 µl dispase solution. Incubate at 37 °C for 1–5 min, until the fragments of hES-cell colonies lift from the surface of the dish *en bloc*.

5. Pick up the colonies with a micropipette or Pasteur pipette, wash twice by passing through 1 ml PBS$^+$, and plate 8–10 pieces onto a feeder layer in a 60 mm organ-culture dish, in SSM.

helpful to use culture vessels that provide easy access to the cells, such as organ-culture dishes, four-well cluster plates, slide culture chambers, or small (12.5 cm^2) tissue-culture flasks. The latter two types of vessels can be filled with medium and used to transport cultures.

3 Human ES cell maintenance: serum-supplemented *versus* serum-free media

Spontaneous differentiation is suppressed in hES cells cultured in serum-free medium, or serum-replacement medium (SRM), in the presence of FGF2, so it is not generally necessary to select undifferentiated colonies of hES cells away from differentiated derivatives under microscopic guidance. Thus, once established, hES cells may be harvested by using enzymes or low-Ca^{2+} buffer solutions, rather than mechanical dissection.

However, we and others have observed that prolonged culture of hES cells in SRM, using either enzymatic dissociation or cell dissociation buffer to disperse cells, can lead to emergence of cell populations bearing karyotypic abnormalities (9–11). These abnormal cell populations can eventually overtake the culture – they often show faster growth rates and a reduced tendency to differentiate compared with euploid cells. The factors that predispose to the development of karyotypic abnormalities in hES cell cultures are unknown, but in our experience karyotypically abnormal cells appear to emerge far less frequently in cultures maintained in serum-containing medium (i.e. SSM) and passaged by mechanical dissection of colonies. We therefore still recommend the latter technique for use in derivation of cell lines. All hES cell cultures should be subjected to karyotype analysis on a regular basis, particularly if changes in growth rate or spontaneous differentiation are observed. We use standard G-banding techniques to karyotype our cells, but it is important to remember that submicroscopic genetic lesions, such as small amplifications, deletions, rearrangements, or point mutations, will escape detection by this method.

In our laboratory we continue to maintain stock cultures using mechanical dissection of colonies, and use this method to seed larger scale cultures grown in SRM; but we do not maintain the latter beyond 15–20 passages. *Figure 2* illustrates typical morphologies of hES-cell colonies grown in the two systems. *Protocol 4* describes culture of hES cells in SRM, and is based upon Thomson *et al.* (Current Protocols in Stem Cell Biology; course conducted at the Jackson Laboratory Bar Harbor, 2003).

4 Cryopreservation of human ES cells

New hES cell lines should be cryopreserved as soon as practicable following their derivation, and at regular intervals thereafter during expansion and characterization. It is possible to cryopreserve hES cells using standard slow-freezing methodology (i.e. 1 °C/min cooling rate in 90% FCS plus 10% dimethylsulphoxide), but we have found that subsequent recovery of viable cells is rather variable; which signifies that the method is particularly unreliable when only small amounts of cells are available for cryopreservation, for example when preserving low-passage cultures during cell-line establishment. *Protocol 5*, which describes cryopreservation by vitrification and is designed for preservation of small numbers of cells (12), is time-consuming but reliably gives a very high recovery of viable cells. (An alternative method for the cryopreservation of primate ES cells is given in Chapter 12, *Protocol 3*.)

5 Characterization of hES cells

The current definition of the hES-cell phenotype relies on immunocytochemical characterization combined with analysis of gene expression, and biological assays to demonstrate pluripotentiality. In addition to these criteria, regular

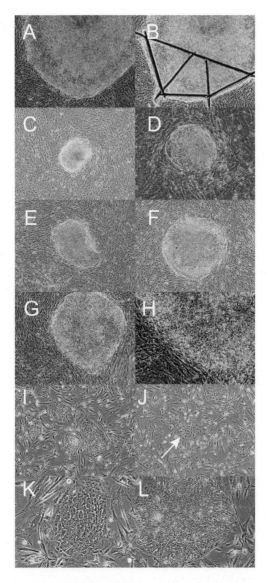

Figure 2 Subculture and growth of hES-cell colonies. (A–H) Colonies grown in serum-supplemented medium with MEF feeder-layer support, at sequential stages following transfer by mechanical dissection. (A) Colony at day 7 of culture, ready for subculture. (B) Colony in (A) scored mechanically for transfer, avoiding the central, differentiated area (denoted by flattened morphology). (C) Colony at day 1 of culture following transfer. (D) Colony at day 3. (E) Colony at day 3 with differentiated area at the top left. (F) Colony at day 4. (G, H) Colony at day 5 with typical smoothened morphology in centre and jagged, defined cell borders at the periphery. (I-L) colonies grown in medium supplemented with proprietary serum replacement and FGF2, with MEF feeder-layer support, and following transfer by enzymatic dissociation. (Feeders are at one-third the density used in serum-based cultures, and the hES colonies are less compact.) (I, J) Colonies at day 1 of culture following transfer, the colony in centre of (J) is differentiated (arrow). (K) Colony at day 3. (L) Colony at day 5. Magnifications: (C, E, F, G, I, and J) as shown in (G); (A, B, H, K, and L) as shown in (H).

Protocol 4

Subculture of hES cells under serum-free conditions, with feeder cell support

Reagents and equipment

- Culture of hES cells 7 d after initial seeding of the flask, or at 80–90% confluence
- Feeder layers in tissue-culture flasks, prepared with freshly inactivated MEFs plated at $2 \times 10^4/\text{cm}^2$ in SRM supplemented with 4 ng/ml FGF2
- Dissociation Solution (Cat. No. C5914, Sigma) *or* freshly prepared collagenase solution: collagenase Type 1A (Cat. No. S3P6854, Worthington Biochemical Company) at a concentration of 4 mg/ml in DMEM/F-12
- Serum-replacement hES-cell medium (SRM): for 500 ml combine 389 ml DMEM/F12, 100 ml Knockout™ Serum Replacement (to give 20% KnockOut™SR; Cat. No. 10828-028, Invitrogen), 5 ml of 10 mM non-essential amino acids (0.1 mM final concentration; Cat. No. 11140-050, Invitrogen), 5 ml of 200 mM L-glutamine solution (2 mM final concentration), and 0.9 ml of 55 mM β-mercaptoethanol solution (90 μM final concentration)
- 40 μl FGF2 stock solution (50 μg/ml prepared in accordance with manufacturer's instructions; final concentration 4 ng/ml): if less than 500 ml medium is required we prepare the medium without FGF2, then add the factor just prior to use
- Sterile, 50 ml and 15 ml centrifuge tubes
- Sterile, 5 ml pipette

Methods

Harvesting hES cells with Dissociation Solution:

1 Aspirate medium from bulk culture flask, and wash cells with 8–10 ml PBS⁻.

2 Aspirate PBS⁻ and add 5 ml Dissociation Solution. Incubate the flask at 37 °C for 3–4 min, and tap the flask until clumps of cells detach leaving only feeder cells and differentiated hES cells attached.

3 Dilute the Dissociation Solution immediately with 10 ml of SRM (*without* added FGF2), and collect the cell suspension into a 50 ml centrifuge tube.

4 Allow clumps to settle under gravity and discard the supernatant fraction.[a]

Or harvesting hES cells with collagenase solution:

1 Aspirate medium from bulk culture flask, add 1 ml collagenase solution per 25 cm² culture surface area, and incubate for 10–15 min at 37 °C.

2 Tap the flask to dislodge feeder cells, leaving only hES-cell colonies attached. Flush hES-cell colonies from the flask surface with 5 ml SRM (*without* added FGF2) using a 5 ml pipette, and collect into a 15 ml centrifuge tube.[b]

3 Dissociate the larger clumps of cell by pipetting up and down several times.[a]

4 Allow the clumps to settle under gravity and discard the supernatant.

Protocol 4 continued

Subculture:

5 Harvest the cell suspension by centrifugation at 500 g for 2 min, and resuspend the pellet of cells in 5–6 ml of SRM supplemented with 4 ng/ml FGF2. The split ratio will depend on a number of factors such as confluence and extent of differentiation, morphology and growth rate, which characteristics may vary from week to week. When initiating serum-replacement cultures from cell stocks grown in serum-containing medium, begin with a 1:2 split ratio.

[a]With either dissociation procedure, occasionally large clumps of hES cells are harvested that are difficult to break up. In this case the cell suspension can be filtered using a cell strainer with 70 μm mesh size (Cat. No. 353350, BD Biosciences) to avoid the transfer of these large clumps, which if not already differentiating will differentiate after subculture.

[b]Some colonies may require gentle scraping using the pipette to detach from the flask surface. Where many differentiated cells are present in the culture, these may be removed by scraping them gently with a blunt-ended Pasteur pipette under microscopic guidance and then aspirating the medium, prior to harvest of undifferentiated colonies.

Protocol 5
Vitrification of hES cells and recovery from cryopreservation (freezing and thawing)

Usually, 8–10 fragments of hES colonies are frozen in each straw. Although this does not represent a large number of cells, recovery is very efficient and re-establishment of the line is usually successful.

Reagents and equipment

- Culture of undifferentiated hES cells [a]
- DMEM–Hepes medium: combine 0.5 ml of 1 M Hepes solution (Cat. No. 15630–080, Invitrogen) and 20 ml DMEM; store at 4 °C for up to 1 week
- Serum supplemented-Hepes medium: combine 8 ml DMEM–Hepes Medium with 2 ml FCS; filter sterilize using a 0.22 μm syringe filter, and store at 4 °C for up to 1 week
- 1 M sucrose solution: dissolve 3.42 g sucrose in 10 ml serum-supplemented Hepes medium; sterilize the solution using a 0.22 μm syringe filter, and store at 4 °C for up to 1 week
- 10% vitrification solution: combine 2 ml serum-supplemented Hepes medium, 0.25 ml ethylene glycol, and 0.25 ml DMSO, and use on day of preparation
- 20% vitrification solution: combine 0.75 ml serum-supplemented Hepes medium, 0.75 ml of 1 M sucrose solution, 0.5 ml ethylene glycol, and 0.5 ml DMSO; store at 4 °C and use on day of preparation
- 0.2 M sucrose solution: combine 4 ml serum supplemented-Hepes medium and 1 ml of 1 M sucrose solution, and store at 4 °C for up to 1 d

Protocol 5 continued

- 0.1 M sucrose solution: combine 4.5 ml serum supplemented-Hepes medium and 0.5 ml of 1 M sucrose solution; and store at 4 °C for up to 1 d
- 5 ml cryovial, labelled with all necessary details (i.e. cell line, passage number, date, and number of straws), and with a hole punched on the upper section (using a flame-heated 18 Gauge hypodermic needle) attached to a labelled metal cane
- Two 20 µl micropipettes
- 4-well multidishes
- straws

Methods

Vitrification and storage:

1 Prepare a 4-well cluster dish by dispensing 1 ml serum-supplemented Hepes medium, 1 ml 10% vitrification solution, and 1 ml 20% vitrification solution each into separate wells, and prewarm to 37 °C.

2 Harvest hES cells for freezing using mechanical dissection of the colony, but cutting pieces slightly larger than described in *Protocol 3*.

3 Using a 20 µl micropipette transfer 8–10 colony pieces to the first well containing serum-supplemented Hepes medium, then into the second well containing 10% vitrification solution for 1 min, ensuring that all colonies settle to the bottom of the well. During incubation in 10% vitrification solution, take a fresh micropipette tip and deposit a 20 µl drop of the 20% vitrification solution onto the inside of the 4-well plate lid (make a fresh drop for each straw to be frozen down).

4 Transfer the hES-cell colony pieces to the third well containing 20% vitrification solution, and incubate for 25 s. Pick up the pieces in the least possible volume and transfer into the 20 µl droplet of 20% vitrification solution.

5 Use a second micropipette (set at 3 µl) to pick up the colony pieces and eject them in a 3 µl droplet onto the lid. Immediately touch the narrow end of the vitrification straw to the side of the droplet at a 45° angle to the plane of the lid. The droplet should be drawn up by capillary action to make a 1 mm column of fluid and cells in the straw.

6 Wearing safety goggles, plunge the straw at a 45° angle into a bucket of liquid nitrogen, transfer the frozen straw into the cryovial, and store the vial in a liquid-nitrogen storage tank.

Recovery of vitrified cells:

1 Prepare a 4-well cluster dish with 1 ml of 0.2 M sucrose solution and 1 ml of 0.1 M sucrose solution, each in separate wells, and 1 ml serum-supplemented Hepes medium in each of two wells, and prewarm to 37 °C.

Protocol 5 continued

2. Collect the cryovial containing those straws required for thawing from liquid nitrogen: carefully, remove one straw at a time with forceps, then (*wearing protective gloves*) hold the straw in one hand, between the thumb and middle finger with the index finger blocking the end of the straw that does not contain cells.

3. Submerge the narrow end of the straw (containing cells) immediately into the first well containing 0.2 M sucrose solution. As the straw warms, the hES-cell colony pieces should be expelled from the straw and into the sucrose solution.

4. After 1 min at room temperature, transfer the colony pieces into the second well containing 0.1 M sucrose solution for 5 min.

5. Finally, transfer the pieces into the third, then fourth, wells containing serum supplemented-Hepes medium for 5 min each time, and plate pieces onto an organ-culture dish prepared as for routine subculture (*Protocol 3*) with fresh MEF.[a]

[a] We recommend that cells recovered from cryopreservation be grown in vessels that provide appropriate access for the subsequent mechanical dissection of colonies. It is possible that some of the thawed colonies may differentiate, and it may be necessary to recover the undifferentiated hES cells by dissection under microscopic guidance.

examination of G-banded karyotype should be performed (by a cytogenetics laboratory) to confirm maintenance of euploidy in hES cells during propagation *in vitro*. These tests should be performed soon after establishment of cell lines, and at regular intervals thereafter.

5.1 Immunochemical characterization of hES cells

An expanding range of immunochemical markers is available with which to characterize hES cells (*Table 1*). The monoclonal antibodies most commonly used in hES-cell characterization at present are those that recognize: the cell surface glycolipids, SSEA-1 (CD15 or Lewis X antigen, which is absent from undifferentiated hES cells), SSEA-3, and SSEA-4 (which are present in undifferentiated hES cells); epitopes on a pericellular matrix of chondroitin sulphate/keratan sulphate proteoglycan, found on the surface of primate pluripotential stem cells (antibodies TRA-1-60, TRA-1-81, GCTM-2); the transcription factor, Oct-4; and the tetraspannin membrane protein, CD9. Additionally, hES cells express the CD34-related sialomucin podocalyxin, and the glycosylphosphatidylinositol-linked membrane protein, CD24. Some workers have reported expression of the CD133 antigen and flk-1 in hES cells.

Detection of these antigens is carried out by indirect immunofluorescence or other immunochemical assays on fixed cells, or by flow cytometry for quantitation. It is important to recognize that none of these markers is absolutely specific for hES cells, and that even well-maintained cultures, but particularly those undergoing a degree of spontaneous differentiation, will be heterogenous in their expression of these antigens with subpopulations of cells expressing some but not all of the hES-cell markers.

Table 1 Monoclonal antibodies reactive with epitopes found on hES cells

Hybridoma	Class and subclass	Target antigen	Supplier[e]	Reference
MC-631	M IgM	SSEA-3[a]	A	(14–16)
MC-813-70	M IgG3	SSEA-4[b]	A	(14–16)
TRA-1-60	M IgM	KSPG[c]	B	(14–19)
TRA-1-81	M IgM	KSPG[c]	B	(14–19)
GCTM-2	M IgM	KSPG[d]	C	(14–19)
TG343	M IgM	KSPG[d]	C	(20)
TG30	MIgG2a	CD9	C	Pera, unpublished
OCT-4 (C-10)	MIgG2b	Oct-4	D	–
P1/33/2	M IgG1	CD 9	E	–
ML5	MIgG2a	CD24	F	–
PHM-5	MIgG1	podocalyxin	G	(21)

[a,b] Globo-series glycolipid epitope.

[c] 200 kDa cell-surface/pericellular matrix keratan sulphate/chondroitin sulphate proteoglycan containing extensive O-linked carbohydrate. These antibodies react with carbohydrate epitopes.

[d] These antibodies react with the core protein of the proteoglycan.

[e] Sources: A, Developmental Studies Hybridoma Bank, University of Iowa; B, Chemicon International; C, this laboratory; D, Santa Cruz Biotechnology Inc.; E, Dako Corporation; F, BD Biosciences Pharmingen; G, Chemicon International

Protocol 6
Immunocytochemical examination of hES cells using indirect immunofluorescence

Indirect immunofluorescence provides high-resolution imaging of the intracellular localization of antigens, and facilitates discrimination between non-specific background binding and genuine reactivity. The interpretation of immunocytochemical staining requires some experience. Each of the surface antigens listed in *Table 1* shows a distinct pattern of staining on hES cells under the fluorescence microscope. Consult the published literature for examples of images of staining of hES cells with particular antibodies (see *Table 1*).

Reagents and equipment

- Culture of hES cells, harvested according to *Protocol 3* or *4*
- Fixative solutions: acetone/water 9:1 (v/v); methanol/acetone 1:1 (v/v); 4% paraformaldehyde in PBS⁻ (see Chapter 9, *Protocol 6*); or absolute ethanol
- PBS : PBS without Ca^{2+} or Mg^{2+} (Chapter 2, *Protocol 1*)
- Primary antibodies (*Table 1*)
- Secondary antibodies either conjugated to appropriate fluorochromes *or* conjugated to alkaline phosphatase for immunohistochemical detection with Fast Red TR[a]
- Mayer's haematoxylin, for counterstaining
- Hoechst 33258 solution: 1 µg/ml Hoechst 33258 in PBS⁻, for fluorescent counterstaining

Protocol 6 continued

- Antifade mountant, such as Vectashield (Cat. No. H1000, Vector Laboratories)
- Chamber-well or multiwell slides (with 8, 12, or 24 wells): these are gelatinized and placed in square or rectangular culture dishes; MEF feeder cells may be added where required, prior to addition of hES cells[b]

Method

1. If feeder layers are required, seed mitotically inactivated MEFs onto chamber well slides, multiwell slides or coverslips at least 1 d and not more than 5 d prior to hES-cell addition.

2. Transfer the hES cells as intact clumps or clusters of cells onto chamber well slides, multiwell slides or coverslips, and culture for any length of time between 1 d (for analysis of the undifferentiated stem cell phenotype) to 3–4 weeks (for analysis of differentiation).

3. Wash slides in PBS⁻ and fix cells. For cell-surface glycolipid antigens use acetone/water for 5 min followed by air-drying. For intracellular antigens, use either methanol/acetone or paraformaldehyde solution, both at r.t. for 5 min. Methanol/acetone-treated slides should be air-dried directly after fixation, whilst paraformaldehyde fixed slides should be rinsed with water prior to air-drying. Alternatively, for many antigens absolute ethanol provides good preservation. Cells should be rinsed in PBS⁻, incubated in ethanol for 5 min, then air dried.[c]

4. Apply antibodies to the slides and incubate in a humidified atmosphere (e.g. a covered culture dish containing moistened paper towel or tissue) for 30 min, wash with PBS⁻, and add secondary detection reagents conjugated to fluorochromes or enzymes. For immunohistochemical detection, use antibodies conjugated to alkaline phosphatase and detection with Fast Red followed by counterstaining with Mayer's haematoxylin to provide high-contrast staining with red.

5. After the detection reagent has been applied, wash the slides again in PBS⁻.[d]

6. Counterstain the nuclei with a DNA-binding dye such as Hoechst 33258 solution for 30 s to help localize cells and discriminate between human cells and MEFs.[e]

7. Cover cells with antifade mountant. For cells fixed on slides apply a coverslip, and for cells fixed on coverslips invert onto a slide, in preparation for microscopic analysis.

[a] The Fast Red TR staining kit (from Sigma) includes levamisole to inactivate endogenous alkaline phosphatase.

[b] Standard coverslips can be used, but are more difficult to handle.

[c] Slides can be stored for at least 6 months at −20 °C.

[d] Proprietary mounts such as Vectashield (Vector Laboratories) contain antifade compounds to stop bleaching of staining during examination. This is a particular problem with high numerical aperture lenses. Although we recommend use of mounts with antiphotobleaching additives, the performance of certain fluorochromes such as Alexa Fluour 350 may be adversely affected.

[e] Mouse nuclei stain differently to human nuclei with Hoechst 33258, with a more granular appearance compared with homogenous staining of human nuclei.

Protocol 7
Multiplex FACS® analysis of hES-cell antigen expression

Reagents and equipment

- Culture of hES cells
- PBS⁻: PBS without Ca^{2+} or Mg^{2+} (Chapter 2, *Protocol 1*)
- Trypsin/EDTA solution: 0.05% trypsin and 200 μ EDTA in PBS⁻
- Serum-supplemented hES-cell medium (SSM) (*Protocol 1*)
- Primary antibodies: mouse GCTM-2 (IgM) or TRA-1-60 (IgM) antibody against hES cell pericellular matrix proteoglycan; mouse TG30 (IgG2a) or similar antibody recognizing CD9; mouse Oct-4 (C-10) (IgG2b) (1:50 dilution); and class-matched negative control antibodies (Dako)
- Alexa Fluor® 488 goat-anti-mouse IgG_{2a} (Cat. No. A-21131, Invitrogen), Alexa Fluor® 647 goat anti-mouse IgG_{2b} (Cat. No. A-21242, Invitrogen) and streptavidin-phycoerythrin (PE; Cat. No. 554061 BD Biosciences Pharmingen)
- Biotinylated rabbit anti-mouse IgM (1:125 dilution)
- Microfuge
- Cell strainer with 70 μm mesh size (Cat. No. 353350, BD Biosciences)
- Fluorescence-activated cell sorter

Method

1 Harvest hES cell colonies using dispase as described above (*Protocol 3*). Wash colonies with PBS⁻, and dissociate using trypsin/EDTA solution into a single cell suspension.

2 Stop trypsin action with 1 ml SSM, and harvest cells by microcentrifugation at 500 g for 2 min. Resuspend cells in 50 μl SSM.

3 Fix cells by transferring them into 5 ml ice-cold, 100% methanol, and incubate for 30 min on ice.[a]

4 Centrifuge cells for 2 min at 500 g, and wash with 500 μl of SSM.

5 Resuspend cells in 300 μl mixture of mouse monoclonal antibody supernatants: GCTM-2 IgM (1:2 dilution), mouse TG30 IgG2a (1:2 dilution), and anti-Oct-4 IgG2b (1:50 dilution), or a mixture of class-matched negative-control antibodies, for 30 min on ice.

6 Centrifuge cells for 2 min at 500 g, and wash with 1 ml SSM.

7 Pellet the cells again by centrifugation at 500 g for 2 min, and remove supernatant medium. Resuspend the cells in 300 μl biotinylated rabbit anti-mouse IgM (1:125 dilution), and incubate for 30 min on ice.

8 Centrifuge cells for 2 min at 500 g, and wash in 500 μl SSM before resuspending in a mixture of goat anti-mouse IgG2a-AF488 (1:1000 dilution), goat anti-mouse IgG2b AF647 (1:1000 dilution), and streptavidin-PE (1:1000 dilution) for 30 min on ice.

9 Wash the cells again in 1 ml SSM prior to resuspending in 500 μl of the same. Filter cells through a cell strainer to remove clumps.

Protocol 7 continued

10 Analyse cells on a flow cytometer with initial gating using forward and right-angle light scatter; and collect AF488, AF647, and PE fluorescence signals. Human ES cells analysed by this method should be compared to single colour controls for TG30, GCTM-2, and Oct-4.

[a] Live, non-fixed hES cells also may be analysed for the presence of the cell surface markers, KSPG and CD9, recognized by antibodies GCTM-2 and TG30.

In *Protocol 7* we describe a rapid and convenient technique for estimating the proportion of undifferentiated hES cells in a culture, by double labelling for cell-surface epitopes KSPG and CD9, combined with fluorescence-activated cell sorting (FACS®) analysis.

5.2 FACS® analysis of hES-cell antigen expression

Flow cytometry provides quantitative information on the proportion of hES cells in a culture expressing particular surface markers. Because the technique requires dissociation of hES-cell colonies into single cell suspensions, which is only poorly tolerated at best, it is important to monitor cell viability and to bear in mind the possibility that some subpopulations of cell types within a hES cell culture may be more susceptible to dying following dissociation than others.

Protocol 7 describes the method of simultaneous immunostaining cells in single-cell suspension for the stem-cell markers KSPG, CD9, and Oct-4, and subsequent quantitative analysis by flow cytometry.

5.3 Immunomagnetic isolation of viable hES cells

Flow cytometry requires dissociation of hES colonies into single cells, a procedure that results in loss of viability and differentiation. Therefore, isolation of viable stem cells by fluorescence-activated cell sorting (FACS®) is problematic. An alternative approach is to immunoisolate small clusters of cells using magnetic beads. In the case of the surface marker defined by antibody GCTM-2, positive cells and negative cells generally appear to segregate spatially within a colony and, as a consequence, the clusters of cells isolated using this method are usually homogeneous with respect to antigen expression and show much better viability than single cells.

5.4 Gene expression in hES cells

Examination of transcription in mouse and human ES cells using conventional methods and microarray profiling have supported the concept of a network of genes characteristic of pluripotent stem cells in both species (3, 13). Molecular embryology in the mouse has also defined many genes that are characteristically expressed in cells that have undergone commitment to specific differentiation lineages, and of course there are vast numbers of genes expressed in particular

Protocol 8
Immunomagnetic isolation of viable hES cells using antibody GCTM-2

For this isolation procedure, hES cells should be harvested as described in *Protocol 3* or *4* above, and dispersed into small clumps.

Reagents and equipment

- Culture of hES cells
- Rat anti-mouse IgM beads (Cat. No. 110.15, Dynal)
- Mouse monoclonal antibody GCTM-2 IgM, reactive with the hES cell-surface pericellular matrix proteoglycan
- Magnetic particle concentrator (Dynal)
- Incubation buffer: Hepes-buffered DMEM (Cat. No. 10315, Invitrogen) supplemented with 1% FCS
- PBS$^-$ (*Chapter 2, Protocol 1*)
- Sterile, 1.5 ml microfuge tubes

Method

1. Resuspend the beads in 1 ml PBS$^-$ (allowing 25 µl bead suspension for up to 4×10^7 cells in a 1 ml sample) and transfer into a microfuge tube. Place the tube in a Dynal magnetic particle concentrator for 1 min.

2. Remove the buffer, wash the beads in 1.5 ml of PBS$^-$, and return to the magnet for 1 min.

3. Aspirate the buffer, and resuspend the beads in 50 µl of PBS$^-$. Remove PBS$^-$, add 500 µl of neat GCTM-2 supernatant, and incubate at r.t. for 30 min.

4. Wash beads several times in PBS$^-$, three times in incubation buffer, and resuspend in 50 µl of incubation buffer. The beads are then ready for addition of cells.

5. Harvest cells from serum-containing cultures using dispase (*Protocol 3*), or from serum-free cultures using collagenase (*Protocol 4*). Disperse the cells into small clumps by trituration, and harvest the clumps by microcentrifugation at 250 g for 4 min.

6. Resuspend the cell pellet in 500 µl incubation buffer. Reserve a small aliquot (50 µl) of cells for antibody staining, to assess the proportion of GCTM-2-positive cells prior to separation.

7. Add a 50 µl aliquot of GCTM-2-coated beads to the 450 µl cell suspension, and incubate at 4 °C for 30 min with occasional gentle agitation.

8. Place the tube in the magnet for 1 min. Retain the liquid phase, which represents the unbound fraction, for later staining by immunofluorescence.

9. Wash beads twice in incubation buffer and resuspend in 500 µl incubation buffer.[a]

Protocol 8 continued

10 Examine the starting material, the unbound fraction, and the bound fraction by indirect immunofluorescence (*Protocol 6*) or by flow cytometry (*Protocol 7*), to assess the efficacy of the magnetic separation.

[a] To propagate the bound fraction it is unnecessary to remove the beads from the cells as they can reattach and grow readily even when initially attached to beads.

Table 2 Primers used in RT-PCR studies on hES cells

Gene	Primer name	Primer sequence(5'–3',F$_w$/R$_v$)	T$_m$°(C)	Product size (bp)
Cripto	CRIPTOF484 CRIPTOR668	CAG AAC CTG CTG CCT GAA TG GTA GAA ATG CCT GAG GAA ACG	55	185
GDF3	GDF3F773 GDF3R958	ATG GAG AGT GTC CCT CCT C CCC ACA TTC ATC GAC TAC CAT G	55	185
FoxD3	FoxD3F1 FoxD3F1429	ATG ACC CTC TCC GGC GGC CCG GCC ATT TGG CTT GAG	60–65	1428
Actin	Act$_w$ bAct R$_v$	FCGC ACC ACT GGC ATT GTC AT TTC TTC TTG ATG TCA CGC AC	57	200
OCT-4	hOCT-4 F$_w$ hOCT-4 Rv	CGA CCA TCT GCC GCT TTG AG CCC CCT GTC CCC CAT TCC TA	57	540
Nanog	hNanog F$_w$ hNanog R$_v$	CAG AAG GCC TCA GCA CCT AC CTG TTC CAG GCC TGA TTG TT	55	216
CD9	hCD9F$_w$ hCD9 R$_v$	TGC ATC TGT ATC CAG CGC CA CTC AGG GAT GTA AGC TGA CT	60	800

types of mature, fully differentiated cells. Although microarrays are now commonly used to study hES gene expression, they are still too expensive for routine studies, and many groups still rely on RT-PCR (*Table 2*). The use of semiautomated, multiwell plate format assays for quantitative RT-PCR (see Chapter 6) will in future probably play an important role in the rapid profiling of hES cell cultures for expression of a standard set of stem cell markers and lineage-specific differentiation markers. Any of these methods will be subject to the limitation that cultures grown under standard conditions are in fact heterogeneous. Flow cytometry protocols, such as that outlined above (*Protocol 7*) provide a more defined starting population for analysis of gene expression.

Gene expression analysis in hES cells often is focused on genes with known patterns of expression and defined roles in the mouse embryo. Although this is a sound starting point, it is important to remember that assumptions about conservation of expression patterns and gene function between the two species are largely speculative, given the lack of information on gene expression during human peri-implantation development. Also, it may be difficult to obtain human tissue containing transcripts for certain genes expressed in early embryos, for use as positive controls in PCR. Where possible, this problem

Protocol 9

Preparation of cDNA by reverse transcription of mRNA from hES cells

Here, RNA isolation from hES cells and derivative cells (for example neural progenitor cells or neurospheres) is carried out using the Dynalbeads mRNA DIRECT kit, according to the manufacturer's instructions.

Reagents and equipment

- Culture of hES cells or their differentiated derivatives
- DynalBeads mRNA DIRECT kit (Cat. No. 610.02, Dynal)
- Cell wash buffer: 0.1 M NaCl, 10 mM Tris-HCl, 1 mM EDTA, pH 7.4
- Lysis/binding buffer: 100 mM Tris-HCl pH 7.5, 500 mM LiCl, 10 mM EDTA, 5 mM dithiothreitol, 1% lithium dodecyl sulphate (LDS)
- Bead wash buffer: 10 mM Tris-HCl pH 7.5, 0.15 M LiCl, 1 mM EDTA, 0.1% LDS

- Reverse transcriptase and reaction buffer supplied by manufacturer, e.g. Superscript (Invitrogen), dNTPs
- RNAase inhibitor, e.g. RNAase OUT (Invitrogen)
- Magnetic particle concentrator or magnet (Dynal)
- Four-well plates or organ-culture dishes

Method

1. Using the methods described above (*Protocol 3*) harvest hES cells (1–8 colonies or equivalent), and wash twice in warmed PBS⁻ and twice in 1 ml cell wash buffer in four-well plates or organ-culture dishes.

2. Lyse cells by resuspension in 300 μl lysis/binding buffer in a microfuge tube, and centrifuge at 15 000 g to remove debris. Remove the supernatant.[a]

3. Prior to use, wash oligo dT-conjugated magnetic beads with lysis/binding buffer; and incubate the lysates with 20 μl washed beads for 10 min at r.t.

4. Remove unbound material by placing the sample tube in a magnetic particle concentrator. Wash beads twice, first with 400 μl bead wash buffer containing LDS and second with 200 μl bead wash buffer without detergent.

5. Resuspend samples in 100 μl of wash buffer without detergent, but supplemented with RNAase inhibitor.

6. Carry out reverse transcription on mRNA conjugated to magnetic beads in a 20 μl reaction, according to the manufacturer's instructions. Omit reverse transcriptase from one reaction tube for use as control in the PCR.

7. Following reverse transcription, remove the enzyme and buffer by washing the beads with water containing RNase inhibitor.

8. Resuspend the beads in 20 μl of water containing RNase Out.

Protocol 9 continued

9. Carry out PCR amplification on 2–5 µl of the suspension of beads that now bear a cDNA library from the hES cells. Perform amplifications according to gene-specific protocols; generally, it is important not to exceed 30 cycles of amplification, and to include appropriate housekeeping gene controls as well as negative controls omitting reverse transcriptase. Complementary DNA templates may be titred with a housekeeping promoter to ensure equal input into assays prior to analysis.

[a] Lysates may be frozen and stored at −80 °C.

Protocol 10
Formation of teratomas by implantation of hES cells into the testis capsule of SCID mice

Reagents and equipment

- Suspension of hES cells as intact colonies or clumps of cells in 50–100 µl of hES cell medium: an inoculum of 5–10 colonies of hES cells (20 000–50 000 cells) injected into one testis is generally sufficient to yield a tumour[a]
- Male SCID mice 5–6 weeks of age
- Sterile surgical instruments
- Injectable, general anesthetic, e.g. avertin
- 26 Gauge hypodermic needle, and either a 25 Gauge hypodermic needle or a drawn-out glass capilllary
- Dissecting microscope with overhead illumination
- 6-0 silk suture
- Laminar flow hood

Methods

Preparation of the testis:

1. Anaesthetize the mouse with avertin or any suitable general anaesthetic.
2. Transfer the mouse into a laminar flow hood, swab the abdomen with alcohol, and using aseptic technique make a small longitudinal incision through the skin and abdominal wall, just below the level of the origin of the hind limb.
3. Place the mouse under a dissecting microscope. Squeeze the scrotum of the mouse to bring the testes into the abdominal cavity. Locate a piece of fat inside the incision, towards the caudal part of the mouse and slightly lateral to the midline; draw this tissue outside the abdomen with forceps, and the vas deferens and testis will follow.

Injection of hES cells:

4. Use a small artery clamp to place gentle traction on the testis. Make a small hole in a region of the testis capsule free of major blood vessels using a 26 Gauge hypodermic

> **Protocol 10 continued**
>
> needle. Insert either a 25 Gauge syringe needle or a drawn-out capillary containing the hES cells through the hole, about halfway into the testis; and inject up to 50 μl of cell suspension.[b]
>
> 5 Replace the first testis inside the abdominal cavity, then repeat the procedure with the other testis.
>
> 6 Close the internal and external incisions separately with 6-0 silk.
>
> 7 Monitor the mouse until recovery from anaesthesia is complete, and check its condition the following day.
>
> *Tumour formation and analysis:*
>
> 8 Monitor animals weekly, beginning at around 4 weeks, for tumour development. Bring the testis down into the scrotum and palpate it. Sometimes, the testis will not come down if it is enlarged or adherent to surrounding tissue, in which case palpate the organ in the abdomen. Lesions usually become apparent as swellings by about 5 weeks.
>
> 9 Remove the tumours, fix in formalin, and send for routine histological processing. Immunocytochemical staining with antibodies showing specificity for human housekeeping genes (e.g. human mitochondrial proteins or ribonuclear proteins) may be carried out on frozen sections, with detection as described in *Protocol 6* to confirm the human origin of the differentiated tissue.
>
> [a] The cells should be prepared for inoculation by another worker so that they are ready for injection when the animal is ready to receive them.
>
> [b] A few seminiferous tubules may herniate out from the incision, but do not attempt to repair the incision.

may be circumvented by the design of primers based on sequences conserved in mouse and human.

5.5 Biological assay of hES-cell pluripotentiality

The defining feature of a pluripotent stem cell is its ability to differentiate into a wide variety of cells and tissues. Protocols for the *in vitro* differentiation of hES cells are described in Chapter 11 of this volume. However, the ability of an hES cell line to form teratomas is the best test of pluripotentiality presently available, as a much greater variety of tissues are usually represented in a single graft than are found during differentiation *in vitro*. Grafting of hES cells under the testis capsule of SCID mice gives good take rates from small number of cells, and provides for simple monitoring of tumour formation and growth. An alternative method for raising teratomas in the limb muscle of SCID mice is provided in Chapter 11.

Teratomas should contain a variety of types of tissues, such as neural tissue, gut, respiratory epithelium, glandular epithelium, and muscle. Definitive identification may be difficult if the tissue has not fully matured; reference to atlases of histology of embryonic tissues may aid in analysis. No undifferentiated hES cells should be present. For those with no previous experience in the evaluation of teratoma histology, it will be helpful to seek the advice of a consultant histopathologist in interpretation of the slides.

Acknowledgements

Work in our laboratory has been supported by the National Health and Medical Research Council, the Juvenile Diabetes Research Foundation, and the National Institutes of Health. We thank all members of our laboratory for their input into this chapter.

References

1. Thomson, J.A., Itskovitz-Eldor, J., Shapiro, S.S., Waknitz, M.A., Swiergiel, J.J., Marshall, V.S., and Jones, J.M. (1998). *Science*, **282**, 1145-7.
2. Pera, M. and Trounson, A.O. (2004). *Development*, **131**, 5515-25.
3. Pera, M.F., Reubinoff, B., and Trounson, A. (2000). *J Cell Sci.*, **113**, 5-10.
4. Wei, C.L., Miura, T., Robson, P., Lim, S.K., Xu, X.Q., Lee, M.Y., et al. (2005). *Stem Cells*, **23**, 166-85.
5. Cheng, L., Hammond, H., Ye, Z., Zhan, X., and Dravid, G. (2003). *Stem Cells*, **21**, 131-42.
6. Cowan, C.A., Klimanskaya, I., McMahon, J., Atienza, J., Witmyer, J., Zucker, J.P., et al. (2004). *N. Engl. J. Med.*, **350**, 1353-6.
7. Richards, M., Tan, S., Fong, C.Y., Biswas, A., Chan, W.K., and Bongso, A. (2003). *Stem Cells*, **21**, 546-56.
8. Richards, M., Fong, C.Y., Chan, W.K., Wong, P.C., and Bongso, A. (2002). *Nat. Biotechnol.*, **20**, 933-6.
9. Draper, J.S., Smith, K., Gokhale, P., Moore, H.D., Maltby, E., Johnson, J., et al. (2004). *Nat. Biotechnol.*, **22**, 53-4.
10. Buzzard, J.J., Gough, N.M., Crook, J.M., and Colman, A. (2004). *Nat. Biotechnol.*, **22**, 381-2; author reply 82.
11. Mitalipova, M.M., Rao, R.R., Hoyer, D.M., Johnson, J.A., Meisner, L.F., Jones, K.L., Dalton, S., and Stice, S.L. (2005). *Nat. Biotechnol.*, **23**, 19-20.
12. Reubinoff, B.E., Pera, M.F., Vajta, G., and Trounson, A.O. (2001). *Hum. Reprod.*, **16**, 2187-94.
13. Fougerousse, F., Bullen, P., Herasse, M., Lindsay, S., Richard, I., Wilson, D., et al. (2000). *Hum. Mol. Genet.*, **9**, 165-73.
14. Reubinoff, B.E., Pera, M.F., Fong, C.Y., Trounson, A., and Bongso, A. (2000). *Nat. Biotechnol.*, **18**, 399-404.
15. Andrews, P.W., Casper, J., Damjanov, I., Duggan-Keen, M., Giwercman, A., Hata, J., et al. (1996). *Int. J. Cancer*, **66**, 806-16.
16. Thomson, J.A., Itskovitz-Eldor, J., Shapiro, S.S., Waknitz, M.A., Swiergiel, J.J., Marshall, V.S., and Jones, J.M. (1998). *Science*, **282**, 1145-7.
17. Henderson, J.K., Draper, J.S., Baillie, H.S., Fishel, S., Thomson, J.A., Moore, H., and Andrews, P.W. (2002). *Stem Cells*, **20**, 329-37.
18. Badcock, G., Pigott, C., Goepel, J., and Andrews, P.W. (1999). *Cancer Res.*, **59**, 4715-9.
19. Cooper, S., Pera, M.F., Bennett, W., and Finch, J.T. (1992). *Biochem. J.*, **286**, 959-66.
20. Cooper, S., Bennett, W., Andrade, J., Reubinoff, B.E., Thomson, J., and Pera, M.F. (2002). *J. Anat.*, **200**, 259-65.
21. Kerjaschki, D., Poczewski, H., Dekan, G., Horvat, R., Balzar, E., Kraft, N., and Atkins, R.C. (1986). *J. Clin. Invest.*, **78**, 1142-9.

Chapter 11
Differentiation of human ES cells

Sharon Gerecht-Nir[1] and Joseph Itskovitz-Eldor

Department of Obstetrics and Gynecology, Rambam Medical Center, The Bruce Rappaport Faculty of Medicine, Technion–Israel Institute of Technology, Haifa, Israel

[1]Current address: Harvard–M.I.T. Division of Health Sciences and Technology, Massachusetts Institute of Technology, Cambridge, Massachusetts 02139, USA.

1 Introduction

Although the establishment of ES cell lines from human embryos was first accomplished relatively recently (1; and see Chapter 10), procedures already elaborated for the *in vitro* differentiation of human ES (hES) cells confirm the pluripotency of this cell type, and the power of the ES cell-based methodology: hES cells hold great promise for the study of human embryonic development, and for transplantation therapies.

For hES cells, as for mouse ES (mES) cells, differentiation processes follow reproducible, temporal patterns that recapitulate early embryogenesis in many ways; and hES cells are endowed with the capability to differentiate into derivatives of all three primary germ layers, both *in vitro* and *in vivo*. Furthermore, and in striking contrast to mES cells, hES cells have the capacity to form cells of the extraembryonic tissues, which normally arise in the early embryo prior to gastrulation. Therefore, hES cells provide a valuable tool for the study of early events in human embryogenesis that otherwise may be inaccessible for analysis.

At the present time, a major focus of research is to define procedures for directing hES-cell differentiation towards a single cell fate. This is in order to provide a source of normal, tissue-specific cells with a view to therapeutic interventions, for correcting tissue damage and dysfunction. So far, protocols have been developed for the derivation from hES cells of various lineages, including neurons (2–8), cardiomyocytes (9–14), endothelial cells (15, 16), smooth-muscle cells (16), haematopoietic cells (17–21), osteogenic cells (22), epidermis (23), hepatocytes (24, 25), insulin-producing cells (26, 27), and trophoblast (28, 29). In this chapter we present a range of protocols for the differentiation of hES cells; and, in particular, methods for inducing haematopoietic differentiation and vascular development are highlighted. Techniques for controlled, mass culture allowing the scaling up of production of differentiated hES-cell derivatives also are described.

2 Maintenance of hES cell lines

Whilst the derivation and culture of hES cell lines are described in detail in Chapter 10, our methods for the maintenance of undifferentiated hES cells are provided in *Protocol 1*. All incubations of hES cells and their differentiated derivatives may be conducted in a humidified incubator at 37 °C, gassed with 5% CO_2 in air; however, recent evidence suggests that hypoxic conditions (i.e. 5% O_2 and 5% CO_2) may be beneficial to the proliferation of undifferentiated hES cells (30, 31). To maximize retention of pluripotency by hES cells (see Section 2.1) and to ensure reproducibility of *in vitro* differentiation experiments, we advise that prescribed culture conditions be followed closely.

As a major aim of research into hES cells is to derive cells for the replacement of diseased or damaged tissues by xenograft transplantation, implications immediately arise for hES cell culture methodologies; most especially, the potential hazard of zoonosis transmitted via feeder cells of animal origin. Therefore extensive research is ongoing into the feasibility of prolonged culture – and even derivation – of hES cells on a wide spectrum of human feeder-cell types, ranging from fetal (32) to adult (33–36). Other potential sources of zoonoses are medium components and additives such as serum, growth factors, and matrices that originate from non-human sources. The ideal is therefore to use reagents of either human or synthetic origin. Moreover, the projected use of hES cells for therapeutic applications demands that culture media and conditions be precisely defined, to enable controlled derivation, maintenance, and scaling up of culture (see Chapter 5 for a commentary on defined culture systems). Initial progress has been made to date in regard to some of these issues:

- Undifferentiated hES cells can be maintained in the absence of feeder cells, on an extracellular matrix (ECM) of Matrigel™ or laminin and with feeder cell-conditioned medium (37).
- Both feeder cells and conditioned medium have been eliminated from a culture system for hES cells that provides instead an ECM of human fibronectin, and supplementation with basic fibroblast growth factor (bFGF) and tumour growth factor β1 (TGFβ1) (38).
- Noggin, an antagonist of bone morphogenetic protein (BMP), is critical in preventing differentiation of hES cells in culture; and the combination of Noggin and bFGF is sufficient to maintain the proliferation of undifferentiated hES cells (39).
- Furthermore, hES cells maintained in medium containing high concentrations of bFGF (24–36 ng/ml), or bFGF in combination with other factors, show characteristics similar to cultures maintained with feeder cell-conditioned medium (40, 41).

Protocol 1
Maintenance of undifferentiated hES cells

Reagents and equipment

- Culture of undifferentiated hES cells at ~80% confluence of colonies, on MEF feeder layers in 6-well culture dishes (Chapter 10, *Protocol 2*)
- Feeder cells for hES cell culture, and for harvesting conditioned medium: prepare a suspension of mitomycin C-inactivated MEFs in MEF medium, according to Chapter 2, *Protocols 2 and 3*[a]
- Human ES-cell medium: 80% KnockOut™ Dulbecco's modified Eagle's medium (KnockOut™ D-MEM; Cat. No.10829–018, Invitrogen) supplemented with 20% (v/v) KnockOut™ serum replacement (KnockOut™ SR; Cat. No. 10828–028, Invitrogen), 4 ng/ml basic fibroblast growth factor (bFGF; Cat. No.13256–029, Invitrogen), 1 mM L-glutamine (Cat. No.25030–024, Invitrogen), 0.1 mM β-mercaptoethanol (Cat. No. 31350–010, Invitrogen), and 1% stock solution of non-essential amino acids (Cat. No. 11140–035, Invitrogen)
- MEF medium (Chapter 2, *Protocol 1*)
- Sterile, tissue culture-grade water (Cat. No. W-3500, Sigma)
- Gelatin solution: add 0.1% (w/v) gelatin powder, Type A from porcine skin (Cat. No. G-1890, Sigma), to sterile water; and dissolve and sterilize by autoclaving[b]
- Matrigel™ (Cat. No. 354234, BD Biosciences): aliquot the supplied solution on ice before storing frozen at $-20\,°C$, to avoid repeated freezing/thawing
- Collagenase solution: dissolve collagenase Type IV (Cat. No. 17104–019, Invitrogen) at 1 mg/ml in KnockOut™ D-MEM, filter sterilize using a 22 µm pore-size filtration unit (e.g. Stericup, Millipore), store at $4\,°C$ and use within 2 weeks of preparation
- 6-well tissue-culture dishes (Cat. No. 140685, Nunc)
- Cell scraper
- Sterile Pasteur pipettes
- 1000 µl micropipette
- Sterile, 5 ml and 10 ml transfer pipettes
- Sterile, 15 ml conical centrifuge tubes
- Haemocytometer
- 23–25 Gauge hypodermic needles
- Benchtop refrigerated centrifuge
- Inverted phase-contrast microscope

Methods

Gelatin-coating of wells:

1 Dispense 2 ml gelatin solution per well of a 6-well culture dish, and incubate the dish overnight in a humidified incubator or other sterile environment.

2 Aspirate residual gelatin from wells prior to seeding MEFs.

Passaging hES cells on feeder layers:

3 Seed $3–4 \times 10^5$ feeder cells in 2 ml MEF medium per well, in gelatin-coated 6-well dishes, and incubate overnight. Next day, replace MEF medium with hES-cell medium and continue incubation. Use 2 ml medium per well throughout the culture.

4 Passage hES cells when they reach ~80% confluence: aspirate medium, add 0.5 ml collagenase solution per well, and incubate the dish for 30 min at $37\,°C$.

Protocol 1 continued

5. Halt enzymatic digestion by addition of hES-cell medium (1 ml per well), harvest cells by gentle scraping, and using a transfer pipette collect the cell suspension into a conical tube. This treatment allows hES cells to remain adherent in small clumps.[c]

6. Pellet cells by centrifugation for 3 min at 800 rpm and (preferably) 4 °C.

7. Aspirate medium and resuspend the cell pellet in hES-cell medium. To preserve the cells in small clumps we recommend pipetting the pellet *gently* using either a 5 ml transfer pipette or a 1000 μl micropipette.

8. Seed hES-cell clumps onto fresh feeder layers at a split ratio of 1:3.[d]

9. Incubate cultures, and change medium daily.

Preparation of MEF-conditioned medium: We recommended that feeder layers be used in the preparation of conditioned medium for up to 1 week only after mitotic inactivation.

1. Seed $5.5-6 \times 10^5$ mitotically inactivated MEF in 2 ml MEF medium per well of 6-well dishes, and incubate overnight.[e]

2. Aspirate the MEF medium and replace with 2 ml hES-cell medium per well, and incubate for 24 h.

3. Collect the MEF-conditioned medium, and store for up to 2 weeks at 4 °C or 1 month at −20 °C.

4. Supplement MEF-conditioned medium with 4 ng/ml bFGF immediately prior to use in hES-cell culture.

Matrigel™-coating of wells:

1. Prepare a 1:30 dilution of Matrigel™ in KnockOut™ D-MEM.[f]

2. Dispense 2 ml diluted Matrigel™ solution per well of a 6-well culture dish, and incubate for 2 h at r.t.

3. Aspirate residual Matrigel™ solution prior to cell seeding.

Feeder-free passaging of hES cells:[g]

1. Seed hES-cell clumps onto Matrigel™-coated plates at a split ratio of 1:3, in MEF-conditioned medium.

2. Change conditioned medium daily.

3. Using a phase-contrast microscope, locate and then mechanically remove spontaneously differentiated areas from the cultures using the tip of a hypodermic needle or similar implement, as described in Chapter 10, *Protocol 4*. Repeat this procedure over several passages, until a stable culture is obtained.

[a] The ICR strain of mouse (Harlan) is recommended for MEF derivation.
[b] Sterile gelatin solution may be stored at r.t.

> **Protocol 1** continued
>
> ^c Longer periods of collagenase treatment (up to 2 h) will allow the effective separation of hES cells from the feeder layer with gentle pipetting using a 5 ml transfer pipette, while the feeder cells remain attached to the substratum.
>
> ^d If spontaneously differentiating cells are apparent in the hES cell culture, the need arises for their mechanical removal prior to passaging (see above). In this case, adjust the split ratio to compensate for the reduced net number of undifferentiated hES cells.
>
> ^e Gelatinization of wells is not required here.
>
> ^f Use of freshly diluted Matrigel™ solution is recommended, but the working solution can be stored at 2–8 °C for a week.
>
> ^g Owing to the higher incidence of spontaneous differentiation in hES cells cultured under feeder-free conditions, mechanical removal of differentiated derivatives may be required prior to passaging.

- The derivation and culture of hES cells has also been achieved with minimal exposure to animal-derived material, using serum replacement (SR) in the medium and human foreskin fibroblasts as feeder cells (42).

It is anticipated that such culture systems may facilitate, and provide a reliable foundation for, future clinical applications of hES cells. Our *Protocol 1* for the maintenance of undifferentiated hES cells includes modifications to provide feeder-free conditions.

2.1 Spontaneous differentiation of hES cells

Human ES cells continue to proliferate in the undifferentiated state when carefully maintained on a suitable feeder layer or other supportive matrix, in medium supplemented with appropriate cytokines or growth factors. However, removal from culture conditions supporting pluripotency and self-renewal causes their rapid differentiation. Moreover, these cells readily and spontaneously differentiate when culture conditions are suboptimal; most especially when permitted to reach over-confluence, whether cultured on feeder layers or feeder-free, using methods otherwise supporting the undifferentiated state. With continued culture, spontaneously differentiating cells subsequently produce recognizable morphological structures such as epithelial sheets or neuronal networks, whilst fibroblast-like cells frequently become apparent; and derivatives of the extraembryonic tissues may be detected (1).

It is remarkable that established lines of hES cells retain the capacity to differentiate into cells of the extraembryonic lineages spontaneously *in vitro*, and this inherent property is being exploited for the study of trophoblast differentiation and placental morphogenesis. It was shown that addition of bone morphogenetic protein 4 (BMP-4) to feeder-free cultures of hES cells induces their differentiation into trophoblast; and furthermore, when plated at low density the BMP-4-treated cells form syncytia expressing chorionic gonadotrophin, which hormone normally is secreted by trophoblastic cells (28). Thus, the study of cultured hES cells may elucidate some of the earliest differentiation events of human pre- and peri-implantational development (29).

In vivo differentiation of hES cells

For mES cell lines, pluripotency is validated *in utero*, in embryo reconstitution experiments (see Chapters 2 and 3). As such manipulations are prohibited for hES cell lines, their pluripotency can only be proved legitimately using model systems; and until sufficient guided *in vitro* differentiation techniques have been formulated, the raising of teratomas in immunosuppressed mice will remain the method of choice for producing from hES cells a full range of differentiated cell phenotypes (*Protocol 2*). And so the detection of derivatives of the three primary germ layers in teratomas is prerequisite to ascribing the characteristic of pluripotency to any novel hES cell line. *Figure 1* illustrates the range of cell types and complex structures that can be found in mature teratomas.

An added advantage of this experimental, *in vivo* model lies in its possible relevance to the study of human neoplasia; for monitoring tumour growth and invasion, as well as for analysing the angiogenic response developing within the teratoma (43). However, a recent study showed that during teratoma formation, spontaneous vasculogenesis from hES cells occurs but reaches an immature

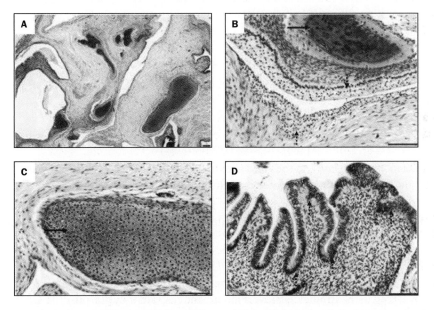

Figure 1 (see Plate 10) Teratoma formation by the hES cell line, H13, in immunocompromised mice. (A–D) Sections of a teratoma 10 weeks after xenografting, stained with haematoxylin and eosin. (A) Low-magnification view of the teratoma. (B–D) Higher magnifications showing tissues derived from all three germ layers: (B) bone formation (mesodermal in origin; solid arrow) near stratified squamous epithelium (ectodermal; dashed arrows); (C) hyaline cartilage development (mesodermal; solid arrow); and (D) columnar heterogeneous epithelium (endodermal; dashed arrows). Scale bars = 100 μm.

Protocol 2
Teratoma formation by xenotransplantation of hES cells into immunocompromised mice

Experimental teratomas raised according to this protocol may be harvested and subjected to RNA or protein extraction, or fixed for histology sectioning using routine procedures. For further details, the reader is referred to Gerecht-Nir et al. (44). All mouse handling and experimental procedures must be performed with the appropriate licenses, and using humane practices.

Reagents and equipment

- 4–8 week-old, male, beige-scid mice (C.B-17/IcrHsd-$Prkcd^{scid}Lyst^{bg}$, Harlan): for the routine, timecourse examination of the pluripotency of an hES cell line, three mice are sufficient for each time point
- Culture of undifferentiated hES cells on feeder layers: approximately $0.5-1 \times 10^7$ hES cells are needed per mouse
- Collagenase solution (Protocol 1)
- PBS, Ca^{2+}- and Mg^{2+}-free (Cat. No. 14190-094, Invitrogen) or saline solution (Cat. No. AWF 7124, Teva Medical) or serum-free culture medium (Protocol 1)
- Cell scraper
- Sterile, 5 ml transfer pipette
- Sterile, 15 ml conical tubes
- Sterile, 1.5 ml Eppendorf tubes
- Benchtop refrigerated centrifuge
- Haemocytometer
- Laminar flow cabinet
- 18–21 Gauge hypodermic needles
- Sterile dissection kit

Methods

Preparation of hES cell suspension:

1. Digest hES-cell colonies with collagenase solution, harvest by gently scraping, and pellet by centrifugation as described in *Protocol 1*, Steps 4–6.
2. Aspirate medium, resuspend the cell pellet in PBS, saline, or serum-free medium, and count the hES cells using a haemocytometer.[a]
3. Adjust the suspension to a density of $2-5 \times 10^7$ hES cells/ml, and dispense aliquots of $0.5-1 \times 10^7$ hES cells in 200–250 µl PBS, saline, or serum-free medium in Eppendorf tubes. Store cells on ice for up to 30 min.

Injection of hES cell suspension into immunocompromised mice:

4. Prepare the mice for injection: transfer them in their cage into a laminar flow cabinet. Working in the cabinet, remove a mouse and restrain it by grasping the scruff of the neck and the tail, using one hand.
5. Using an hypodermic needle, inject into the hind-limb muscle of the mouse 200–250 µl of hES cell suspension – inject *slowly*, to reduce the dissipation of cells into the surrounding tissue.[b]

> **Protocol 2** continued
>
> 6 When all the mice have been injected, return them to their housing and monitor their health regularly.
>
> 7 After ~4 weeks, teratomas become palpable at the site of injection and are normally ~0.5 cm in diameter at this stage. To obtain a well differentiated teratoma, it is recommended to allow at least 7-8 weeks of growth.
>
> 8 For harvesting, sacrifice the mouse humanely and recover the teratoma by dissection and separation from the surrounding tissues.
>
> [a] When counting hES cells in a suspension containing feeder cells using a haemocytometer, hES cells are readily distinguished as being rounder and smaller than feeder cells. And being mitotically inactivated, the presence of a small proportion of feeder cells in the inoculum is inconsequential.
>
> [b] Injection into one leg only of each mouse is advisable, to minimize discomfort to the animal.

state, without expression of mature endothelial markers and other indicators of the sprouting–remodelling processes; and so the experimental teratoma may have only limited value as a model system for vasculogenesis/angiogenesis (44). In this chapter we provide more direct and effective *in vitro* systems for examining early events in human vascular development and neovasculogenesis using hES cells (*Protocols 4, 7,* and *8*).

4 *In vitro* differentiation of hES cells

4.1 Differentiation via the formation of embryoid bodies

Human ES cells, like their mouse and non-human primate counterparts (see Chapters 5, 6, 8, 9, and 12), can be induced to enter a program of *in vitro* differentiation through the primary germ layers via the formation of embryoid bodies (EBs) (45). The basic principles for EB production using mES cells may be applied successfully also to hES cells. For example aggregation may be promoted by seeding hES cells into bacteriological-grade Petri dishes, to which these cells are non-adherent; or the hanging-drop technique may be used, to confer relative control over the number of hES cells per forming aggregate and hence the eventual size and uniformity of EBs produced. By analogy with the mouse system (see Chapter 5) it was anticipated that the use of serum-containing medium would promote mesoderm differentiation and cyst formation, whilst the use of KnockOut™ SR or other serum replacement inhibits those processes, stimulating instead neuronal differentiation (6; and personal observations); and our protocols for inducing mesodermal and haematopoietic differentiation of hES cells consequently utilize fetal calf serum (FCS or FBS), amongst other factors. As yet, techniques for EB differentiation are less well developed for hES cells than for mES cells, in terms of the range of differentiated derivatives obtained.

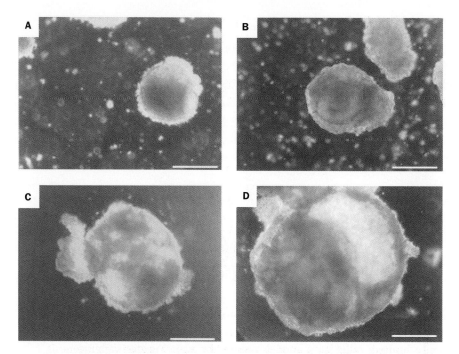

Figure 2 Time-course of development of EBs from hES cells in suspension culture. Photomicrographs of EBs derived using the I9 line of hES cells, demonstrating their gradual growth and morphological development starting with (A) complete aggregation of cells to form simple EBs by day 2, followed by (B) the appearance of distinct layers within EBs on day 3, (C) clear cavitation of EBs on days 7–10, and finally (D) development of fully cystic EBs after 2 weeks of differentiation. Scale bars = 100 μm

The differentiation of hES cells in EBs has been shown to reproduce aspects of early embryogenesis, and occurs in sequential stages (45, 46). *Figure 2* shows a time-course of EB development from seeding a culture of hES cells onto a non-adhesive substratum, where the day of seeding is 'day 0': by day 2 the hES cells have coalesced, resulting in the formation of simple EBs; this is followed by cavitation of some EBs as early as day 3 (and in all EBs between days 7 and 10); and development and growth of cystic EBs between days 7 and 14. Subsequently a more complex, three-dimensional cellular 'architecture' is elaborated within the EBs, where cell–cell and cell–ECM interactions are considered to influence the observed differentiation of the three embryonic germ layers and their derivatives (*Figure 3*). Thus, the hES cell/EB system offers opportunities for the *in vitro* investigation of cellular interactions normally occurring during early embryonic development, and of mechanisms of lineage determination. This potential is illustrated by a large-scale, complementary DNA (cDNA) analysis that was performed on hES cell-derived EBs at different stages of differentiation using DNA microarrays, which revealed sets of temporally expressed genes that could be related to the sequential stages of human embryonic development (47). Using the same technique, it was further demonstrated that many genes known

DIFFERENTIATION OF hES CELLS

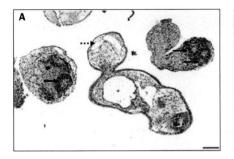
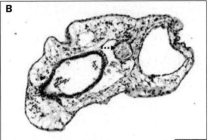
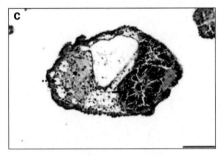

Figure 3 (see Plate 11) Ultrastructure of EBs derived from hES cells. (A–C) Haematoxilin- and eosin-stained sections of EBs cultured in suspension for 1 month from aggregation, demonstrating cells of a variety of morphological types. (A) A group of EBs at lower magnification; (B) and (C), individual EBs at higher magnification. Different cell types with characteristic organization can be observed within the EBs, including: neuronal rosettes (solid arrows), vascular structures (i.e. blood vessels; dashed arrows), an epithelial tube (arrowhead), and connective tissue (asterisks). Scale bars = 100 μm.

to be involved in human vascular development are activated also during the differentiation of EBs (48). Consequently it is anticipated that EBs derived from hES cells may be utilized for determining the cellular basis of certain embryonic defects, and for interpreting knockout phenotypes; and that they may be especially appropriate for the analysis of mutations where comparative studies in the mouse are complicated by early embryonic lethality.

4.2 Vascular morphogenesis within EBs

When hES cell-derived EBs reach a certain state of maturation, they display distinct vascular structures that are formed by differentiated endothelial cells (ECs) and vascular smooth-muscle cells (v-SMC) (15, 16). A convenient way to visualize these three-dimensional structures is by confocal microscopy, which enables inspection of the EB *in toto* (*Figure* 4). When such vascular structures in EBs are stained by immunofluorescence (*Protocol* 4) for the EC markers, CD31 and CD34, and the v-SMC markers, SMA (alpha smooth-muscle actin) and SM-MHC (smooth-muscle myosin heavy chain), it is observed that most of the $CD31^+$ or $SM\text{-}MHC^+$ cells are contained within these three-dimensional structures. The $CD34^+$ population comprises cells that are elongated into an organized structure (*Figure 4A*) and cells that are heaped and rounded,

Protocol 3
Production of EBs from hES cells in suspension culture

EBs are produced by seeding hES cells in clumps onto bacteriological-grade Petri dishes, to provide a non-adhesive substratum, with concomitant culture in suspension. *Figure 3* shows the course of EB growth with time of incubation. EBs produced by this method may be processed for immunocytochemistry, for RNA or protein extraction, or may be fixed for histology sectioning using routine procedures: for detailed methods, the reader is referred to Gerecht-Nir *et al.* (48). Other processing methods based on mES cell-derived EBs (e.g. those provided in Chapters 5, 6, 8, and 9) are adaptable also to hES cell-derived EBs. For RT-PCR analysis of lineage-restricted gene expression, human-specific primers are listed in Table 1.

Reagents and equipment

- Culture of undifferentiated hES cells on feeder layers: approximately $5-6 \times 10^6$ cells are needed to seed each 60 mm bacteriological-grade Petri dish
- Collagenase solution (*Protocol 1*)
- EB medium: 80% KnockOut™ D-MEM (*Protocol 1*) supplemented with 20% 'defined fetal bovine serum' (dFBS; Cat No. SH30070.03, HyClone), 1 mM L-glutamine (Cat No. 25030081, Invitrogen), and 1% non-essential amino acids solution (Cat No. 1114035, Invitrogen)
- 60 mm or 100 mm bacteriological-grade Petri dishes (Cat. No. 628102 and 663102, respectively, Greiner Bio-one)
- Cell scraper
- Sterile, 5 ml transfer pipette or 1000 μl micropipette
- Sterile, 15 ml conical centrifuge tubes
- Benchtop refrigerated centrifuge
- Haemocytometer

Method

1 Digest hES cell colonies with collagenase solution, harvest by gently scraping, and pellet by centrifugation as described in *Protocol 1*, Steps 4–6.

2 Aspirate medium and resuspend the cell pellet in EB medium. To preserve small clumps of hES cells, it is recommended to resuspend the cell pellet gently using either a transfer or a micropipette.

3 Adjust the cell suspension to $0.8-1 \times 10^6$ hES cells/ml in EB medium, and seed a total of 8 ml into a 60 mm Petri dish, or 10 ml into a 100 mm Petri dish, and incubate.

4 Change medium in EB cultures every 3–5 d, the frequency depending upon EB size and density, to prevent acidification as well as to replenish nutrients; this is achieved by gently collecting the EBs and transferring them in their conditioned medium to a (clear) centrifuge tube, and allowing them to settle under gravity (which takes 5–10 min). The conditioned medium can then be safely aspirated and fresh medium gently added, taking care not to break up the EBs. Transfer the contents of the tube into a fresh Petri dish and incubate to the desired stage.

Protocol 4

Immunofluorescence analysis of vascular structures in EBs

EBs are plated and cultured on an adhesive substratum for outgrowth in two dimensions, to facilitate the visualization and analysis of the vascular plexus (see *Figure 4*).

Reagents and equipment

- Day 10–14 EBs, prepared as described in Protocol 3[a]
- EB medium (*Protocol 3*)
- PBS (*Protocol 1*)
- Gelatin solution (*Protocol 1*)
- Paraformaldehyde solution: 4% paraformaldehyde (Cat. No. P6148, Sigma) in PBS (see Chapter 9, *Protocol 6*)
- Blocking solution: dissolve 3% (w/v) bovine serum albumin (BSA; Cat. No. A9418, Sigma) in PBS
- Primary antibodies (*Table 2*)
- Fluorescent, Cy3-conjugated secondary antibody (Cat. No. C2181, Sigma), diluted 1:100 in background-reducing antibody diluent (Cat. No. S3022, Dako)
- TO-PRO solution: TO-PRO3-iodide (Cat. No. T3605, Molecular Probes) diluted 1:500 in PBS

- 6-well culture dish (*Protocol 1*)
- Sterile, 5 ml transfer pipettes
- 1000 µl and 200 µl micropipettes
- Glass microscope slides (e.g. SuperFrostPlus, 76 mm × 26 mm, from Menzel)
- Sterile, glass coverslips of No. 1 thickness (from Marienfeld): both round (13 mm diameter) and rectangular (24 mm × 50 mm) coverslips are required
- Silicon pen (Cat. No. S2002, Dako)
- Humidified chamber
- Fluorescence mounting medium (Cat. No. S3023, Dako)
- Nail polish, or other sealant
- Confocal laser scanning system, e.g. Nikon Eclipse E600 microscope and Bio-Rad Radiance 2000 scanning system

Method

1. Spread coverslips over the surface of wells of a 6-well culture dish, cover them with gelatin solution, and incubate the dish at 37 °C for at least 2 h. Aspirate residual gelatin solution.[b]

2. Plate EBs onto the gelatin-coated coverslips in culture dishes. All of the EBs produced from one 60 mm bacteriological-grade Petri dish may be plated onto a maximum of three wells, in 2 ml EB medium per well. Incubate cultures for 1–3 d, during which time EBs will adhere to the coverslips and form outgrowths.

3. Once adherent, wash EBs twice with PBS, and fix by incubation with 0.5 ml paraformaldehyde solution per well for 20–25 min at r.t.

4. Aspirate fixation solution, wash fixed cells gently with PBS twice, and incubate with 0.5 ml blocking solution for 15 min at r.t.

Protocol 4 continued

5   Transfer individual coverslips (cell-side up!) onto microscope slides, and delineate the circumferences of the coverslips on the slides using a silicon pen.[c]

6   Wash the samples gently with PBS, apply primary antibodies at the appropriate dilution, and incubate slides in a humidified chamber for 1 h at r.t.

7   Wash the samples gently with PBS twice, apply Cy3-conjugated secondary antibody, and incubate in a humidified chamber for 30 min at r.t.

8   Wash the samples gently with PBS twice, and once with TO-PRO solution.

9   Apply a small quantity of mounting medium to the samples and cover with a rectangular coverslip.

10   Allow the mounting medium to air-dry, and then seal the edges of the rectangular coverslips onto the slides with nail polish.

11   Using confocal microscopy, locate and examine the depth of organization of vessel-like structures by scanning different layers within the EBs.

[a] This protocol was developed using the following hES cell lines: H9, H9.2, H13, I6, and I9.

[b] Alternatively, for immunofluorescence analysis EBs may be plated directly onto, and processed in, a 4-well chamber slide system (Cat. No. 177399, Nunc). These slides are similarly precoated with gelatin.

[c] This will allow the application of a minimal volume of antibody solution.

representing endothelial and haematopoietic derivatives, respectively. The SMA^+ population includes epithelial cells that are organized as sheets in different locations within the EBs, flattened cells with particularly intensive SMA expression, and more elongated cells forming wide tubes (*Figure 4B*). Together, these tubular channels and cords carry the hallmarks of a primitive vascular network.

4.3 Cardiac differentiation in EBs

When EBs are cultured in suspension to the stage when they become cystic (at day 8–14), a minority of them display rhythmic pulsing due to the spontaneous contractile activity of cardiac-muscle cells lining their central cavities (9, 45). These contractile areas within EBs display properties consistent with an early-stage cardiac phenotype, including expression of structural proteins such as actin, as well as proteins of gap junctions such as Cx40 and Cx43 (9, 10). The plating of contractile EBs onto an adhesive substratum to induce attachment and spreading facilitates observation of such areas, and enables their functional characterization. In particular, the culture of contractile EBs on microelectrode array systems (MEA; *Protocol 5*) proves both convenient and highly informative (see also Chapter 6, Section 6.3). These systems comprise 60 microelectrodes with diameters of 10–30 μm, and were developed originally to: (a) record spatio-temporal parameters of electrophysiological activity propagating through

Table 1 Human-specific primers for RT-PCR analysis of lineage-restricted gene expression (highlighting vascular genes), and PCR conditions

Cell type	Gene transcript	Primer sequences (5' to 3', F_w/R_v)	Product (bp)	Cycles	Annealing temp. (°C)	[MgCl$_2$] (mM)
Neuron	Neurofilament 68 kD	GAGTGAAATGGCACGATACCTA TTTCCTCCTCTTCACCTTC	473	35	60	1.5
Neuron	Dopamine β Hydroxylase (DβH)	CACGTACTGGTGCTACATTAAGGAGC AATGGCCATCACTGGCGTGTACACC	440	35	55	1.5
Keratynocyte	CK10	GCTGACCTGGAGATGCAAATTGAGAGCC GGGCAGCAGCATTCATTTCCACATTCACATCAC	129	35	65	1.5
Cardiac muscle	α-cardiac actin	GGAGTTATGGTGGGTATGGGTC AGTGGTGACAAAGGAGTAGCCA	468	35	65	2.0
Cartilage	CMP	ATGACTGTGAGCAGGTGTGC GTCCAGCGTATCCACGATCT	224	35	55	1.5
Hepatocyte	Albumin	TGCTTGAATGTGCTGATGACAGGG AAGGCAAGTCAGCAGCCATCTCAT	157	35	60	1.5
Hepatocyte	α-fetoprotein	GCTGATTGTCTGCAGGATGGGAA TCCCCTGAAGAAAATTGGTTAAAAT	216	32	60	1.5
Early endoderm/ Cardic muscle	GATA 4	AGACATCGCACTGACTGAGAAC GACGGGTCACTATCTGTGCAAC	475	35	60	1.0
Kidney	Renin	GTGTCTGTGGGGTCATCC ATCAAACAGCCTCTTCTTGGC	142	35	65	1.5
Endothelial	CD31	CAACGAGAAAATGTCAGA GGAGCCTTCCGTTCTAGAGT	260	32	60	1.5
Endothelial	CD34	TGAAGCCTAGCCTGTCACCT CGCACAGCTGGAGGTCTTAT	200	32	60	1.5
Haemato-vascular	Tal-1	ATGGTGCAGCTGAGTCCTCC TCTCATTCTTGCTGAGCTTC	331	40	55	1.5
Vascular	Ang1	GGGGGAGGTTGGACTGTAAT AGGGCACATTTGCACATACA	362	35	60	1.5
Vascular	Ang2	GGATCTGGGGAGAGGAAC CTCTGCACCGAGTCATCGTA	535	35	60	1.5

Table 1 (Continued)

Cell type	Gene transcript	Primer sequences (5′ to 3′, F_w/R_v)	Product (bp)	Conditions Cycles	Annealing temp. (°C)	[MgCl$_2$] (mM)
Endothelial	Tie2	ATCCCATTTGCAAAGCTTCTGGCTGGC TGTGAAGCGTCTCACAGGTCCAGGATG	512	35	60	1.5
Endothelial	VE-cad	ACGGGATGACCAAGTACAGC ACACACTTTGGGCTGGTAGG	596	35	60	1.5
Endothelial	KDR	CTGGCATGGTCTTCTGTGAAGCA AATACCAGTGGATGTGATGGCGG	790	35	60	1.5
Endothelial	Von Willebrand factor (vWF)	ATGTTGTGGGAGATGTTTGC GCAGATAAAGAGCTCAGCCTT	656	40	55	1.0
Vascular	[a]VEGF outer	GGGCAGAATCATCACGA CCGCCTCGGCTTGTCACA	534 483	35	55	1.5
	[a]VEGF inner	ATCGAGACCCTGGTGGACA CCGCCTCGGCTTGTCACA	411 351 279			
Endothelial	AC133	CAGTCTGACCAGCGTGAAAA GGCCATCCAAATCTGTCCTA	200	32	60	1.5
Endothelial	VCAM	GAAGCCAGCTTCCACATAAC AGTGGTGGCCTCGTGAATGG	700	35	60	1.5
Smooth muscle	SMA	CCAGCTATGTGAAGAAGAAGAGG GTGATCTCCTTCTGCATTCGGT	965	35	60	1.5
Smooth muscle	Calponin	GAGTGTGCAGAACGGAACTTCAGCC GTCTGTGCCCAACTTGGGGTC	671	35	60	1.0
Smooth muscle	Caldesmon	AACAACCTGAAAGCCAGGAGG GCTGCTTGTTACGTTTCTGC	530	35	60	1.5
Smooth muscle	SM-MHC	AAGCCAAGAGCTTGGAAGC TCCTCCTCAGAACCATCTGC	179	35	62	1.0
Household gene	GAPDH	AGCCACATCGCTCAGACACC GTACTCAGCGCCAGCATCG	302	27	60	1.5

[a] In order to detect the expression of different isoforms of VEGF, nested PCR should be performed, i.e. initial PCR amplification using 'outer-set' primers followed by additional PCR amplification of the product using 'inner-set' primers.

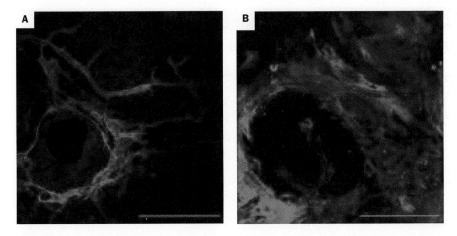

Figure 4 (see Plate 12) Vascular development within EBs. Immunofluorescence staining and confocal analysis of day 10–13 EBs for CD34 and SM-MHC reveal: (A) distinct areas with complex vascular structure composed of endothelial (CD34$^+$) cells; and (B) smooth-muscle (SM-MHC$^+$) cells forming wide tubes, rarely with network formation. Scale bars = 50 μm.

Table 2 Human-specific monoclonal antibodies for immunostaining of vascular and cardiac-muscle cells derived from differentiated EBs

Mouse anti-human antibodies	Supplier	Cat. No.	Dilution
Endothelial and smooth muscle markers:			
CD34	Dako	M7165	1:20
CD31	Dako	M0823	1:20
vWF	Dako	M0616	1:50
Smooth-muscle myosin heavy chain (SM-MHC)	Dako	M3558	1:20
Smooth-muscle alpha actin (SMA)	Dako	M0851	1:100
Calponin	Dako	M3556	1:50
Cardiac markers:			
Sarcomeric α-actinin	Sigma	A5044	1:500
Troponin I	Chemicon	MAB3438	1:5000
Myosin ventricular heavy chain alpha/beta	Chemicon	MAB1552	1:10
Atrial natriuretic peptide (ANP)	Chemicon	CBL66	1:100
Isotype control: IgG$_1$	R&D	MAB002	1:20

nervous or heart-muscle tissue, and (b) stimulate tissues electrically. Contractile hES cell-derivatives may be dissected from EBs and cultured on MEAs for days whilst their activity is recorded non-invasively. High-resolution activation maps are thereby generated that may demonstrate the presence of a functional syncytium with stable, focal activation and conduction properties (10). In addition, after plating intact, contractile EBs onto MEAs, the effects of cardiotropic drugs or other treatments may be evaluated (9, 10).

Protocol 5
Preparation of contractile EBs for analysis

Reagents and equipment

- Day 8–10 EBs, prepared as described in Protocol 3
- EB medium (Protocol 3)
- Anti-human monoclonal antibodies (Table 2)
- 18–21 Gauge hypodermic needle
- 24-well dishes (Cat. No. 142475, Nunc), precoated with gelatin (Protocol 1)
- Inverted phase-contrast microscope
- Multiarray electrode systems (MEAs; Multi Channel Systems MCS GmbH)

Method

1. Plate one to three EBs per well of a gelatinized 24-well dish, in 0.5 ml EB medium per well, and incubate to allow EB attachment and outgrowth, changing medium every second day.

2. Conduct daily microscopic observation to detect EBs with spontaneously contracting areas, and determine the beating rate.

3. Mechanically dissect contractile areas of cells using a hypodermic needle. For analysis, pulsating cells may be: (a) transferred onto MEAs precoated with gelatin (as detailed for dishes in Protocol 1) and cultured, and electrophysiological measurements recorded according to the manufacturer's instructions; (b) transferred onto gelatin-coated coverslips and cultured, followed by immunofluorescence analysis (Protocol 4) using cardiac-specific antibodies (Table 2); or (c) collected and processed for RNA or protein extraction using routine procedures.[a]

[a] For RNA extraction and RT-PCR assays, the reader is referred to Chapter 6, Sections 5.1–5.2 and Protocols 8 and 9. For protein extraction and Western blot analysis, see Gerecht-Nir et al. (48).

5 Guided differentiation

With the newly identified, potential therapeutic uses of hES-cell derivatives in medicine, much attention currently is being focused on methods for achieving guided differentiation to generate highly purified populations of the desired cell types. The success of these particular differentiation systems ultimately will rely on a high level of understanding of the wide range of biochemical and physical cues that are involved during lineage specification in human embryogenesis. For the interim, methods are available to guide differentiation towards preferred pathways.

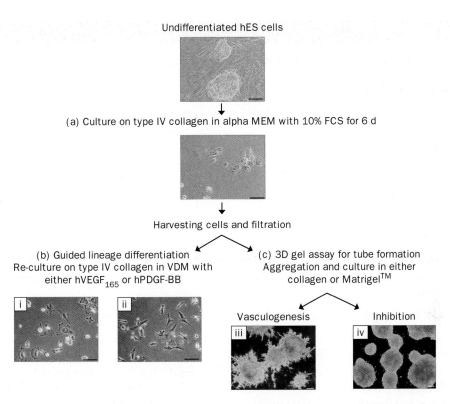

Figure 5 Experimental strategy for guided, vascular differentiation of hES cells. By this two-step differentiation strategy (*Protocol 7*), a culture of undifferentiated hES cells is first induced to undergo mesodermal differentiation in monolayer culture, in step (a). In step (b), the mesodermal progenitor cells are guided towards maturation into v-SMCs and ECs by further incubation in VDM supplemented with hVEGF$_{165}$ (i) or hPDGF (ii), respectively. In step (c), mesodermal progenitor cells are assayed for vasculogenesis and angiogenesis by aggregation and culture in collagen or Matrigel™ (*Protocol 8*): aggregates in Matrigel™ show vigorous sprouting of vascular structures (iii), which process is inhibited in the presence of anti-VE-cad monoclonal antibody (iv), which was previously suggested to be a potent inhibitor of angiogenesis (67, 68).

5.1 Haematopoietic differentiation

The induction of haematopoiesis in hES cells is a guided differentiation system currently under investigation (*Protocol 6*): here, single-cell suspensions of differentiated hES cells are cultured in either liquid or semisolid medium supplemented with specific cytokines, resulting in further differentiation into haematopoietic colony forming units (CFU) (17, 18).

5.2 Non-EB-mediated hES-cell differentiation: vascular development

For murine and primate ES cells, differentiation into certain cell types can now be achieved in monolayer culture–without recourse to EB formation–by

Protocol 6
Haematopoietic differentiation of hES cells

Reagents and equipment

- Day 12–15 EBs in suspension culture, produced as described in *Protocol 3*
- Trypsin/EDTA solution: 0.05% trypsin/ 0.53 mM EDTA (Cat. No. 15400054, Invitrogen)
- EB medium (*Protocol 3*)
- MethoCult™ GF+ medium (Cat. No. H4435, Stem Cell Technologies): this complete, methylcellulose medium is supplied supplemented with recombinant human cytokines for growth of CFU-GM, CFU-G, and CFU-M
- Water bath set at 37 °C
- Sterile, 15 ml conical centrifuge tube
- Sterile, 5 ml transfer pipette, cut to have a wide-bore tip if necessary
- Benchtop refrigerated centrifuge
- 1000 µl micropipette
- Haemocytometer
- 35 mm tissue-culture dishes (Cat. No. 627102, Greiner)

Method

1. Transfer EBs using a wide-bore pipette into a 15 ml conical tube and allow them to settle. Aspirate supernatant medium carefully (also using a wide-bore pipette), add 2 ml of trypsin/EDTA solution to the EBs, and incubate the tube in a water bath for 5–10 min at 37 °C.

2. Halt trypsinization by addition of 5 ml EB culture medium, and pellet cells by centrifugation for 5 min at 1200 rpm and 4 °C.

3. Resuspend the cells in MethoCult™ GF+ medium, count the cells and adjust the density to $1-2 \times 10^5$ cells/ml. Dispense 1 ml of this mixture per 35 mm dish, and incubate.

4. CFUs can be scored by their morphology ~10 d after plating, according to established criteria (see *Atlas of Human Hematopoietic Colonies*, Stem Cell Technologies Inc.).[a]

5. The differentiated cells can be assessed further (as described in refs 17–19) using FACS® analysis or immunofluorescence[b]

[a] Using this protocol and system of identification, human CFUs were generated from hES cells and identified as primitive erythroid (CFU-Ep), granulocyte–erythrocyte–macrophage–megakaryocyte (CFU-GEMM), granulocyte–macrophage (CFU-GM), macrophage (CFU-M), granulocyte (CFU-G), and erythroid burst-forming units (BFU-E) (17).

[b] In particular, cells cultured from day 15 EBs have been shown to express low levels of CD45 (1.4% ± 0.7%). Most of the CD45+ cells also coexpress CD34 (1.2% ± 0.5), which phenotype is similar to that of the first definitive haematopoietic cells to be detected within the wall of the dorsal aorta of human embryos (18).

providing a differentiation-inducing layer in combination with differentiation-guiding medium. Depending on the desired course of differentiation, the layer may consist of either a specific ECM (such as poly-L-lysine, fibronectin, or collagen) or feeder cells (such as stromal cell lines derived from the relevant tissue). Use of a mouse bone-marrow stromal cell line to induce haematopoietic differentiation in mouse and non-human primate ES cells is described in Chapters 7 and 12, respectively.

Here we describe our method for inducing vascular differentiation of hES cells in monolayer culture using such a conditioning environment. The overall strategy (depicted in *Figure 5*) involves a two-step process (*Protocol 7*): (a) induction of differentiation into mesodermal progenitor cells, followed by (b) subculture of those mesodermal cells in growth factor-supplemented medium, to further direct differentiation into either ECs or v-SMCs.

The preliminary differentiation step (a) involves culture of single-cell suspensions in the absence of feeder cells, on collagen Type IV-coated dishes, and produces mainly two morphologically distinct cell types of mesodermal origin: (i) smaller cells with high nuclear/cytoplasmic ratios, and (ii) larger, flattened cells with an intracytoplasmic fibre arrangement (*Figure 6A*). Most of the type (i) cells are proliferating (*Figure 6B*) and express markers of endothelial progenitor cells (16); and it is these cells that give rise to vascular structures *in vitro* (see below). The larger cells express smooth muscle actin (SMA) as well as other SMC markers, and further studies are required to determine their exact nature and origin.

For the second differentiation step (b), mesodermal progenitors are induced further to differentiate and mature into either ECs or v-SMCs by their consecutive reculture on collagen Type IV-coated dishes, in medium supplemented with either human vascular endothelial growth factor (hVEGF) or human platelet-derived growth factor BB (hPDGF-BB), respectively (16). Under these conditions,

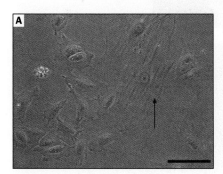 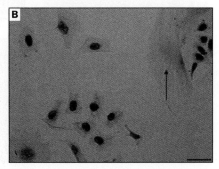

Figure 6 Guided, vascular differentiation of hES cells: step (a), mesoderm differentiation in monolayer culture. Phase-contrast micrographs of hES cells cultured for 6 d on type IV collagen-coated dishes for mesodermal differentiation, showing: (A) two derivative cell types comprising large, flattened cells with intracellular fibers (arrow) and smaller cells with higher nuclear/cytoplasmic ratios; and (B) the proliferative ability of the smaller cells, most of which become labelled with BrdU (dark nuclear stain), while the larger cells (arrow) fail to incorporate. Scale bars = 100 μm. (Adapted from Ref. 16.)

vascular plexuses with ring- or cord-like arrangements can be observed forming from the maturing cells (16). Supplementation of medium with hVEGF results specifically in extensive uptake of Dil-acetylated low-density lipoprotein, with approximately 20% of cells producing von Willebrand factor and expressing CD31, denoting ECs (*Figure 7A*). Supplementation with hPDGF-BB promotes differentiation into v-SMC, with approximately 80% of cells expressing the v-SMC markers, SMA, calponin, and SM-MHC (16; *Figure 7B*).

Alternatively, mesodermal progenitor cells may be subjected to a three-dimensional gel assay, in step (c), with either collagen Type I or Matrigel™ (*Figure 8A*), to determine their capacity for *in vitro* vasculogenesis and the sprouting angiogenic process (*Protocol 8*). (These two ECMs are known to promote the formation of three-dimensional vessel-like structures from ECs *in vitro* (15, 16).) Transverse sections of the mesodermal cultures in Matrigel™ reveal penetration of the cells into the gel and formation of tube-like structures (*Figure 8B*). Ultrastructural observation by transmission electron microscopy also reveals typical, tube-like arrangements of elongated vascular ECs within the Matrigel™, as well as intracellular Weibel–Palade bodies, lipoprotein capsules, and glycogen deposits in the developing ECs (7; *Figure 8C*).

Taking into consideration the limited means available for studying vascular development (comprising vasculogensis and angiogenesis) during human embryogensis *in vivo*, this sequential system for vascular-guided hES-cell differentiation offers a useful model in which to examine the mechanisms of vascular lineage

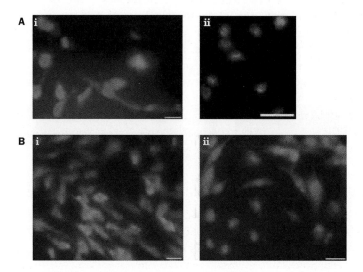

Figure 7 (see Plate 13) Guided, vascular differentiation of hES cells: step (b), maturation of mesodermal progenitor cells as analysed by immunofluorescence. (Panel A) Immunofluorescence anlayis of mesodermal cells cultured in the presence of hVEGF$_{165}$ reveals expression of EC markers such as: (i) CD31 (red stain; nuclei counterstained in blue); (ii) Dil-Ac-LDL (red) and von Willebrand Factor (green). (Panel B) Similar analysis of cells cultured in the presence of hPDGF-BB reveals expression of v-SMCs markers such as: (i) SMA (red; nuclei in blue); (ii) calponin (red; nuclei in blue). Scale bars = 100 μm. (Adapted from ref. 16)

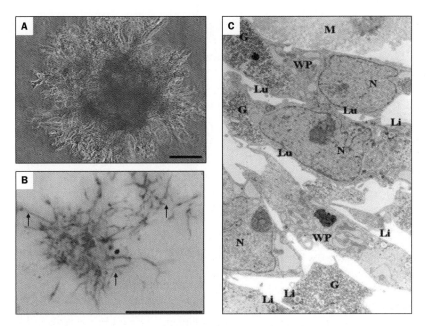

Figure 8 Directed, vascular differentiation of hES cells: step (c), vasculogenesis and angiogenesis in a three-dimensional gel composed of Matrigel™. (A) Inverted phase-contrast microscopy demonstrates extensive vasculogenesis with sprouting (angiogenesis) by a mesodermal aggregate on Matrigel™ in the presence of hVEGF. (B) Sections of vascular structures stained with toluidine blue reveal cord-like structures (*arrows*) penetrating the gel. (C) Transmission electron micrograph showing the arrangement of developing ECs (N = nucleus) within the Matrigel™ matrix (M): here the cells contain lipoprotein capsules (Li) at the lumen surfaces (Lu), and intracellular Weibel–Palade bodies (WP) and deposits of glycogen (G), the endogenous energy source for this cell type. Scale bars = 100 μm.

commitment and early angiogenesis. Furthermore, as considerable interest is currently centred on the inhibition of new vascular growth to treat the spread of cancer, hES cells differentiated in this manner may readily be employed for an *in vitro* evaluation of antiangiogenesis agents.

6 Scale-up procedures

While hES cells and their differentiated derivatives hold tremendous potential for many experimental and therapeutic applications, their eventual utility is dependent on availability in sufficient quantities. This in turn requires an increase in the scale of production of undifferentiated hES cells and/or the target cells derived from them. It then becomes necessary to consider, in addition to the protein microenvironment (in terms of growth factors and cytokines), the broad physicochemical parameters (such as pH, oxygen tension, and glucose availability) that have a significant effect on the overall performance of stem cell cultures. Consequently, to facilitate the future applications of hES cells, the development of a controlled and scaleable bioprocess is essential.

Protocol 7
Vascular differentiation of hES cells in monolayer culture

Reagents and equipment

- Culture of undifferentiated hES cells: $3-4 \times 10^6$ cells are needed to seed each 6-well culture dish
- EDTA solution: 5 mM EDTA in PBS (Cat. No. V4231, Promega), supplemented with 1% (v/v) dFBS (Protocol 3)
- Human ES-cell medium (Protocol 1)
- Basal vascular-differentiation medium (VDM): alpha Minimum Essential Medium (alpha MEM; Cat. No. 12571–063, Invitrogen) supplemented with 10% dFBS (Protocol 3), and 0.1 mM β-mercaptoethanol (Protocol 1)
- Growth factor-supplemented VDM: VDM containing, for endothelial differentiation, 50 ng/ml $hVEGF_{165}$ (Cat. No. 293-VE, R&D Systems); or for v-SMC differentiation, 10 ng/ml hPDGF-BB (Cat. No. 220-BB, R&D Systems)
- Cell scraper
- Benchtop refrigerated centrifuge
- 1000 μl micropipette
- Cell strainer, with 40 μm mesh size (Cat. No. 352340, BD Biosciences)
- Haemocytometer
- Collagen Type IV-coated 6-well (Cat. No. 354428, BD Biosciences) and 24-well (Cat. No. 354430, BD Biosciences) culture dishes

Methods

(a) Induction of mesodermal differentiation:

1. Wash hES cell culture once with PBS, add 0.5 ml EDTA solution per well and incubate for 25–30 min at 37 °C, to dissociate the colonies. Add 1 ml hES-cell medium per well, and harvest the hES cells by scraping.
2. Collect and transfer the cell suspension into a conical tube, and centrifuge for 5 min at 1200 rpm and 4 °C.
3. Aspirate medium, and resuspend the cell pellet in VDM. Draw the cells repeatedly up and down a 1000 μl micropipette tip to obtain a single-cell suspension.
4. Filter the suspension through a cell strainer, and count the filtered single cells.[a]
5. Plate $0.5-0.7 \times 10^6$ hES cells per well in collagen Type IV-coated 6-well dishes.
6. Add VDM to a final volume of 2 ml per well and incubate dishes, changing medium daily. Well-differentiated populations of mesodermal cells are obtained by day 6, where the day of seeding is designated 'day 0'.

(b) Induction of maturation into ECs or v-SMCs:

7. At day 6 of mesodermal differentiation, aspirate medium and incubate cells with 0.5 ml EDTA solution per well for 25–30 min at 37 °C.

Protocol 7 continued

8 Add 1 ml hES-cell culture medium per well, and harvest the cells by scraping. Collect and transfer the cell suspension into a conical tube, and centrifuge for 5 min at 1200 rpm and 4 °C.

9 Aspirate medium and resuspend the cell pellet in VDM. Draw the cells repeatedly up and down using a 1000 µl micropipette tip to obtain a single-cell suspension.

10 Filter the cell suspension through a cell strainer and count the filtered single cells, as before.

11 Seed 5×10^4 cells per well in collagen Type IV-coated, 24-well dishes.

12 Add VDM supplemented with either hVEGF or PDGF-BB, to a final volume of 1 ml per well; and incubate dishes, changing medium daily. The presence of either ECs or v-SMC can be observed after 10–12 d of culture with the appropriate growth factor (i.e. between day $6+10$ and $6+12$ of differentiation).

[a] This filtration step is essential to obtain a single-cell suspension of cells that is free of clumps, lysed cells, and other debris.

Protocol 8

Three-dimensional gel assays for vasculogenesis by mesodermal hES-cell derivatives

Reagents and equipment

- Mesodermal progenitor cells in collagen Type IV-coated 6-well dishes, obtained by differentiation of hES cells as described in *Protocol 7*, steps 1–6, and at day 6 of differentiation
- EDTA solution (*Protocol 7*)
- Cell strainer (*Protocol 7*)
- VDM supplemented with hVEGF (*Protocol 7*)
- Uncoated, 35 mm bacteriological-grade Petri dishes (Cat. No. 351008, BD Biosciences)
- 2 × alpha MEM: reconstitute powdered alpha MEM (Cat. No. 11900-024, Invitrogen) according to the manufacturer's instructions but at double the prescribed concentration
- Collagen Type I solution: dissolve collagen Type I (Cat. No. 1179179, Roche) at 3 mg/ml in sterile 0.3% acetic acid, according to the manufacturer's instructions
- 24-well (Cat. No. 142475, Nunc) or 4-well (Cat. No. 176740, Nunc) culture dishes
- Matrigel™ solution, undiluted (*Protocol 1*)
- Sterile, 15 ml conical centrifuge tube
- Sterile, 5 ml transfer pipette, cut to have a wide-bore tip if necessary
- Haemocytometer

Protocol 8 continued

Method

1. Dissociate the monolayers of differentiated hES cells in wells using EDTA solution (as described in *Protocol 7*, Steps 7-9); filter the cell suspension, and count the filtered single cells (Step 10).

2. Resuspend the cells at $3\text{-}5 \times 10^5$ cells/ml in hVEGF-supplemented VDM seed onto 35 mm bacteriological-grade Petri dishes using 6 ml per dish, and incubate for 12-24 h to induce cell aggregation in suspension.

Collagen assay:[b]

3. Collect the cell aggregates from Step 2, centrifuge for 3 min at 1500 rpm and 4 °C, and resuspend aggregates from each 35 mm dish in 1 ml of a mixture of equal volumes of $2 \times$ alpha MEM and collagen Type I solution, precooled on ice.[a]

4. Dispense 250 μl of the cell suspension per well into either 24-well or 4-well dishes, and incubate for 15 min at 37 °C to allow gelling to occur.

5. Supplement the cultures with 500 μl hVEGF-supplemented VDM per well and incubate. Sprouting and network formation should appear ~72 h after seeding.

Matrigel™ assay:[b]

3. Thaw the undiluted Matrigel™ solution overnight at 4 °C, dispense 380 μl per well of either 24-well or 4-well dishes, and incubate at 37 °C for 30 min to allow gelling to occur.

4. Collect the cell aggregates from Step 2, centrifuge for 1 min at 800 rpm and 4 °C, and resuspend aggregates from each 35 mm dish in 2 ml hVEGF-supplemented VDM.[c]

5. Seed 250 μl cell suspension onto the Matrigel™ layer and incubate. Vascular sprouting and network formation should appear 24-48 h after seeding.

6. The extent of vascular network formation, and the properties of those networks, can be studied using various qualitative and quantitative approaches: microscopic measurements may be taken to monitor the kinetics of sprouting and branching, and total capillary tube length; and transmission electron microscopy and immunostaining may be used for morphological and cellular characterization.

[a] It is important to perform this procedure on ice, to preserve a liquid-gel status.

[b] For both assays, vascular initial sprouting and tube-like structures can be observed to form within 48 h of culture, using an inverted phase-contrast microscope (*Figure 8A*).

[c] The reduced time and speed of centrifugation are to preserve the integrity of the aggregates.

6.1 Dynamic systems for EB production

At the present time, the aggregation of multiple hES cells is an obligatory process to initiate efficiently EB formation. Established methods for producing EBs (45, 46) include culture of hES cells in hanging drops, suspension culture in liquid medium using non-adhesive substrata, and culture in semisolid medium containing methylcellulose (MC). These systems all involve 'static' culture in Petri dishes, are laborious and time-consuming, are not controllable, and therefore are impractical for industrial scale up. And so we have introduced 'dynamic' systems for the large-scale production of EBs, which are collectively termed 'dynamic' because of the motion applied to maintain and equilibrate the internal environment. Our prototype systems are described below.

Any dynamic system for EB production must strike a balance between allowing ES cells to aggregate, which is necessary for EB formation, and causing EBs to adhere and to agglomerate into large masses, which is detrimental to their uniform growth and differentiation (49). Our initial attempts at producing differentiating EBs from mouse and human ES cells cultured in suspension in spinner flasks resulted in the formation of large clumps of cells within a few days, which is indicative of a significant tendency of both mES cells (50) and hES cells (personal observations) to aggregate under these conditions. However, the high rate of stirring that was required to avoid EB agglomeration was accompanied by massive hydrodynamic damage to the cells from turbulence and collision between the fragile, incipient EBs and the apparatus. Therefore, to circumvent these problems we employed a static system for an initial aggregation period of 4 d, followed by a period in dynamic culture in spinner flasks, to successfully achieve the bulk production of cardiomyocytes from differentiating mES cells (51).

Another, more technically advanced, dynamic approach that is highly effective for hES cells is to generate and culture EBs within rotating cell-culture systems (RCCS^{TM}; manufactured by SYNTHECON^{TM} Inc.) originally developed by NASA for tissue culture in zero gravity (52; *Figure 9A*). These bioreactors provide exceptionally supportive environments for the three-dimensional organization and differentiation of human tissue (further details and information are available on the website, http:/www.synthecom.com). In the RCCS^{TM} the operating principles are:

(a) *Whole-body rotation around a horizontal axis, with cells or aggregates in permanent free-fall*. The resulting flow pattern in the RCCS^{TM} is laminar with mild fluid mixing, as the vessel rotation is slow. The settling of cell clusters, which is associated with oscillation and tumbling, generates the fluid mixing. The net outcome is a very low-shear, dynamic environment allowing efficient mass transfer, and from which turbulence and the collision of cells are absent.

(b) *Oxygenation by active or passive diffusion across a membrane to the exclusion of all but dissolved gasses from the reactor chamber*. Crucially, the vessel is devoid of gas bubbles and gas/fluid interfaces, which are otherwise deleterious. An added advantage of the RCCS^{TM} is that the vessels are geometrically designed so that the membrane-area to volume-of-medium ratio is high, thus facilitating efficient gas exchange.

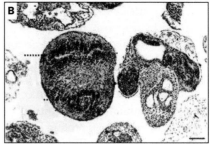
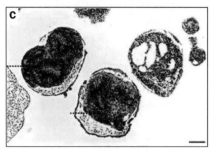
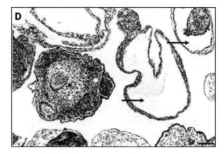

Figure 9 (see Plate 14) Mass production of EBs by a rotating cell-culture system. (A) Cross-sectional diagram of the rotating bioreactor (STLV^{TM}) used for the formation of small EBs from hES cells (*Biotechnology and Bioengineering* copyright (2004); copyright owner as specified in that Journal). As the vessel rotates slowly around its horizontal axis, hES cells aggregate in suspension with the minimum of shearing force. Note the absence of a medium-to-air interface found in conventional bioreactors. (B–D) Haematoxilin- and eosin-stained sections of EBs generated after 1 month in rotation culture, showing a variety of cell types including: epithelial neuronal tubes (dashed arrows), blood vessels (arrowheads), connective tissues, and cyst formation (solid arrows). Note that these dynamically produced EBs are more uniform in size and shape than those obtained in static suspension culture, as demonstrated in *Figure 3*. Scale bars = 100 μm.

With regard to EB formation, culture, and differentiation in particular, these bioreactors offer: an optimized environment for hES cell aggregation; minimal hydrodynamic damage to incipient EBs; reduced opportunity for EB fusion and agglomeration (the speed of rotation may be increased according to EB size, to this end); and the uniform growth and differentiation of EBs in three dimensions, as they oscillate and rotate evenly (see below). We recommend the Slow Turning Lateral Vessel^{TM} (STLV^{TM}) for this purpose.

Human ES cells cultured within the STLV^{TM} aggregate by 12 h, in contrast to 48 h for hanging-drop or suspension culture. The average diameter of dynamically formed EBs after 1 week of culture is about 20 μm, and from day 7 onwards the total number of EBs does not change significantly but their size increases with time, reaching 400–800 μm after 1 month of culture: this compares with 400–1500 μm for EBs in static culture. Furthermore, the use of a STLV^{TM} has a significant impact on the overall process of EB formation from hES cells: in addition to their smaller size, the STLV^{TM}-borne EBs exhibit evenly rounded shape and uniform size, with minimal agglomeration, and three-fold

Protocol 9
Dynamic formation of EBs from hES cells using the STLV™

All aspects of vessel preparation, conditioning, and maintenance are described in detail in the RCCS™ user manual, available on the website, http://www.SYNTHECON.COM. EBs produced in this way may be collected and processed for RNA or protein extraction, or fixed for histology sections using routine procedures (see *Protocol 3*). Analysis of EB-conditioned medium provides metabolic indices such as medium pH, and rates of glucose consumption and lactic acid production, to confirm efficient cell expansion and differentiation in these dynamic cultures (52).

Reagents and equipment

- A 55 ml volume, Slow Turning Lateral Vessel (STLV™; Synthecon™ Inc.)
- Culture of undifferentiated hES cells on feeder cells: 2-3.5×10^7 hES cells are needed to seed a 55 ml STLV™
- EDTA solution (*Protocol 7*)
- EB culture medium (*Protocol 3*)
- Cell scraper (*Protocol 1*)
- Sterile, 10 ml and 5 ml transfer pipettes
- Sterile, 15 ml conical tubes
- Benchtop refrigerated centrifuge
- 1000 µl micropipette
- Cell strainer (*Protocol 7*)

Method

1. Harvest hES colonies using EDTA solution, as described in *Protocol 7*, Steps 1-2. Resuspend cells in EB medium at a density of 3-7×10^5 hES cells/ml.
2. Transfer 55 ml cell suspension into the STLV™ through the 'fill port' and ensure that the vessel is devoid of gas bubbles, according the manufacturer's instructions.
3. Place the vessel on its rotator base inside an incubator, connect the base to an external power source, and set the rotation of the vessel at 15-20 rounds/min.[a]
4. Continue rotation undisturbed for 5 d. Every further 3-5 d, replace 70% of the conditioned medium with fresh medium, as follows.

Changing medium:

5. Stop the vessel's rotation, and place it upright in a laminar flow hood to allow the cell aggregates to settle.
6. Open the 'fill port' and remove 35 ml medium using a 10 ml transfer pipette. Be sure not to aspirate or disturb any small aggregates.
7. Add 30 ml fresh medium, close the port, complete the addition of a further 5 ml medium through a 'syringe port', and continue the incubation and rotation of the vessel.

> **Protocol 9 continued**
>
> *Medium analysis:*
>
> 8 Remove triplicate 1 ml samples from the conditioned medium, and immediately analyse for pO_2, pCO_2, pH, glucose, and lactic acid levels using a Blood Gas Analyzer (StatProfile M, Nova Biomedical); or store at -70 °C for subsequent analysis of lactate dehydrogenase (LDH) activity using a Chemistry Analyzer (Roche Hitachi Modular P800).
>
> [a] At this speed the suspended cell aggregates remain close to a stationary point within the bioreactor vessel.

enhancement in their yield compared to EBs produced by static methods. Dynamically formed EBs exhibit steady and progressive differentiation, with cyst formation and elaboration of complex structures such as neuroepithelial tubes, blood vessels, and glands (13; *Figure 9B–D*) as observed in statically formed EBs. Overall, we consider this to be the system of choice for controlling the aggregation and EB-mediated differentiation of hES cells, and for producing relatively large numbers of EBs in a reproducible manner, with minimal handling involved.

6.2 Production of hES cell-derived EBs within three-dimentional matrices

Another approach to guiding the differentiation of hES cells of relevance to scale-up procedures involves manipulating EBs by imposing physical constraints on their formation and growth; cells in EBs are responsive to physical, in addition to chemical, cues. Such confinement may be achieved by the encapsulation of hES cells in agarose hydrogel capsules (46), which permits culture at high cell density, enables EB formation, and allows differentiation into haematopoietic cell lineages. Recently, this system was successfully applied to enhance cardiomyocyte derivation from mES cells under hypoxic conditions (53).

Another confining environment is the three-dimensional porous scaffold; culture of hES cells within an alginate scaffold of this kind enables efficient formation of EBs with a relatively high degree of cell proliferation and differentiation (54). Under these conditions, EBs form mainly within the scaffold pores, and become distributed evenly over the entire scaffold volume. Most probably, the relatively small pore size together with the hydrophilicity of the alginate scaffold facilitate the generation and dispersal of EBs in this way. Scanning electron micrographs of the hES cell-seeded scaffolds recorded after 1 month of culture reveal that the forming EBs occupy the entire pore volume (*Figure 10A*), whilst some cells form a lining along the scaffold fibres (*Figure 10B*). Furthermore, this confining environment induces vasculogenesis in the differentiating EBs to a higher degree than observed in either static, suspension cultures, or in dynamic, rotating cultures (54).

Figure 10 EB formation within three-dimensional scaffolds. Scanning electron micrographs of LF120 alginate scaffolds seeded with hES cells and cultured for 1 month: (A) EBs (dashed arrows) develop mainly within pores confined by the scaffold fibres (solid arrows), whilst (B) other cells (dashed arrows) become disposed along the scaffold fibres (solid arrows).

Polymer scaffolds also have been shown to provide a conducive and supportive environment for the organization of early differentiating hES cells into tissue-like structures: cells dissociated from day 8 EBs, seeded on synthetic scaffolds and treated with appropriate cytokines subsequently undergo organization into complex three-dimensional structures resembling, and with features characteristic of, embryonic tissues (55, 56).

7 Epilogue: future clinical applications and limitations

There are a number of different potential sources of cells for tissue regeneration and repair. These include mature, somatic cells and adult stem cells, which may be obtained from the patient him/herself, as well as the embryonic cells represented by hES cells and fetal germ cells. As somatic cells are committed to a particular cell lineage and commonly have a low proliferative potential, the orthodox view is that their ability to repair tissue will be low. And as the plasticity of adult stem cells is currently under dispute, theoretically hES cells may serve best as a source of any cell type for transplantation. For example hES cell-derived mesovascular progenitor cells have the characteristic of a high proliferative activity (*Protocol 7* and *Figure 6*), which may render them a good potential source of cells for building functional blood vessels, *ex vivo* and *in vivo*, for the treatment of ischaemia and cardiovascular disease.

However, two main obstacles hinder the clinical development of hES cell-based replacement therapy at this time. The first is the difficulty associated with selection and expansion of pure populations of desired cell types. A widely adopted strategy is to genetically manipulate hES cells, to enable the production (by enhancement or selection) and propagation of specific cell populations. Different techniques for gene 'knock-in' and 'knock-out' in undifferentiated hES cells have been established, and include their transfection with plasmid-borne transgenes (57, 58), the generation of homologous recombination events (57), and the introduction of transgenes using self-inactivated lentiviruses (59, 60) or adenoviral and adenoassociated viral vectors (61). Another potentially important genetic manipulation, from the point of view of transplantation, is the incorporation into hES cells of 'suicide' genes to provide the facility of ablating the cells or their derivatives, should the need arise (62).

The second hindrance to hES cell-based clinical applications, and a major obstacle for successful organ transplantation, is immunological intolerance for allogenic cells. As long-term immunosuppressive therapy is generally detrimental, the creation of immunological tolerance would be greatly advantageous to stem cell-based transplantation therapy. To this end, potential strategies are currently under review and development and include: (i) the establishment of banks of hES cell lines, large enough to represent the majority of tissue types; (ii) the modification of the cells to carry patient-specific nuclear genomes (63); (iii) the creation of a 'universal cell' suitable for all patients, by manipulating the major histocompatibility complex (MHC) (64); and (iv) the generation of haematopoietic chimerism, to create the required tolerance for derived tissues or cells (65). The feasibility of the last strategy has been demonstrated using rat ES-like cells, where their injection into fully MHC-mismatched rats resulted in permanent engraftment; this enabled further long-term graft acceptance of second-set transplantation, via mixed chimerism (66).

Once these obstacles are overcome, the contribution of hES cells to medicine is likely to be enormous.

Acknowledgments

Funding was provided by NIH (grant 1RO1HL073798-01 and 1R24RR018405-0 to J I-E), grant from the Israel Ministry of Science (to J. I.-E.), and the Juvenile Diabetes Research Foundation (fellowship to S. G.-N.).

References

1. Thomson, J.A., Itskovitz-Eldor, J., Shapiro, S.S., Waknitz, M.A., Swiergiel, J.J., Marshall, V.S., and Jones, J.M. (1998). *Science*, **282**, 1145-7.
2. Carpenter, M.K., Inokuma, M.S., Denham, J., Mujtaba, T., Chiu, C.P., and Rao, M.S. (2001). *Exp Neurol.*, **172**, 383-97.
3. Reubinoff, B.E., Itsykson, P., Turetsky, T., Pera, M.F., Reinhartz, E., Itzik, A., and Ben-Hur, T. (2001). *Nat. Biotechnol.*, **19**, 1134-40.
4. Schuldiner, M., Eiges, R., Eden, A., Yanuka, O., Itskovitz-Eldor, J., Goldstein, R.S., and Benvenisty, N. (2001). *Brain Res.*, **913**, 201-5.
5. Zhang, S.C., Wernig, M., Duncan, I.D., Brustle, O., and Thomson, J.A. (2001). *Nat Biotechnol.*, **19**, 1129-33.
6. Schulz, T.C., Palmarini, G.M., Noggle, S.A., Weilero, D.A., Mitalipova, M.M., and Condie, B.G. (2003). *BMC Neurosci.*, **4**, 27.
7. Perrier, A.L., Tabar, V., Barberi, T., Rubio, M.E., Bruses, J., Topf, N., Harrison, N.L., and Studer, L. (2004). *Proc. Natl. Acad. Sci. USA*, **101**, 12543-8.
8. Buytaert-Hoefen, K.A., Alvarez, E., and Freed, C.R. (2004). *Stem Cells*, **22**, 669-74.
9. Kehat, I., Kenyagin-Karsenti, D., Snir, M., Segev, H., Amit, M., Gepstein, A., Livne, E., Binah, O., Itskovitz-Eldor, J., and Gepstein, L. (2001). *J. Clin. Invest.*, **108**, 407-14.

10. Kehat, I., Gepstein, A., Spira, A., Itskovitz-Eldor, J., and Gepstein, L. (2002). *Circ. Res.*, **91**, 659–61.
11. Xu, R.H., Chen, X., Li, D.S., Li, R., Addicks, G.C., Glennon, C., Zwaka, T.P., and Thomson, J.A. (2002). *Nat. Biotechnol.*, **20**, 1261–4.
12. Mummery, C., Ward, D., van den Brink, C.E., Bird, S.D., Doevendans, P.A., Opthof, T., Brutel de la Riviere, A., Tertoolen, L., van der Heyden, M., and Pera, M. (2002). Cardiomyocyte differentiation of mouse and human embryonic stem cells. *J. Anat.*, **200**, 233–42.
13. He, J.Q., Ma, Y., Lee, Y., Thomson, J.A., and Kamp, T.J. (2003). *Circ Res.*, **93**, 32–9.
14. Mummery, C., Ward-van Oostwaard, D., Doevendans, P., Spijker, R., van den Brink, S., Hassink, R., van der Heyden, M., Opthof, T., Pera, M., de la Riviere, A.B., Passier, R., and Tertoolen, L. (2003). *Circulation*, **107**, 2733–40.
15. Levenberg, S., Golub, J.S., Amit, M., Itskovits-Eldor, J., and Langer, R. (2002). *Proc. Natl. Acad. Sci. USA*, **99**, 4391–6.
16. Gerecht-Nir, S., Ziskind, A., Cohen, S., and Itskovitz-Eldor, J. (2003). *Lab. Invest.*, **83**, 1811–20.
17. Kaufman, D.S., Hanson, E.T., Lewis, R.L., Auerbach, R., and Thomson, J.A. (2001). *Proc. Natl. Acad. Sci. USA*, **98**, 10716–21.
18. Chadwick, K., Wang, L., Li, L., Menendez, P., Murdoch, B., Rouleau, A., and Bhatia, M. (2003). *Blood*, **102**, 906–15.
19. Cerdan, C., Rouleau, A., and Bhatia, M. (2004). *Blood*, **103**, 2504–12.
20. Wang, L., Li, L., Shojaei, F., Levac, K., Cerdan, C., Menendez, P., Martin, T., Rouleau, A., and Bhatia, M. (2004). *Immunity*, **21**, 31–41.
21. Vodyanik, M.A., Bork, J.A., Thomson, J.A., and Slukvin, I.I. (2004). *Blood*, **105**, 617–26.
22. Sottile, V., Thomson, A., and McWhir, J. (2003). *Cloning Stem Cells*, **5**, 149–55.
23. Green, H., Easley, K., and Iuchi, S. (2003). *Proc. Natl. Acad. Sci. USA*, **100**, 15625–30.
24. Rambhatla, L., Chiu, C.P., Kundu, P., Peng, Y., and Carpenter, M.K. (2003). *Cell Transplant.*, **12**, 1–11.
25. Lavon, N., Yanuka, O., and Benvenisty, N. (2004). *Differentiation*, **72**, 230–8.
26. Assady, S., Maor, G., Amit, M., Itskovitz-Eldor, J., Skorecki, K.L., and Tzukerman, M. (2001). *Diabetes*, **50**, 1691–7.
27. Segev, H., Fishman, B., Ziskind, A., Shulman, M., and Itskovitz-Eldor, J. (2004). *Stem Cells*, **22**, 265–74.
28. Xu, R.H., Chen, X., Li, D.S., Li, R., Addicks, G.C., Glennon, C., Zwaka, T.P., and Thomson, J.A. (2002). *Nat. Biotechnol.*, **20**, 1261–4.
29. Gerami-Naini, B., Dovzhenko, O.V., Durning, M., Wegner, F.H., Thomson, J.A., and Golos, T.G. (2004). *Endocrinology*, **145**, 1517–24.
30. Ginis, I., Luo, Y., Miura, T., Thies, S., Brandenberger, R., Gerecht-Nir, S., Amit, M., Hoke, A., Carpenter, M.K., Itskovitz-Eldor, J., and Rao, M.S. (2004). *Dev. Biol.*, **269**, 360–80.
31. Ezashi, T., Das, P., and Roberts, R.M. (2005). *Proc. Natl. Acad. Sci. USA*, **102**, 4783–8.
32. Richards, M., Fong, C.Y., Chan, W.K., Wong, P.C., and Bongso, A. (2002). *Nat. Biotechnol.*, **20**, 933–6.
33. Amit, M., Margulets, V., Segev, H., Shariki, K., Laevsky, I., Coleman, R., and Itskovitz-Eldor, J. (2003). *Biol. Reprod.*, **68**, 2150–6.
34. Cheng, L., Hammond, H., Ye, Z., Zhan, X., and Dravid, G. (2003). *Stem Cells*, **21**, 131–42.
35. Hovatta, O., Mikola, M., Gertow, K., Stromberg, A.M., Inzunza, J., Hreinsson, J., Rozell, B., Blennow, E., Andang, M., and Ahrlund-Richter, L. (2003). *Hum. Reprod.*, **18**, 1404–9.
36. Richards, M., Tan, S., Fong, C.Y., Biswas, A., Chan, W.K., and Bongso, A. (2003). *Stem Cells*, **21**, 546–56.

37. Xu, C., Inokuma, M.S., Denham, J., Golds, K., Kundu, P., Gold, J.D., and Carpenter, M.K. (2001). *Nat. Biotechnol.*, **19**, 971–4.
38. Amit, M., Shariki, C., Margulets, V., and Itskovitz-Eldor, J. (2004). *Biol. Reprod.*, **70**, 837–45.
39. Xu, R.H., Peck, R.M., Li, D.S., Feng, X., Ludwig, T., and Thomson, J.A. (2005). *Nat. Methods*, **2**, 185–90.
40. Wang, L., Li, L., Menendez, P., Cerdan, C., and Bhatia, M. (2005). *Blood*, **105**, 4598–603.
41. Xu, C., Rosler, E., Jiang, J., Lebkowski, J.S., Gold, J.D., O'Sullivan, C., Delavan-Boorsma, K., Mok, M., Bronstein, A., and Carpenter, M.K. (2005). *Stem Cells*, **23**, 315–23.
42. Inzunza, J., Gertow, K., Stromberg, M.A., Matilainen, E., Blennow, E., Skottman, H., Wolbank, S., Ahrlund-Richter, L., and Hovatta, O. (2005). *Stem Cells*, **23**, 544–9.
43. Tzukerman, M., Rosenberg, T., Ravel, Y., Reiter, I., Coleman, R., and Skorecki, K. (2003). *Proc. Natl. Acad. Sci. USA*, **100**, 13507–12.
44. Gerecht-Nir, S., Osenberg, S., Nevo, O., Ziskind, A., Coleman, R., and Itskovitz-Eldor, J. (2004). *Biol. Reprod.*, **71**, 2029–36.
45. Itskovitz-Eldor, J., Schuldiner, M., Karsenti, D., Eden, A., Yanuka, O., Amit, M., Soreq, H., and Benvenisty, N. (2000). *Mol. Med.*, **6**, 88–95.
46. Dang, S.M., Gerecht-Nir, S., Chen, J., Itskovitz-Eldor, J., and Zandstra, P.W. (2004). *Stem Cells*, **22**, 275–82.
47. Dvash, T., Mayshar, Y., Darr, H., McElhaney, M., Barker, D., Yanuka, O., Kotkow, K.J., Rubin, L.L., Benvenisty, N., and Eiges, R. (2004). *Hum. Reprod.*, **19**, 2875–83.
48. Gerecht-Nir, S., Dazard, J.-E., Golan-Mashiach, M., Osenberg, S., Botvinnik, A., Amariglio, N., Domany, E., Rechavi, G., Givol, D., and Itskovitz-Eldor, J. (2005). *Dev. Dyn.*, **232**, 488–98.
49. Dang, S.M., Kyba, M., Perlingeiro, R., Daley, G.Q., and Zandstra, P.W. (2002). *Biotechnol. Bioeng.*, **78**, 442–53.
50. Wartenberg, M., Donmez, F., Ling, F.C., Acker, H., Hescheler, J., and Sauer, H. (2001). *FASEB J.*, **15**, 995–1005.
51. Zandstra, P.W., Bauwens, C., Yin, T., Liu, Q., Schiller, H., Zweigerdt, R., Pasumarthi, K.B.S., and Field, L.J. (2003). *Tissue Eng.*, **9**, 767–78.
52. Gerecht-Nir, S., Cohen, S., and Itskovitz-Eldor, J. (2004). *Biotechnol. Bioeng.*, **86**, 493–502.
53. Bauwens, C., Yin, T., Dang, S., Peerani, R., and Zandstra, P.W. (2005). *Biotechnol. Bioeng.*, **90**, 452–61.
54. Gerecht-Nir, S., Cohen, S., Ziskind, A., and Itskovitz-Eldor, J. (2004). *Biotechnol. Bioeng.*, **88**, 313–20.
55. Levenberg, S., Huang, N.F., Lavik, E., Rogers, A.B., Itskovitz-Eldor, J., and Langer, R. (2003). *Proc. Natl. Acad. Sci. USA*, **100**, 12741–6.
56. Levenberg, S., Burdick, J., Kraehenbuehl, T., and Langer, R. (2005). *Tissue Engineering*, **11**, 506–12.
57. Eiges, R., Schuldiner, M., Drukker, M., Yanuka, O., Itskovitz-Eldor, J., and Benvenisty, N. (2000). *Curr. Biol.*, **11**, 514–8.
58. Zwaka, T.P. and Thomson, J.A. (2003). *Nat. Biotechnol.*, **21**, 319–21.
59. Ma, Y., Rmezani, A., Lewis, R., Hawley, R.G., and Thomson, J.A. (2003). *Stem Cells*, **21**, 111–7.
60. Groop, M., Itsykson, P., Singer, O., Ben-Hur, T., Reinhartz, E., Galum, E., and Reubinoff, B.E. (2003). *Mol. Ther.*, **7**, 281–7.
61. Smith-Arica, J.R., Thomson, A.J., Ansell, R., Chiorini, J., Davidson, B., and McWhir, J. (2003). *Cloning Stem Cells*, **5**, 51–62.
62. Schuldiner, M., Itskovitz-Eldor, J., and Benvenisty, N. (2003). *Stem Cells*, **21**, 257–65.
63. Lanza, R.P., Chungm, H.Y., Yoo, J.J., Wettstein, P.J., Blackwell, C., Borson, N., Hofmeister, E., Schuch, G., Soker, S., Moraes, C.T., West, M.D., and Atala, A. (2002). *Nat. Biotechnol.*, **20**, 689–96.

64. Drukker, M., Katz, G., Urbach, A., Schuldiner, M., Markel, G., Itskovitz-Eldor, J., Reubinoff, B., Mandelboim, O., and Benvenisty, N. (2002). *Proc. Natl. Acad. Sci. USA*, **99**, 9864-9.
65. Bradley, J.A., Bolton, E.M., and Pedersen, R.A. (2002). *Nature*, **2**, 859-71.
66. Fandrich, F., Lin, X., Chai, G.X., Schulze, M., Ganten, D., Bader, M., Holle, J., Huang, D.S., Parwaresch, R., Zavazava, N., and Binas, B. (2002). *Nat. Med.*, **8**, 171-8.
67. Corada, M., Zanetta, L., Orsenigo, F., Breviario, F., Lampugnani, M.G., Bernasconi, S., Liao, F., Hicklin, D.J., Bohlen, P., and Dejana, E. (2002). *Blood*, **100**, 905-11.
68. Liao, F., Li, Y., O'Connor, W., Zanetta, L., Bassi, R., Santiago, A., Overholser, J., Hooper, A., Mignatti, P., Dejana, E., Hicklin, D.J., and Bohlen, P. (2000). *Cancer Res.*, **60**, 6805-10.

Chapter 12
ES cell lines from the cynomolgus monkey (*Macaca fascicularis*)

Hirofumi Suemori,[1] Yoshiki Sasai,[2] Katsutsugu Umeda[3] and Norio Nakatsuji[4]

[1]Laboratory of Embryonic Stem Cell Research, Stem Cell Research Center, Institute for Frontier Medical Sciences, Kyoto University, 53 Kawaharacho, Shogoin, Sakyo-ku, Kyoto 606–8507, Japan; [2]Neurogenesis and Organogenesis Group, Riken Center for Developmental Biology, 2–2–3 Minatojima-Minamimachi, Chuo, Kobe, 650–0047, Japan; [3]Department of Pediatrics, Graduate School of Medicine, Kyoto University, 54 Kawahara-cho, Shogoin, Sakyo-ku, Kyoto 606–8507, Japan; [4]Division of Development and Differentiation, Institute for Frontier Medical Sciences, Kyoto University, 53 Kawaharacho, Shogoin, Sakyo-ku, Kyoto 606–8507, Japan.

1 Introduction

Human ES (hES) cells hold great promise as an unlimited source of functional cells and tissues for transplantation therapy and regenerative medicine, owing to their pluripotency and capacity for indefinite proliferation *in vitro*. However, the realization of this goal demands not only robust methods for establishing and maintaining the undifferentiated hES cells, and for inducing their differentiation into specific cell types: animal model systems are both requisite and timely for testing the efficacy and safety of ES cell-based transplantation procedures.

The use of monkey ES cells and transplantation models offers considerable advantages over murine systems:

(a) Strong similarities in the biology of human and monkey ES cells, and clear differences between primate (specifically, human and monkey) and murine ES cells, are emerging.

(b) It is crucial to evaluate the possible risk of immunorejection, and of tumourigenesis (arising from the immortal nature of undifferentiated ES cells), following transplantation of allogenic ES cell-derivatives. Both eventualities can only be adequately assessed using animal models as close to the human as possible with respect to both transplanted cells and host species.

(c) For preclinical research into the cell- or tissue-transplantation process *per se*, it is necessary for the size and structure of the relevant organ or tissue of the model host to approach those in the human. This is especially pertinent to surgical procedures; the absolute depth of structures in the brain, for example, determines in large part the transplantation protocol to be followed. Non-human primates, including monkeys, clearly offer superior models due to their phylogeny, body sizes, and comparative physiology.

(d) Temporal analysis of the functionality of ES cell-derivatives following their transplantation is better addressed in a primate system; observations can be performed in monkeys over several years, compared to 2–3 years at most in mice.

Collectively, these considerations warrant the study of monkey ES cells in the context of an animal model system.

For mouse ES (mES) cells, methods have now been established for the lineage-specific differentiation of mesodermal, endodermal, and ectodermal derivatives (see Chapters 5–9). Certain protocols have recently been transposed (with modification) from the mouse to human and monkey ES cells, and differentiated derivatives successfully produced. Consequently, clinical trials to test the efficacy of hES cell-derivatives in the treatment of a variety of severe diseases are expected to commence imminently. The potential treatment of Parkinson's disease with ES cell-derived dopaminergic neurons is an example of such a therapy in development; dopaminergic neurons can already be produced efficiently from mouse and monkey ES cells (1–3; and see below), and their generation from hES cells is being pursued. As a monkey model of Parkinson's disease has been established (4, 5), it is now feasible to evaluate ES cell-based transplantation strategies in that system before the testing of hES-cell derivatives is conducted in a clinical setting.

Among the monkeys, the macaques (such as rhesus and cynomolgus) are in general most suitable as model, non-human primates. They are commonly bred as experimental animals, widely used for medical research, and various disease models are now available in these species. For these reasons we are investigating ES cells from the cynomolgus monkey (*Macaca fascicularis*), or 'cES cells'; and our current protocols for their derivation and *in vitro* differentiation are provided in this chapter.

1.1 Primate *versus* murine ES cells

ES cell lines were first established from primates (specifically, the rhesus monkey and marmoset) by Thomson and coworkers in 1995, and from the cynomolgus monkey by our group in 2001 (6–9). The methods used to derive these non-human primate and, subsequently, human (10, 11) ES cell lines were broadly similar to those developed for the mouse (12). Features that are common to all of these primate ES-cell types and to mES cells include a high nuclear-to-cytoplasmic ratio, and expression of stem cell-specific genes such as *Oct-3/4* and *Nanog*

Table 1 ES-cell marker expression in monkey, human and mouse ES cells

Marker	Cynomolgus monkey (7, 31)	Rhesus monkey (6)	Human (32)	Mouse (32)
Alkaline phosphatase activity[1]	+	+	+	+
SSEA-1[2]	−	−	−	+
SSEA-3[2]	−	+	+	−
SSEA-4[2]	+	+	+	−
TRA-1–60, TRA-1–81[3]	+	+	+	*
Expression of genes *Oct − 3/4*, *Nanog* and *Rex-1*	+	?	+	+

[1] Detection is using Vector Red substrate (Vector Labs, USA).
[2] Anti-SSEA-1, -3, and -4 monoclonal antibodies, from Developmental Studies Hybridoma Bank (DSHB), University of Iowa, and see *Table 2*.
[3] Anti-TRA-1–60 and TRA-1–81 monoclonal antibodies, from Chemicon International, and see *Table 2*.
*These monocolonal antibodies do not react with mouse antigens.

(which are implicated in self-renewal and the maintenance of pluripotency). However, notable differences exist between murine and primate ES cells:

(a) Regarding morphology, primate ES cells grow in monolayered colonies that are epithelial-like and flatter than the compact, multilayered colonies produced by mES cells.

(b) Accepted ES-cell markers differ between primate and murine species (*Table 1*).

(c) In contrast to mES cells, addition of LIF to the culture medium has no effect on the maintenance of primate ES cells, which depend rather on the presence of feeder cells to sustain their proliferation (8).

(d) Mouse ES cells are able to proliferate after dissociation into single cells with a clonal (or 'plating') efficiency of over 20% (see Chapter 2), whereas for primate ES cells this efficiency is extremely low (13).

(e) Primate ES cells divide more slowly than mES cells, with a cell-cycle time of about 30 h compared with 12 h.

The last two features can be problematical for the maintenance and manipulation of primate ES cells.

2 Establishment and maintenance of cES cell lines

The overall strategy for establishing cES cell lines is similar to that first elaborated for the mouse (see Chapter 2) in that inner cell masses (ICMs) are isolated from blastocysts, cultured on a feeder layer of primary or immortalized mouse embryonic fibroblasts in gelatinized dishes, and outgrowing cES-cell colonies are selected manually from amongst differentiated derivatives, and serially passaged until stable cES cell lines are established. For the cynomolgus system, as for the mouse system, the careful timing of harvesting and passaging of ES cells is crucial to successful cell-line isolation. Using the conditions described here, the efficiency of establishing cES cell lines is 60–80% – an optimized rate that exceeds that typically obtained for mES cell lines (see Chapter 2). As reported for ES cells from other primate species, spontaneous differentiation may be a hindrance to developing and stably maintaining cES cell lines. However, the prevalence of spontaneous differentiation in cES cells is markedly reduced when fetal-calf serum (FCS) in the medium is replaced with KnockOut™ Serum Replacement (KnockOut™ SR; Invitrogen), from which it is inferred that FCS contains differentiation-inducing factor(s) that are absent from KnockOut™ SR (see also Chapter 5 for a commentary on serum replacement). Thus, use of KnockOut™ SR allows cES cells to be maintained in their undifferentiated state for longer periods and *en masse*, that is without periodical, mechanical harvesting of undifferentiated colonies. Also, the possibility of FCS being a source of pathogens precludes its use in hES-cell derivation and culture with a view to potential therapeutic applications. Therefore it is desirable that for cES-cell derivation and culture also, media be formulated that circumvent the use of this component. Consequently, we recommend the use of KnockOut™ SR-supplemented medium, which performs acceptably in the establishment of cES cell lines, despite a detectable decrease in the rate of cES-cell proliferation compared with maintenance using conventional FCS-supplemented medium.

In common with other primate ES cells, cES cells exhibit a very low plating efficiency following enzymatic dissociation into single cells using a trypsin-based solution. Therefore, for effective subculturing a less aggressive treatment is required and only limited dissociation should be carried out (whether mechanical or enzymatic) into clusters of 50–100 cES cells, in order to sustain their proliferation. We have devised an efficient method for routine subculturing, using a dissociation solution containing low concentrations of trypsin and collagenase type IV (dissociation solution I, *Protocol 2*). This method enables well-controlled and reproducible dissociation of cES cell colonies from the entire culture dish or vessel without requirement for mechanical harvesting of undifferentiated colonies, which otherwise is a laborious and time-consuming procedure.

All embryo and cell incubations described below are at 37 °C, in a humidified incubator gassed with 5% CO_2 in air.

2.1 Preparation of feeder layers for cES-cell derivation

Mouse embryonic fibroblasts (MEFs) are highly recommended for use as feeder layers for the derivation and culture of cES cells. (Alternatively, the continuous line of STO mouse embryonic fibroblasts (ATCC CRL 1503) may be used.) As the condition of the feeder layers is very important for establishing and maintaining undifferentiated cES cells, only early passages of MEFs should be used to prepare them, up to passage 5. MEF cultures and derivative, mitotically inactivated feeder layers are prepared as described fully in Chapter 2. (A method for gelatinization of dishes prior to seeding feeder layers is given in Chapter 5, *Protocol 1*). Note that for cES-cell culture, the feeder cells are seeded at a lower density (compared with mES-cell culture, Chapter 2) of $1.5–2 \times 10^4/cm^2$, as higher-density feeder layers tend to induce differentiation of primate ES cells. Feeder layers should be used for cES cell culture within 4 d of their preparation.

2.2 Culture of ICMs from cynomolgus monkey blastocysts

ES cell lines can be established successfully from cynomolgus monkey blastocysts obtained by either *in vitro* fertilization (IVF) or intracytoplasmic sperm injection (ICSI), as described elsewhere (9). Blastocysts produced as a result of natural mating also may be used, but the yield of embryos from a single female is then small.

All animal experiments must be performed with the approval of the institutional research board for animal experiments, and of the relevant licensing authorities.

Protocol 1
Immunosurgical isolation of ICMs from cynomolgus monkey blastocysts

Reagents and equipment

- Expanded, cynomolgus monkey blastocysts at day 7–10 of *in vitro* culture following IVF or ICSI (14), in CMRL 1 Medium-1066 (Cat. No. 11530037, Invitrogen)
- Protease solution: 0.5% protease (Cat. No. P8811, Sigma) in DMEM (Cat. No. D5796, Sigma)
- Anti-monkey antiserum, raised by injecting cynomolgus monkey spleen cells ($\sim 10^8$ cells in total) intravenously into rabbits, three times at fortnightly intervals: antiserum is collected 2 weeks after the last immunization, heat-inactivated at 56 °C for 30 min, and filter-sterilized; and aliquots are stored at −20 °C (15)
- Complement solution: reconstitute lyophilized, guinea pig complement (Cat. No. S1639, Sigma) with water according to the manufacturer's instructions, aliquot, and store at −80 °C
- M16 medium (Cat. No. M7292, Sigma)
- Mineral oil (Cat. No. M8410, Sigma)
- 60 mm plastic Petri dishes, preferably bacteriological grade
- Aspirator tube assembly (Cat. No. A5177, Sigma), for mouth-controlled pipetting
- Glass capillaries: these are finely drawn over the flame of a spirit burner to have an inner diameter of 300 µm for blastocyst handling and transfer, or of 100 µm for ICM isolation
- Stereo dissecting microscope

Method

1. Prepare 1:10 dilutions of anti-monkey antiserum and complement solution in M16 medium; and into a Petri dish dispense single 30 µl microdrops of the diluted antiserum and complement solution, and of protease solution, together with several microdrops of M16 medium (see *Figure 1*). Overlay the microdrops with mineral oil.[a]

Protocol 1 continued

2 Incubate the blastocysts in the microdrop of protease solution for 10 min at r.t., to remove the zonae pellucidae. Check that the embryos are denuded using a dissecting microscope.[b]

3 Wash the zona-free blastocysts by passing serially through three microdrops of M16 medium.

4 Incubate the zona-free blastocysts in the microdrop of antiserum for 1 h at 37 °C.

5 Wash the blastocysts three times in M16 medium as before (Step 3), and incubate in the microdrop of diluted complement solution for 30 min at 37 °C. At the end of this period the trophoblast cells will be observed to have lysed.

6 Remove the surrounding layer of lysed trophoblast cells from each ICM by pipetting using a fine glass capillary, and immediately explant the isolated ICMs into culture (*Protocol 2*).

[a] Alternative arrangements are possible, for example microdrops of protease solution and M16 medium may be dispensed separately into different dishes, if the operator finds this easier overall for embryo handling.

[b] 10–20 blastocysts can be treated at a time.

Protocol 2
Primary ICM culture for cES-cell derivation

Reagents and equipment

- Isolated ICMs from cynomolgus monkey blastocysts (*Protocol 1*)
- Feeder layers in 4-well dishes: seed mitomycin C-treated MEFs or STO cells (Chapter 2) onto gelatin-coated wells (*Protocol 1*, footnote c, Chapter 6) at a density of $1.5–2 \times 10^4$ cells/cm^2
- CEM, or culture medium for cES cells: for 500 ml CEM, supplement 400 ml of a 1:1 mixture of DMEM/F12 (Cat. No. D6421, Sigma) with the following additives; 5 ml of 100 × non-essential amino acids (Cat. No. M7145, Sigma), 6.25 ml of 200 mM L-glutamine (Cat. No. G7513, Sigma), 4 μl of β-mercaptoethanol (final concentration 0.1 mM; Cat. No. 137-06862, WAKO), and 100 ml Knockout™ Serum Replacement (Knockout™ SR; Cat. No. 10828-028, Invitrogen). Mix, and store at 4 °C for up to 2 weeks
- LIF- and bFGF-supplemented CEM: cES-cell medium supplemented with 4 ng/ml recombinant, human basic fibroblast growth factor (bFGF: Cat. No. 01-106, Upstate Biotechnologies) and 10 ng/ml recombinant, human leukaemia inhibitory factor (LIF; Cat. No. LIF1010, Chemicon International)[a]
- PBS, Ca^{2+}- and Mg^{2+}-free (Cat. No. D8537, Sigma)
- Dissociation solution I: to 59 ml PBS add 10 ml of 2.5% trypsin solution (Cat. No. 15090-046, Invitrogen), 10 ml of 10 mg/ml collagenase IV (Cat. No.17104-019, Invitrogen) in PBS, 20 ml of Knockout™ SR, and 1 ml of 100 mM $CaCl_2$. Aliquots should be stored at −20 °C, and thawed only once
- Aspirator tube assembly (*Protocol 1*)
- Inverted phase-contrast microscope

Protocol 2 continued

- Stereodissecting microscope
- 27 Gauge hypodermic needle
- Drawn-out glass capillaries, inner diameter 200–300 μm (see Protocol 1)

Methods

Primary culture:

1. Explant the isolated ICMs singly onto feeder layers in LIF- and bFGF-supplemented CEM, ensuring that each ICM is placed in close contact with the feeder cells.[a,b]
2. Culture ICM explants overnight.
3. Using a dissecting microscope, monitor the attachment of ICMs to the substratum: those ICMs that fail to attach promptly will differentiate or die.[c]
4. Continue incubation of ICM explants. Usually, by 7–10 d of culture cES-cell colonies – typically one colony per ICM – can be discerned, surrounded by differentiating cells. Within these colonies individual cES cells are readily identifiable from their morphology: a bright nucleus, a high nuclear-to-cytoplasmic ratio, and prominent nucleoli. Colonies of cES cells are characteristically epithelial-like, and their typical morphology is shown in *Figure 2A*.[d]

First passage of primary cES-cell culture:

Primary cES-cell colonies are harvested and passaged when they reach about 500 μm in diameter, as follows.

1. Aspirate medium, and wash cells once with PBS.
2. Add 0.5 ml dissociation solution I to each well, and score the primary cES-cell colony into several sections using the tip of a hypodermic needle under a stereo dissecting microscope.
3. Incubate the culture for ~5 min at r.t., until the edges of the colony begin to lift from the substratum; and using a finely drawn capillary connected to an aspirator tube assembly, collect the clusters of cES cells.
4. Carefully wash the clusters twice in CEM in culture dishes, to remove residual dissociation solution I; and transfer the clusters deriving from each primary colony onto a fresh feeder layer in a 4-well dish. Culture overnight.
5. The next day, many dead cells will be observable around the edges of nascent cES-cell colonies. This is normal. Change the culture medium, and continue to do so daily until each colony reaches ~500 μm in diameter. At this stage the culture is ready to be expanded, and is passaged further (as in Steps 5–8) into either a single well of a 4-well dish or a (larger) well of a 12-well dish, depending on cell numbers.[e]

[a] Note that until stable cES cell lines are obtained, LIF- and bFGF-supplemented CEM is used; and that after establishment, these additives are unnecessary.

[b] Residual cell debris from immunosurgery sometimes inhibits ICM attachment.

ES CELL LINES FROM THE CYNOMOLGUS MONKEY

Protocol 2 continued

^c Adherent ICMs may not appear to be actively growing during the following several days, but should commence spreading outwards. Occasionally, an ICM will differentiate and grow rapidly, in which case undifferentiated cES cells cannot subsequently be recovered.

^d Cells within these colonies will become tightly packed if allowed to proliferate.

^e The colonies should be subcultured before they become overly compacted and mounded.

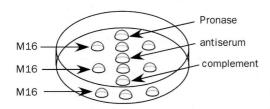

Figure 1 Dispensing microdrops in preparation for microsurgical isolation of ICMs from cynomolgus monkey blastocysts.

2.3 Expansion and routine maintenance of undifferentiated cES cells

Typically, the first three to five passages of cES cells are performed by mechanical disaggregation (as described in Steps 5–8 of *Protocol 2*, for primary explants). The newly isolated cES cells are expanded into increasingly larger dishes, until they reach 50–60% confluence in a 35 mm tissue-culture dish. Thereafter the cES cells can be subcultured *en masse* by enzymatic treatment combined with pipetting, and can usually be stably maintained in this way in an undifferentiated state for more than 1 year. Using conventional, slow-rate cooling methods for cryopreservation, the survival rate of cES cells on thawing is extremely low. Therefore we have developed a simple and effective technique for the cryopreservation of cES cells (and other primate ES cells) based on their vitrification in conventional cryovials (17). Following recovery from vitrification and storage, cES cells show low rates of spontaneous differentiation and their pluripotency is retained.

3 Characterization of cES cell lines

3.1 Expression of cES-cell markers and karyotype analysis

Those ES cell lines established to date from diverse primates have been characterized with respect to cell-surface markers by immunocytochemistry, and alkaline phosphatase activity by chromogenic enzyme detection assay. In common with hES cells, monkey (specifically marmoset, rhesus, and cynomolgus) ES cells express SSEA-4, TRA1-60, and TRA1-81 cell-surface antigens, but not SSEA-1; and whereas human, marmoset, and rhesus monkey ES cells have been reported to express also SSEA-3 at variable levels, we have found that cES cells are negative by immunostaining, which may reflect a species-specific difference. (In contrast mES cells express SSEA-1, but neither SSEA-3 nor SSEA-4.) The expression of genes characteristic of pluripotent cells, such as *Oct-3/4*, *Rex-1*,

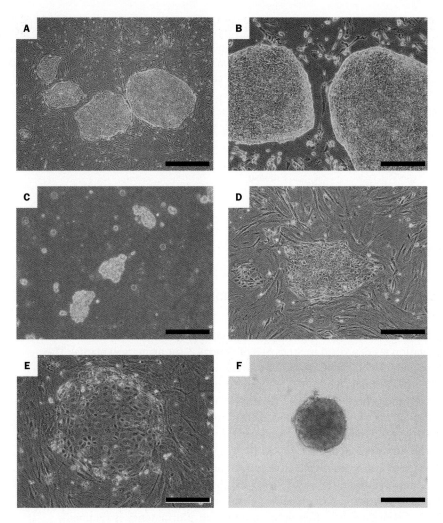

Figure 2 Morphology of cES cells during sequential stages of routine culture on feeder layers, viewed by light microscopy. (A) Sub-confluent culture of cES cells on a feeder layer, with several cES-cell colonies of various sizes. (B) Morphology of cES-cell colonies during treatment with dissociation solution: feeder cells have become rounded and are detaching from the substratum whilst the two cES-cell colonies remain intact, but are beginning to detach. (C) Clusters of cES cells obtained after dissociation of colonies by pipetting, each containing an optimal 50–100 cells. (D) Appearance of cES-cell colonies 1 day after subculturing on a fresh feeder layer. (E) A typical spontaneously differentiated colony of cES cells, showing epithelial morphology and surrounded by feeder cells. (F) A simple embryoid body generated by cES cells in suspension. Scale bars: 750 μm for (A); 300 μm for (B), (C), (D) and (E); and 150 mm for (F).

and *Nanog*, are detectable in undifferentiated cES cells by RT-PCR (unpublished). These ES cell-specific markers are summarized in *Table 1*.

Male and female cES cell lines can be isolated at equal frequencies. Usually, at least 70% of chromosome spreads demonstrate the euploid chromosome

Protocol 3
Routine maintenance of cES cells

Reagents and equipment

- Culture of cES cells, on a feeder layer in a 60 mm culture dish
- PBS (*Protocol 2*)
- Fresh feeder layers in gelatinized, 60 mm culture dishes
- Dissociation solution I (*Protocol 2*)
- CEM (*Protocol 2*)
- DAP freezing medium: 2 M dimethyl-sulphoxide (DMSO), 1 M acetamide, and 3 M propylene glycol in CEM
- 2 ml transfer pipette
- Inverted phase-contrast microscope
- 15 ml conical tubes
- Benchtop centrifuge
- Cryovials of 1 ml capacity (Cat. No. 5000–1012, Nalgene)

Methods

Passaging cES cells:

1. Aspirate medium from the cES-cell culture, and wash cells with PBS.
2. Cover the cells with 0.5 ml dissociation solution I, and incubate for 5–6 min at 37 °C. Examine the cells by inverted microscopy, and confirm that the feeder cells have become rounded and that most cES-cell colonies have partially detached from the substratum (*Figure 2B*).
3. Add 1.5 ml CEM, and dissociate cES-cell colonies into clusters of 50–100 cells with gentle pipetting (*Figure 2C*).[a,b]
4. Transfer the cES-cell suspension into a tube and centrifuge at 1000 rpm for 5 min.
5. Aspirate the supernatant, and resuspend the cell pellet in 6 ml CEM.
6. Dispense 2 ml cell suspension into each of three fresh feeder layers in 35 mm dishes, and incubate. Change medium daily. Following this passage, the cES-cell number will increase about three fold in 3–4 d (*Figure 2D*); and when cultures are 50–60% confluent they are ready to be repassaged.
7. During routine culture, small numbers of differentiated-cell colonies, characterized by flattened and epithelial morphology, are occasionally observed (*Figure 2E*); remove these simply and effectively by scraping and collecting them with a drawn-out glass capillary connected to an aspirator tube assembly.

Vitrification, cryopreservation, and thawing of cES cells (17):
The average cES-cell survival rate by this method is 2–7%, depending on the cell line used.

8. Harvest cells from a confluent culture of cES cells on a feeder layer in a 60 mm to 10 cm dish using dissociation solution I and centrifuge, as in Steps 1–4 above.

Protocol 3 continued

9. Aspirate supernatant and resuspend the cell pellet in 200 μl DAP freezing medium: the final cell density may range from 5×10^5 to 1×10^7 cells/200 μl.[c]

10. Quickly transfer the cell suspension into a cryovial, and *immediately* transfer the vial into liquid nitrogen so that the lower 2/3 of each vial is immersed.[d]

11. Transfer the cyrovials of cells into a −150 °C freezer for long-term storage.

12. To rapidly thaw cryopreserved cES cells, add to the frozen cryovial a small amount (∼0.8 ml) of CEM prewarmed to 37 °C, and pipette carefully.[e]

13. Transfer the cell suspension into a tube containing 10 ml CEM, and centrifuge at 1000 rpm for 5 min.

14. Aspirate medium, resuspend the cells in CEM, reseed the cell suspension onto a fresh feeder layer in a 60 mm dish and incubate.

[a] Excessive pipetting and dissociation will damage the cES cells and decrease their yield.

[b] Feeder cells produce a viscous matrix that is resistant to trypsin digestion in CEM, and consequently sheets or strands of adhering feeder cells will remain at this stage. Do *not* try to dissociate the feeder cells, as this will cause over-dissociation of cES cells and be detrimental to their viability.

[c] This cryopreservation method tolerates a wide range of cell densities.

[d] The interval between cell suspension in DAP freezing medium and freezing in liquid nitrogen should be ∼15 s, to minimize cytotoxicity due to the cryoprotectant.

[e] To avoid explosion on contacting the medium with liquid nitrogen, it is necessary to store vials prior to thawing in a −150 °C freezer, or in vapour-phase liquid nitrogen. Frozen cells should be thawed quickly (within 15 s). The use of a water bath for thawing is *not* recommended.

complement of 42; and even after prolonged culture, cES cell lines demonstrate normal karyotypes. For the analysis of cynomolgus monkey chromosomes, the reader is referred to Borrell *et al.* (18). Other, standard methods for karyotyping mES cells (Chapter 2, *Protocol 10*) may be applied successfully also to cES cells. The nomenclature and G-banded ideogram for the cynomolgus monkey karyotype are according to the International System for Cytogenetic Nomenclature (ISCN 1995).

4 Genetic modification of cES cells

It is anticipated that the genetic modification of non-human primate ES cells will constitute a valuable experimental tool for basic and preclinical research by extrapolation from other systems. For mES cells it has been amply demonstrated that expression of endogenous and exogenous genes may be manipulated to promote differentiation into defined cell types *in vitro*, as exemplified by the generation of dopaminergic neurons and insulin-producing cells (19–21; and see Chapter 9). Also, in animal model systems, it is has proved highly informative to introduce exogenous reporter genes for the *in vivo* tracing of engrafted cells (22; and see Chapter 10), e.g. *GFP* and β-galactosidase. For the human, moreover, modification of specific endogenous genes, such as those in the HLA system, could in theory be applied to produce hES cells whose differentiated derivatives are not recognized by the immune system of the transplanted host, and which therefore escape immune rejection. Thus, non-human primate ES cells offer valuable systems for the development of these broad categories of genetic modification.

Generally, the transformation efficiency of primate ES cells, whether by electroporation or by lipid-mediated gene transfer, is much lower than that of mES cells. Although highly efficient gene transfer into primate ES cells has been achieved using a lentiviral vector (22), this method is less desirable because of (a) exposure to potential biohazards associated with virus-mediated transformation, and (b) difficulties in achieving homologous recombination, owing to the limited size of DNA inserts that the vectors can accommodate (8–13 kb). Therefore, electroporation is currently the preferred method for transforming primate ES cells.

Here we provide a generic protocol for the electroporation of transgenes into cES cells, which was optimized for the cell line CMK6, and is also applicable to other cES and primate ES cell lines. The method allows the pluripotency of the cells to be maintained (23). Using an SV40-*neo* expression vector, over 100 transformed colonies are consistently produced from 10^7 cES cells following (antibiotic G418) drug selection, the precise efficiency being cell-line dependent. By this method we have introduced into cES cells a reporter gene incorporating yellow fluorescent protein with a mitochondrial targeting sequence (*EYFPmito*). In addition, we have compared the transcriptional activities of the *PGK-1*, *SV40*, and *CMV* promoters in cES cells: the *PGK-1* and *SV40* promoters are expressed relatively efficiently, whereas the *CMV* immediate-early promoter displays significantly lower activity (23).

5 Differentiation of cES cells

Cynomolgus monkey ES cells, like ES cells from other species, differentiate *in vitro* spontaneously when permitted to reach confluence. During high-density culture of cES cells a heterogeneous mixture of differentiated derivatives and structures becomes observable, including clusters of neurons and pigment cells, mesenchymal outgrowths from differentiating colonies, and vesicular epithelia resembling visceral endoderm or yolk sac.

For cES cells as for mES (Chapters 6–9) and hES cells (Chapter 11), the formation of embryoid bodies (EBs) is an effective method to induce their differentiation into derivatives of the three primary germ layers. And for cES cells as for hES cells, it is crucial for the efficient production of EBs to commence with cellular aggregates of sufficient size, in order to sustain cellular survival in suspension. Methods for the production and analysis of EBs from cES cells are essentially as described for those from hES cells (9; and see Chapter 11, Protocol 3).

Protocol 4

Transfection of cES cells by electroporation

Equipment and reagents

- Culture of cES cells, on a feeder layer in a 10 cm dish (containing $1-2 \times 10^7$ cES cells)
- PBS (*Protocol 2*)
- CEM (*Protocol 2*)
- Trypsin/EDTA solution I: 0.25% (w/v) trypsin and 0.5 mM EDTA in HBSS (Cat. No. T4049, Sigma), or in PBS
- Linearized plasmid DNA carrying the selectable marker gene, *neo*: this is resuspended at a concentration of 1–2 mg/ml in PBS
- Feeder layer of G418-resistant MEF, prepared from mice carrying the *neo* gene (e.g. the ROSA26 strain), in a 10 cm culture dish
- G418 solution: 50 mg/ml antibiotic G418 (or Geneticin™, disulphate salt solution) in tissue culture-grade water (Cat. No. G7034, Sigma)
- Finely drawn glass capillary
- Feeder layers of (*non*-G418-resistant) MEF in 24-well, tissue-culture plates; these are required about 10 d postelectroporation
- A pulse generator, such as the Gene Pulser (Bio-Rad), with electroporation cuvettes having a 0.4 cm electrode gap (Cat. No. 165-2081EDU, Bio-Rad)

Method

1. Aspirate medium from cES cells, and wash with PBS.
2. Add 2 ml trypsin/EDTA solution I, gently dissociate cES-cell colonies into small clumps of 5–20 cells with gentle pipetting, and transfer into 10 ml CEM in a conical tube.[a]
3. Pellet cES-cell clumps by centrifugation at 1000 rpm for 5 min.
4. Aspirate supernatant, wash cells once by resuspending in PBS and recentrifugation; aspirate supernatant, and resuspend cells in 800 µl PBS.
5. Add 50 µg linearized plasmid DNA to the cell suspension, and incubate on ice for 10 min.[b]
6. Mix the suspension well and transfer into an electroporation cuvette. Deliver one electrical pulse to the cells at 500 µF capacitance and 250 V (i.e. 625 V/cm).[c]
7. Incubate the cuvette on ice for 10 min, then plate the cell suspension onto a layer of *neo*-transformed, G418-resistant feeder cells in a 10 cm dish, and incubate.
8. At 72 h postelectroporation, start selection by adding G418 to the medium at a final concentration of 50 µg/ml, and maintain selection for 7 d in total, changing medium daily.[d]
9. At 10 d postelectroporation (and following 7 d of G418 selection), G418-resistant, *neo*-transformed cES-cell colonies should be apparent. Mechanically divide each of these surviving colonies into clumps of about 50 cells (see Steps 5–8, *Protocol 2*), transfer separately onto fresh feeder layers prepared in 24-well plates in medium without added G418, and incubate. Expand the G418-resistant cES-cell colonies.[e]

Protocol 4 continued

10. Determine the integration pattern of the transgene in expanded, G418-resistant cultures by Southern-blot analysis of their DNA, using appropriate restriction endonucleases and probes.[f]

[a] This reduced clump size is to achieve a balance between efficient electroporation and sustained cell viability.

[b] This amount of DNA usually gives the highest yield of transformants. Transformation efficiency is hardly increased by using more.

[c] The cell survival rate after this pulse is typically 10–20%.

[d] Although this concentration of G418 is much lower than the $\geq 200\,\mu g/ml$ used in the selection of *neo*-transformed mES cells, it effectively eliminates non-transformed cES-cell colonies.

[e] Under these conditions we consistently obtain about 100 transformed, G418-resistant colonies per 10^7 (line CMK6) cES cells (23).

[f] As cES cells are refractory to single-cell cloning, clonal sublines are difficult to isolate. However, under the conditions here described roughly half the cultures expanded from individual G418-resistant colonies are found to contain multiple copies of the transgene; and subcultures derived from them seldom segregate cells with lower copy numbers, with continued passage. Therefore it can be inferred that those original G418-resistant colonies were effectively clonal, deriving from a single, multiple transformant per clump of 5–20 cES cells.

Teratoma formation is a simple and reliable method for analysing the differentiation potency of cES cells *in vivo*. Protocol 2 in Chapter 12 may be followed to raise teratomas using cES cells, injecting about 10^7 cells per SCID mouse (corresponding to a confluent culture in a 10 cm dish). The cells may be injected either subcutaneously or intraperitoneally, and teratoma formation becomes apparent in 2–3 months. Tumours are fixed in Bouin's solution in preparation for haematoxilin and eosin staining, or with paraformaldehyde for immunohistochemistry.

Anti-human-specific antibodies may be used to detect tissue-specific proteins, due to cross-reaction with monkey target homologues; but prescreening of monkey specimens is highly advisable. (Localization of antigens may be conveniently visualized with an HRP-labelled secondary antibody and DAB.) The histological and immunohistochemical examination of teratomas raised in this way can reveal various tissues derived from all three primary germ layers: ectodermal tissues containing neurons, glia, glands, and epithelia are most commonly observed; mesodermal tissues such as muscle, cartilage, and bone are also frequently formed; however, endodermal tissues, such as gut epithelium, are less commonly produced (9).

Protocol 5
Production of EBs from cES cells

Reagents and equipment

- Culture of cES cells, on a 60 mm feeder layer: undifferentiated colonies 500-1000 μm in diameter are required
- PBS (*Protocol 2*)
- Collagenase solution: 1 mg/ml collagenase IV (Cat. No. 17104019, Invitrogen) in PBS
- CEM (*Protocol 2*)
- 15 ml conical tube
- Benchtop centrifuge
- Bacteriological-grade Petri dish, 10 cm diameter (Falcon™ No. 351029, BD Biosciences)

Method

1. Aspirate medium, and wash cells with PBS.
2. Cover cells with 0.5 ml collagenase solution per 35 mm dish, and incubate for 20 min.
3. Add 1.5 ml CEM and detach cES-cell colonies from the feeder layer into the medium by tilting the dish.
4. Transfer the suspension of colonies into a 15 ml conical tube, and allow the colonies to settle to the bottom.
5. Carefully remove supernatant medium, collect and transfer cES-cell colonies into 10-15 ml fresh CEM in a bacteriological-grade Petri dish, and culture. Within a few days the cES-cell aggregates will have formed simple EBs in suspension (*Figure 2F*).[a]
6. Continued culture in suspension leads to apparent cellular differentiation: by 2-3 weeks, beating heart muscle and red blood cells may be observed within the core of the structures.[b]

[a] Alternatively, EBs may be plated from the suspension culture directly onto a gelatinized tissue-culture dish at any time. EBs will then attach to the dish and undergo cellular differentiation into various derivatives, such as neurons and cardiac muscle.

[b] Conditions for inducing neural differentiation from EBs are given in Furuya *et al.* (23).

6 Directed differentiation of cES cells induced by coculture

The wide spectrum of differentiated derivatives obtained from cES cells via teratomas and EBs validates their pluripotency, but the cellular heterogeneity hinders the study of specific phenotypes. Thus we have transposed from the murine to the cynomolgus monkey system methods for haematopoietic and neural lineage-specific differentiation of ES cells, which circumvent the requirements for EB formation, addition of morphogens such as retinoic acid, or positive selection systems. Our methods depend on the coculture of cES cells with two distinct stromal cell types, PA6 and OP9, to yield relatively pure populations of either neural or haematopoietic precursors and derivatives, respectively. In this way, lineage-specific patterns of differentiation are consistently achieved.

6.1 Neural differentiation of cES cells induced by coculture with PA6 cells

We have previously described a strong neuralizing activity present on the cell surface of the PA6 line of stromal cells (originally derived from the bone marrow of mouse skull), termed stromal cell-derived inducing activity, or SDIA (1). When cocultured with PA6 cells under serum-free conditions, mES cells differentiate rapidly and efficiently into neural precursors and neurons (Figure 3): over 90% of the cells become positive for the pan-neural marker, neural cell adhesion molecule (NCAM), within a week of coculture (1, 2); and dopaminergic neurons of the midbrain type are produced at a high frequency of ~30% of induced neurons, and are transplantable (24). Our cES cells demonstrate a similar neural-differentiative capacity in the SDIA-based system (3); 45% of cells become NCAM-positive, and midbrain (tyrosine hydroxylase-positive; TH^+) dopaminergic neurons are produced at a frequency of ~35% of induced neurons. (It is notable that this process of differentiation occurs within 10 d of *in vitro* coculture, compared with 5 weeks during cynomulgus monkey embryogenesis.) Interestingly, dorsal- and not ventral-neural markers predominate in SDIA-treated cES cells, in the absence of patterning signals. (This is in contrast to SDIA-treated mES cells, where *both* dorsal- and ventral-neural markers are induced.) In addition to neural tissues, SDIA-treated cES cells differentiate into eye tissues, a proportion (8%) of differentiating colonies producing patches of ($Pax6^+$, pigmented) retinal epithelium (3, 25; and see *Figure 4*). Using this system, Ooto *et al.* (26) also reported the generation of lens tissues of characteristic transparent appearance (lentoid) in a subpopulation of differentiating cES-cell colonies.

This SDIA-based neural differentiation system has subsequently been refined (by addition of exogenous morphogens and growth factors at defined periods and concentrations, and in combinations) so that the full panoply of major

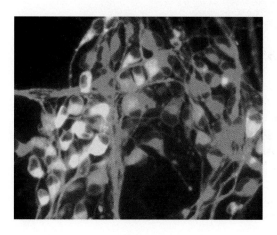

Figure 3 (see Plate 15) Immunostaining of neurons differentiated from cES cells by the SDIA method, after 13 d of coculture with PA6 cells. Green fluorescence represents anti-TH staining (i.e. catecholaminergic neurons), and red, anti-TuJ1 staining (i.e. postmitotic neurons). Cells that are doubly positive appear yellow.

neuroectodermal categories (dorsal and ventral) is produced by cES cells, as well as by mES cells, *in vitro* (24). Thus, it is inferred that in the absence of exogenous patterning signals, SDIA-treated cES cells generate naïve precursors having the potential to differentiate into the full dorsal–ventral range of neuroectodermal derivatives, (24). The implications and relevance of this differentiation system for neurodegenerative diseases (such as Parkinson's disease, and retinitis pigmentosa) are highlighted.

The PA6 stromal cell line requires careful maintenance, avoiding specifically over-confluence, to preserve their neural-inducing activity. Prior to coculture, mitotic inactivation of PA6 cells may be conducted by mitomycin-C treatment (see Chapter 2, for MEF inactivation) but is not advised, as this may result in less efficient neural inducing activity.

The efficiency of neural differentiation of cES cells is affected by the quality and activity of KnockOut™ SR. Therefore batch testing and titration is required to determine an optimal lot and concentration of KnockOut™ SR, which may range from 5–15%. Also, inappropriate concentrations of β-mercaptoethanol cause diminished yields of neural cells.

Due to the low cloning efficiency of cES cells, partially dissociated clumps of 10–50 cells are plated onto PA6 cells, in differentiation medium. For small-scale differentiation experiments, and where cES cell cultures display spontaneous differentiation, undifferentiated colonies may be harvested manually for coculture with PA6 cells. (Unusually, large cES-cell colonies tend to differentiate into non-neural derivatives, and should be avoided where possible.) For large-scale differentiation experiments it is appropriate to use enzymatic disaggregation of cES cells. After 10 d of coculture with PA6 cells, cES-cell colonies display extensive neurite projections, and are often surrounded by mesenchyme-like cells.

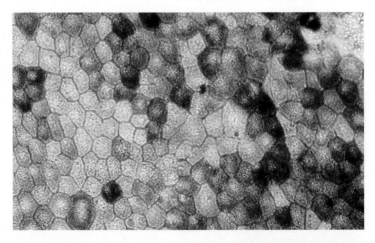

Figure 4 (see Plate 16) A high magnification view of retinal pigment epithelium induced from cES cells by the SDIA method, after 24 d of coculture with PA6 cells. The characteristic polygonal cells are arranged in a cobblestone fashion. These cells accumulate more pigments after an additional week of culture.

Protocol 6
Maintenance of PA6 cells

Reagents and equipment

- Culture of PA6 cells (C3T3-G2/PA6; Cat. No. RCB1127, Riken Cell Bank, Japan; http://www.brc.riken.go.jp/Eng/index.html/)
- PA6-cell medium: alpha MEM (Cat. No. 11900-024, Invitrogen) supplemented with 50 U/ml penicillin and 50 µg/ml streptomycin (Cat. No. 15140-122, Invitrogen), and 10% FCS (Cat. No. 12103-78P, JRH)
- PBS (Protocol 2)
- Trypsin/EDTA solution II: 0.05% trypsin and 0.53 mM EDTA in HBSS (Cat. No. 14175-095, Invitrogen)
- 10 cm tissue-culture dishes (e.g. Cat. No. 353003, BD Biosciences)

Methods

Thawing cryopreserved PA6 cells:
To revive a vial of frozen PA6 cells, follow steps 1–4 of *Protocol 1* in Chapter 7 for OP9 cells, substituting PA6-cell medium for OP9-cell medium, and seeding a thawed aliquot at 10^5 cells/ml. The PA6-cell culture is ready to passage when near confluence.

Expansion of PA6 cells:

1. Aspirate culture medium, and wash PA6 cells with PBS.
2. Cover the cells with 3 ml trypsin/EDTA solution II, and incubate for 5 min at 37 °C.
3. Add 5 ml PA6-cell medium, and break the cell aggregates into a single-cell suspension with gentle pipetting.
4. Collect the suspension, centrifuge the cells at 1000 rpm for 5 min, and resuspend the cells in 5 ml PA6-cell medium.
5. Dispense the cells at a split ratio of 1:5 into 10 cm dishes, using 10 ml medium per dish, and culture.[a]
6. Passage the cells every third day, or when the density reaches near confluence, when a total of $0.5–1.0 \times 10^7$ cells/10 cm dish is attained.

[a] Precoating of dishes is not required.

6.1.1 Evaluation, and time course, of neural and neuronal differentiation

To monitor neural differentiation of cES cells, immunostaining may be performed using antibodies recognizing specific subtypes of neural derivatives. A full description of the time course of SDIA-induced neural differentiation is given elsewhere (3).

Table 2 lists a selection of primary antibodies that are effective for cES-cell derivatives. The neuronal markers, class III β-tubulin (TuJ) and NeuN, are useful

Protocol 7
SDIA-induced neural differentiation of cES cells

Reagents and equipment

- Cynomolgus monkey ES-cell culture in a 10 cm dish. For differentiation studies we use lines, 9 and 6.4 (available from N. Nakatsuji).
- Subconfluent culture of PA6 cells in a 10 cm dish (Protocol 6)
- PA6-cell medium (Protocol 6)
- Trypsin/EDTA solution II (Protocol 6)
- Differentiation medium I: Glasgow Minimum Essential Medium (G-MEM: Cat. No. 11710-035, Invitrogen) supplemented with 10% KnockOut™ SR (Cat. No. 10828-028, Invitrogen); 0.1 mM non-essential amino acids (Cat. No. 11140-050, Invitrogen), 1 mM sodium pyruvate (Cat. No. S-8636, Sigma), 0.1 mM β-mercaptoethanol (Protocol 2)[a]
- Collagen type I solution: prepare a solution of 25–50 µg/ml collagen type I (Cat. No. 354236, BD Biosciences) in 0.02 N acetic acid, and use 5 µg/cm^2 for coating a culture plate or slide; or 1% gelatin solution (Chapter 6, Protocol 1)
- Dissociation solution II: 0.25% trypsin in PBS supplemented with 1 mM CaCl$_2$ and 20% KnockOut™ SR
- Collagenase B solution: (Cat. No. 1-088-807, Roche) 1 mg/ml in Hanks' balanced salt solution (HBSS; Cat. No. 24020-117, Invitrogen)
- KCl-supplemented HBSS (contains 5.33 mM KCl; Cat. No. 24020-117, Invitrogen)
- BSA (Cat. No. A-3156, Sigma)
- Biocoat™ culture slides with 8 wells (Cat. No. 354630, BD Biosciences), or 10 cm tissue-culture dish (Cat. No. 353003, BD Biosciences) containing a 22 × 22 mm glass coverslip[b]
- Inverted phase-contrast microscope
- Haemocytometer
- Conical centrifuge tubes
- Benchtop centrifuge
- Cell strainer with 70 µm mesh size (Cat. No. 352350, BD Biosciences)

Methods

Preparation of PA6-cell monolayers for coculture:

1. Precoat BioCoat™ culture slides with collagen type I solution, and aspirate. Alternatively, pretreat slides with a 0.1% solution of gelatin for 1 h (see Chapter 6, Protocol 1).[c,d]

2. Harvest a 10 cm dish of PA6 cells by trypsinization (Protocol 6, Steps 1–4), and resuspend cells in 25 ml PA6-cell medium giving 2–4 × 10^5 cells/ml.

3. For 8-well chamber slides, dispense 500 µl of the PA6-cell suspension (i.e. 1–2 × 10^5 cells) into each well and for 10 cm dishes, 2–4 × 10^6 cells in 10 ml. Incubate cultures at overnight 37 °C.

4. The next day, monitor that a confluent PA6-cell monolayer has been produced. As long as this is achieved, small differences in cell density will not adversely affect the differentiation-inducing efficiency. Confluent cultures may be incubated for up to 2 d, prior to experimentation.

Protocol 7 continued

Setting up cES/PA6-cell cocultures:

5 Prior to coculture, aspirate medium and wash PA6-cell monolayers with differentiation medium I several times to ensure removal of residual FCS, which strongly inhibits neural differentiation.

6 Prepare clusters of 10–50 undifferentiated cES cells either mechanically, or *en masse* with dissociation solution II, and transfer into 10 ml CEM in a conical tube.

7 Centrifuge at 1000 rpm for 2 min; aspirate the supernatant and resuspend clumps in 6 ml differentiation medium I.

8 Repeat Step 7 twice.

9 Count cES-cell clusters using a haemocytometer and inverted microscope, and plate onto PA6-cell monolayers.[e]

10 Incubate cocultures, changing medium every third day.

Dopamine assay:

1 After 3 weeks of coculture, incubate cell monolayers for 15 min in KCl-supplemented HBSS, using 3 ml per 10 cm dish, to induce depolarization of neurons.

2 Collect the solution, stabilize by addition of EDTA to a final concentration of 0.1 mM EDTA, and store at $-80\,°C$. Analyse the presence of dopamine using HPLC with fluorescence detection (27).

Harvesting neural cells:

(i) To prepare cells for FACS® analysis (from a 10 cm dish), dissociate with 4 ml trypsin/EDTA dissociation solution II for 10 min at 37 °C, add 4 ml of medium, and filter through a cell strainer to remove cell clumps. Pellet cells at 1000 rpm for 5 min, and resuspend in HBSS supplemented with 1% BSA. Cells are now ready for incubation with appropriate antibodies. A selection of suitable markers for sorting neural-precursor cells are given elsewhere (3), and include CD56.

(ii) For subculture, or for transplantation experiments, detach the neural cells by treatment with collagenase B solution (1 mg/ml in HBSS) supplemented with DNase I (25 µg/ml; Cat. No. D-5025, Sigma) for 5 min at r.t., and resuspend in fresh differentiation medium I. The cell suspension can now be either replated, or used as an inoculum for grafting (3).

Directed differentiation of SDIA-induced neural progenitor cells:

With the day of initiation of coculture of cES cells with PA6 cells designated as 'day 0', the following treatments result in the generation of dorsal and ventral neural lineages:

(i) Addition of 0.5–5 nM human bone morphogenetic protein 4 (hBMP4; Cat. No. 314-BP-010, R&D Systems) on days 7–12, or 7–13, promotes differentiation of dorsal-most CNS cells and neural crest derivatives, such as $Brn3a^+/Peri^+$ (sensory lineage) and $TH^+/Phox2b^+/Peri^+$ (autonomic lineage) neurons.

> **Protocol 7** continued
>
> (ii) Addition of sonic hedgehog (Shh; Cat. No. 461-SH-025, R&D Systems) at concentrations of 30–300 nM on days 3–10 suppresses differentiation of neural crest lineage cells, and promotes that of ventral CNS tissues, such as Is11/2$^+$ and HB9$^-$ brain stem-type motor neurons, and HNF3β$^+$ floor plate cells.
>
> (iii) Combined treatment with Shh and 0.2 μM all-*trans* retinoic acid (RA; Cat No. R 2625, Sigma) on days 3–13 inducs differentiation of Is11/2$^+$ and HB9$^+$ spinal cord-type motor neurons.
>
> *Generation of pigmented retinal epithelium and lens tissue:*
>
> (i) After 3 weeks of coculture on PA6 cells, large patches of Pax6$^+$ retinal pigment cells with characteristic, polygonal morphology (*Figure 4*) are observable in one to several percent of the colonies. These cells are easily isolated under a dissecting microscope by scraping with a 0.2 ml Pipetman tip, and expanded (without feeder cells) by plating on a collagen I or Matrigel® (Cat. No. 356234, BD Biosciences) coated culture dish, in DMEM supplemented with 10% FCS and 20 ng/ml bFGF (3, 25). Pigmented cells may be subcultured in the same way.
>
> (ii) Lentoids (26) positive for alpha A-crystallin and Pax6 appear after days 14–16 of SDIA-induction, increasing in number until day 40. Lentoid production may be increased further to 35–40% of colonies by pretreating cES cells with 8 ng/ml bFGF for 3–4 d prior to coculture, and by using a high initial colony density (200–300 colonies/10 cm dish).
>
> [a] To maximize yields of neuronal cells, store medium at 4 °C for no longer than 4 weeks.
>
> [b] The coverslip is later removed for immunocytochemistry, to assess the efficiency of cellular differentiation.
>
> [c] These coatings prevent detachment of PA6 cells during the differentiation period.
>
> [d] Alternatively, if large numbers of differentiated colonies are required, precoat a 10 cm tissue-culture dish containing a glass coverslip.
>
> [e] Do not exceed 500 clusters/10 cm dish, or 20 clusters per chamber well as neural differentiation is inhibited when cES cells are plated at higher density.

for identifying postmitotic neurons. The onsets of NCAM and TuJ expression are typically at 5 d and 7 d of induction, respectively, for SDIA-treated cES cells. Usually 40–50% of differentiated cES cells are found to be NCAM-positive neural cells, and 20–30% are TuJ-positive postmitotic neurons.

TH$^+$ cells are the most predominant neuron type, and first appear in colonies following 10 d of SDIA-treatment. After 2 weeks of induction, 80–90% of colonies contain TH$^+$ cells, and 30–35% of TuJ$^+$ postmitotic neurons are TH$^+$ at the cellular level. The SDIA-treated cES cells release about 1.0 pmol dopamine/10^7 cells in response to high K$^+$ stimuli, indicating the presence of a significant number of functional dopaminergic neurons. In addition, norepinephrine is detectable at about 0.5 pmol/10^7 cells. The SDIA-induced cES cells also gave

Table 2 Antibodies for immunostaining undifferentiated cES cells, and neural and neuronal derivatives

Antibody	Cell specificity	Supplier	Cat. No.	Host	Dilution
SSEA-4	Undifferentiated cES cells	DSHB	MC-813–70	Mouse	1:50
TRA-1–60	Undifferentiated cES cells	Chemicon International	MAB4360	Mouse	1:100
TRA-1–81	Undifferentiated cES cells	Chemicon International	MAB4381	Mouse	1:100
AP2	Neural crest derivatives	DSHB	5EA	Mouse	1:40
Brn3a	Sensory neurons	Chemicon International	MAB1585	Mouse	1:600
ChAT	Neurons	Chemicon International	AB144	Goat	1:20
DBH	Noradrenalinergic neurons	Chemicon International	AB1585	Rabbit	1:200
DA transporter	Dopaminergic neurons	Chemicon International	MAB369	Rat	1:1000
E-cadherin	Undifferentiated cES cells, epidermal cells	Takara	M108	Rat	1:50
GAD	GABAergic neurons	Santa Cruz	SC7513	Rabbit	1:10
HNF3β	Floor plate cells	DSHB	4C7	Mouse	1:100
Human nuclei	Monkey-cell nuclei	Chemicon International	MAB1281	Mouse	1:50
Isl1	Motor neurons	DSHB	39.4D5	Mouse	1:200
NCAM	Neural cells	Chemicon International	AB5032	Rabbit	1:300
NeuN	Postmitotic neurons	Chemicon International	MAB377	Mouse	1:100
MNR2	Motor neurons	DSHB	81.5C10	Mouse	1:10
Pax6	Ventral neural tube cells	DSHB	Pax6	Mouse	1:200
Pax6	Ventral neural tube cells	BAbCO	PRB-278P	Rabbit	1:500
Pax7	Dorsal neural tube cells	DSHB	Pax7	Mouse	1:200
Peripherin	Peripheral neurons	Chemicon International	AB1530	Rabbit	1:100
Peripherin	Peripheral neurons	Santa Cruz	sc7604	Goat	1:100
Serotonin	Serotonin	DiaSorin	20079	Goat	1:1000
TH	Dopaminergic neurons	Chemicon International	AB152	Rabbit	1:200
TH	Dopaminergic neurons	Chemicon International	AB1542	Sheep	1:100
TuJ	Postmitotic neurons	BAbCO	PRB-435P	Rabbit	1:600

rise to neurons that are GAD$^+$ (GABAergic), ChAT$^+$ (cholinergic), and 5HT$^+$ (serotonergic) at efficiencies of 20%, 5%, and 1%, respectively (3). Addition of hBMP4 (0.5 nM) efficiently suppresses neuronal differentiation in cES cells, as in mES cells. BMP4 treatment at an appropriate period can induce specification of the neural crest lineage in SDIA-treated cells. Markers for pigmented retinal epithelium include Pax6, RPE65, CRALBP, and Zo-1; and for lentoid bodies specifically, alpha A-crystallin (26).

Protocol 8
Immunostaining of cES-cell derivatives for neural markers

Reagents and equipment

- Differentiated cES cells on BioCoat™ slides, or on coverslips (*Protocol 7*)
- Paraformaldehyde solution: 4% paraformaldehyde in PBS (see *Protocol 9*, Chapter 9)
- Permeabilization solution: 0.3% Triton X-100 in PBS
- Blocking solution: either 2% skim milk (Cat. No. 232100, BD Biosciences) in PBS, or 10% goat serum (Cat. No. 16210-064, Invitrogen) in PBS
- Primary antibodies (*Table 2*)
- Fluorescent dye-conjugated secondary antibodies
- PBS (*Protocol 2*)

Method

1 Fix cells on coverslips or Biocoat™ slides by incubation with paraformaldehyde solution for 1 h at r.t.[a]

2 Incubate the fixed cells with permeabilization solution for 15 min at r.t., replenishing the solution three times. Wash the samples three times with PBS.

3 Incubate samples for 1 h with blocking solution, to remove non-specific staining.

4 Incubate samples with primary antibodies, diluted appropriately in the blocking solution (see *Table 2*), for 24 h at 4 °C.

5 Wash cells with PBS, and incubate with secondary antibodies conjugated with fluorescent dyes according to either *Protocol 13* in Chapter 6, or *Protocol 10* in Chapter 10.

[a] A detailed procedure for paraformaldehyde fixation that is applicable to cES cells is given in *Protocol 9*, Chapter 10.

6.2 Haematopoietic differentiation of cES cells induced by coculture with OP9 cells

Both primitive and definitive haematopoiesis can be induced sequentially in cES cells by coculture with OP9 stromal cells (28) using a similar method to that described previously for mES cells (see Chapter 7). This cES/OP9-cell coculture system is not only valuable for analysing pathways of primitive and definitive haematopoiesis in primates, but also may serve a role in the provision of haematopoietic derivatives for use in transplantation models.

7 Conclusion

Directed differentiation into a variety of neural and non-neural cell types is now feasible with cES cells. Among the many cell types generated, dopaminergic

Protocol 9
Induction of haematopoetic differentiation in cES cells by coculture with OP9 cells

Detailed methods for the maintenance of OP9 cells, and for their preparation for coculture, are provided in Section 2 of Chapter 7. For the characterization of primitive and definitive haematopoietic cells derived from cES cells according to this Protocol, the reader is referred to Umeda et al. (28).

Reagents and equipment

- Confluent monolayers of OP9 stromal cells, cultured in 6-well dishes
- Culture of undifferentiated cES cells
- Trypsin/EDTA solution III: 0.25% trypsin and 1 mM EDTA in HBSS (Cat No. 25200-072, Invitrogen)
- Cell dissociation buffer (enzyme-free in HBSS; Cat. No.13150-015, Invitrogen)
- Differentiation medium II: alpha MEM supplemented with 10% FCS (Cat. No. 12103-78P, JRH) and 50 mM β-mercaptoethanol (Protocol 2)
- Cell strainer with 70 μm mesh size (Cat. No. 352350, BD Biosciences)
- Recombinant, human vascular endothelial cell growth factor (rhVEGF; Cat. No. 293-VE, R&D Systems Inc.)
- 10 U/ml human erythropoietin (hEPO; 286-EP-250, R&D Systems)

Method

1. Harvest cES cells (as described in *Protocol 3*) using trypsin/EDTA solution III, pellet by centrifugation, and resuspend in differentiation medium II; reseed onto confluent monolayers of OP9 cells using 4×10^3 cES cells/well, and coculture for 6 d.[a]

2. On day 6 of differentiation, harvest the cells using dissociation buffer, resuspend in differentiation medium II, and filter through a cell strainer to remove OP9 cells.

3. Collect the filtrate, count suspended cells, and reseed 1×10^5 cells per well of a fresh, 6-well dish of OP9 cells. Continue coculture, changing medium every 2-3 d.

4. Nucleated cells, corresponding to primitive erythrocytes (eryP), are released into the culture medium on day 8 of differentiation; and smaller erythrocytes corresponding to definitive erythrocytes (eryD) on days 16-18. These non-adherent fractions may be harvested and processed for May-Grünwald/Giemsa staining, immunostaining, or RT-PCR; or may be subjected to FACS® analysis (28; and see Chapter 7).

5. Adherent clusters of haematopoietic progenitor cells with characteristic 'cobblestone' appearance become apparent at days 8-10. Colony-forming assays for progenitor cells are described in Umeda et al. (28).

[a] To induce eryD development, supplement differentiation medium II with 10 U/ml hEPO throughout; and to stimulate both primitive and definitive haematopoiesis (increasing the yield of floating haematopoietic cells and of adherent progenitor cells), add 20 ng/ml rhVEGF from days 0-6.

neurons and retinal pigment epithelium are now directly applicable to preclinical studies into cell therapy for Parkinson's disease and retinal degeneration, respectively. This has already been demonstrated by a group at Kyoto University Hospital, who reported functional motor recovery in drug-induced Parkinson's disease-model monkeys by striatal implantation of dopaminergic neurons induced from cES cells by the SDIA method (29). Thus, the nature and versatility of cES cells, coupled with the suitability of the cynomolgus monkey as an animal model, signify that the value of this species for the development of cell therapies is likely to be significant.

Acknowledgements

This work was supported by a grant from the Japan Society for the Promotion of Science. We thank Drs Kenji Mizuseki, Kiichi Watanabe, and Morio Ueno for their useful advice on neural differentiation protocols.

References

1. Kawasaki, H., Mizuseki, K., Nishikawa, S., Kaneko, S., Kuwana, Y., Nakanishi, S., Nishikawa, S.-I., and Sasai, Y. (2000). *Neuron*, **28**, 31–40.
2. Kawasaki, H., Mizuseki, K., and Sasai, Y. (2001). *Methods Mol. Biol.*, **185**, 217–28.
3. Kawasaki, H., Suemori, H., Mizuseki, K., Watanabe, K., Urano, F., Ichinose, H., Haruta, M., Takahashi, M., Yoshikawa, K., Nishikawa, S., Nakatsuji, N., and Sasai, Y. (2002). *Proc. Natl. Acad. Sci. USA*, **99**, 1580–5.
4. Akai, T., Yamaguchi, M., Mizuta, E., and Kuno, S. (1993). *Ann Neurol.*, **33**, 507–11.
5. Jenner, P. (2003). *Parkinsonism Relat. Disord.*, **9**, 131–7.
6. Thomson, J.A., Kalishman, J., Golos, T.G., Durning, M., Harris, C.P., Becker, R.A., and Hearn, J.P. (1995). *Proc. Natl. Acad. Sci. USA*, **92**, 7844–8.
7. Thomson, J.A., Kalishmanm, J., Golos, T.G., Durning, M., Harris, C.P., and Hearn, J.P. (1996). *Biol. Reprod.*, **55**, 254–9.
8. Thomson, J.A. and Marshall, V.S. (1998). *Curr. Top. Dev. Biol.*, **38**, 133–65.
9. Suemori, H., Tada, T., Torii, R., Hosoi, Y., Kobayashi, K., Imahie, H., Kondo, Y., Iritani, A. and Nakatsuji, N. (2001). *Dev. Dynamics*, **222**, 273–9.
10. Thomson, J.A., Itskovitz-Eldor, J., Shapiro, S.S., Waknitz, M.A., Swiergiel, J.J., Marshall, V. S., and Jones, J.M. (1998). *Science*, **282**, 1145–7.
11. Reubinoff, B.E., Pera, M.F., Fong, C.Y., Trounson, A., and Bongso, A. (2000). *Nat. Biotechnol.*, **18**, 399–404.
12. Nagy, A., Gertsenstein, M., Vinterstein, K., and Behringer, R. (2003). Isolation and culture of blastocyst-derived stem cell lines. In *Manipulating the Mouse Embryo*, 3rd edn (eds A. Nagy, M. Gertsenstein, K. Vinterstein, R. Behringer), pp. 359–97, Cold Spring Harbor Laboratory Press, New York.
13. Amit, M., Carpenter, M.K., Inokuma, S., Chiu, C.P., Harris, C.P., Waknitz, M.A., Itskovitz-Eldor, J., and Thomson, J.A. (2000). *Dev. Biol.*, **227**, 271–8.
14. Torii, R., Hosoi, Y., Masuda, Y., Iritani, A., and Nigi, H. (2000). *Primates*, **41**, 39–47.
15. Solter, D. and Knowles, B.B. (1975). *Proc. Nat. Acad. Sci. USA*, **72**, 5099–102.
16. Reubinoff, B.E., Pera, M.F., Vajta, G., and Trounson, A.O. (2001). *Hum. Reprod.*, **16**, 2187–94.
17. Fujioka, T., Yasuchika, K., Nakamura, Y., Nakatsuji, N., and Suemori, H. (2004). *Int. J. Dev. Biol.*, **48**, 1149–54.

18. Borrell, A., Ponsà, M., Egozcue, J., Rubio, A., and Garcia, M. (1998). *Mutat. Res.*, **403**, 185-98.
19. Chung, S., Sonntag, K.C., Andersson, T., Bjorklund, L.M., Park, J.J., Kim, D.W., Kang, U.J., Isacson, O., and Kim, K.S. (2002). *Eur. J. Neurosci.*, **16**, 1829-38.
20. Kim, J.H., Auerbach, J.M., Rodriguez-Gomez, J.A., Velasco, I., Gavin, D., Lumelsky, N., Lee, S.H., Nguyen, J., Sanchez-Pernaute, R., Bankiewicz, K., and McKay, R.D. (2002). *Nature*, **418**, 50-6.
21. Blyszczuk, P., Czyz, J., Kania, G., Wagner, M., Roll, U., St-Onge, L., and Wobus, A.M. (2003). *Proc. Natl. Acad. Sci. USA*, **100**, 998-1003.
22. Asano, T., Hnazono, Y., Ueda, Y., Muramatsu, S., Kume, A., Suemori, H., Suzuki, Y., Kondo, Y., Harii, K., Hasagawa, M., Nakatsuji, N., and Qzawa, K. (2002). *Mol. Therapy*, **6**, 162-8.
23. Furuya, M., Yasuchika, K., Yoshimura, Y., Nakatsuji, N., and Suemori, H. (2003). *Genesis*, **37**, 180-7.
24. Mizuseki, K., Sakamoto, T., Watanabe, K., Muguruma, K., Ikeya, M., Nishiyama, A., Arakawa, A., Suemori, H., Nakatsuji, N., Kawasaki, H., Murakami, F., and Sasai, Y. (2003). *Proc. Natl. Acad. Sci. USA*, **100**, 5828-33.
25. Haruta, M., Sasai, Y., Kawasaki, H., Amemiya, K., Ooto, S., Kitada, M., Suemori, H., Nakatsuji, N., Ide, C., Honda, Y., and Takahashi, M. (2004). *Invest. Ophthalmol. Vis. Sci.*, **45**, 1020-5.
26. Ooto, S., Haruta, M., Honda, Y., Kawasaki, H., Sasai, Y., and Takahashi, M. (2003). *Invest. Ophthalmol. Vis. Sci.*, **44**, 2689-93.
27. Mitsui, A., Nohta, H., and Ohkura, Y. (1985). *J. Chromatogr.*, **344**, 61-70.
28. Umeda, K., Heike, T., Yoshimoto, M., Shiota, M., Suemori, H., Luo, H.Y., Chui, D.H., Torii, R., Shibuya, M., Nakatsuji, N., and Nakahata, T. (2004). *Development*, **131**, 1869-79.
29. Takagi,Y., Takahashi, J., Saiki, H., Morizane, A., Hayashi, T., Kishi, Y., Fukuda, H., Okamoto, Y., Koyanagi, M., Ideguchi, M., Hayashi, H., Imazato, T., Kawasaki, H., Suemori, H., Omachi, S., Iida, H., Itoh, N., Nakatsuji, N., Sasai, Y., and Hashimoto, N. (2005). *J. Clin. Invest.*, **115**, 102-9.
30. Sumi, T., Fujimoto, Y., Nakatsuji, N., and Suemori, H. (2004). *Stem Cells*, **22**, 861-72.
31. Cibelli, J.B, Grant, K.A., Chapman, K.B., Cunniff, K., Worst, T., Green, H.L., Walker, S.J., Gutin, P.H., Vilner, L., Tabar, V., Dominko, T., Kane, J., Wettstein, P.J., Lanza, R.P., Studer, L., Vrana, K.E., and West, M.D. (2002). *Science*, **295**, 819.
32. Ginis, I., Luo, Y., Miura, T., Thies, S., Brandenberger, R., Gerecht-Nir, S., Amit, M., Hoke, A., Carpenter, M.K., Itskovitz-Eldor, J., and Rao, M.S. (2004). *Dev. Biol.*, **269**, 360-80.

List of suppliers

Abbott Laboratories, Abbott Park (IL), USA.
http://www.abbott.com/

Abcam Ltd, Cambridge, UK.
http://www.abcam.com/

American Type Culture Collection (ATCC), Manassas (VA), USA.
http://www.atcc.org/

Amersham Biosciences UK Ltd, Little Chalfont (Bucks.), UK.
http://www1.amershambiosciences.com/

Amicon, part of **Millipore Corporation**, Billerica (MA), USA.
http://www.millipore.com/

Applied Biosystems, Foster City (CA), USA.
http://www.appliedbiosystems.com/

Aska Pharmaceutical Company Ltd, Tokyo, Japan.
http://www.aska.co.jp/

Axon Instruments, part of **Molecular Devices Corporation**, Sunnyvale (CA), USA.
http://www.moleculardevices.com

BabCO, part of **Covance Research Products, Inc.**, Berkeley (CA), USA.
http://www.crpinc.com/

Bacto Laboratories Pty Ltd, Liverpool (NSW), Australia.
http://www.bacto.com.au/main.htm

Bandelin Electronic GmbH & Co. KG, Berlin, Germany.
http://www.bandelin.com/

BD Biosciences (Pharmingen), San Jose (CA), USA.
http://www.bdbiosciences.com/index.shtml

Bellco Glass, Inc., Vineland (NJ), USA.
http://www.bellcoglass.com/

Bethyl Laboratories, Inc., Montgomery (TX), USA.
http://www.bethyl.com/

Biochemical Laboratory Service Ltd, Budapest, Hungary.
http://www.bls-ltd.com/

Biometra G.m.b.H. i.L., Goettingen, Germany.
http://www.biometra.de/

Bioptechs, Inc., Butler (PA), USA.
http://www.bioptechs.com/

Bio-Rad Laboratories, Inc., Hercules (CA), USA.
http://www.bio-rad.com/

BLS Ltd, see **Biochemical Laboratory Service Ltd BTX Technologies, Inc.**, Hawthorne (NY), USA.
http://www.btx.com/

LIST OF SUPPLIERS

CALBIOCHEM, brand of **EMD Biosciences, Inc.**, San Diego (CA), USA.
http://www.emdbiosciences.com/html/CBC/home.html

Cambrex Corporation, East Rutherford (NJ), USA.
http://www.cambrex.com/

Campden Instruments Ltd, Leicester, UK.
http://www.campden-inst.com/

Carl Zeiss A.G., Oberkochen, Germany.
http://www.zeiss.com/

Cascade Biologics Inc., Portland (OR), USA.
http://www.cascadebio.com/

CellSystems® Biotechnologie Vertrieb GmbH, St. Katharinen, Germany.
http://www.cellsystems.de/

Charles River Laboratories International, Inc., Wilmington (MA), USA
http://www.criver.com/

Chemicon International, Inc., Temecula (CA), USA.
http://www.chemicon.com/

CLEA Japan, Inc., Tokyo, Japan.
http://www.clea-japan.com/

Clontech International, Inc., Mountain View (CA), USA.
http://www.clontech.com/

Colorado Video, Inc., Boulder (CO), U.S.A.
http://www.colorado-video.com/

Columbus Instruments, Inc., Columbus (OH), USA.
http://www.colinst.com/

Corning, Inc., Corning (NY), USA.
http://www.corning.com/lifesciences/

DakoCytomation California Inc., Carpinteria (CA), USA
http://www.dakocytomation.us/

Delaware Diamond Knives, Inc., Wilmington (DE), USA.
http://www.ddk.com/

Developmental Studies Hybridoma Bank, University of Iowa, USA.
http://www.uiowa.edu/~dshbwww/

DiaSorin, Inc., Stillwater (MN), USA.
http://www.diasorin.com/

Difco, Inc., see **BD and Company**, Franklin Drive (NJ), USA.
http://www.bd.com/

Dynal Biotech, Oslo, Norway.
http://www.dynalbiotech.com/ *and see* http://www.invitrogen.com/DYNAL

Electron Microscopy Sciences, Hatfield (PA), USA.
http://www.emsdiasum.com/default.htm

Engelbrecht Medizin- und Labortechnik GmbH, Edermünde, Germany.
http://www.engelbrecht.de/

Eppendorf AG, Hamburg, Germany.
http://www.eppendorf.com/

Fine Science Tools, Inc., North Vancouver, Canada.
http://www.finescience.com/

Fisher Scientific International, Inc., Hampton (NH), USA.
https://www1.fishersci.com/

Fluka GmbH, Buchs, Switzerland.
http://www.sigmaaldrich.com/fluka

FST, see **Fine Science Tools, Inc.**
Greiner Bio-One International AG, Kremsmuenster, Austria
http://www.gbo.com/

Harlan Bioproducts for Science Inc., Indianapolis (IN), USA.
http://www.harlan.com/

LIST OF SUPPLIERS

Helena Laboratories Corporation,
Beaumont (TX), USA.
http://www.helena.com/

Heraeus Holdings GmbH, Hanau,
Germany.
http://www.heraeus-instruments.de/

HyClone, Logan (UT), USA.
http://www.hyclone.com/

Invitrogen Corporation, Carlsbad (CA), USA.
http://www.invitrogen.com/

Iwaki Glass Company Ltd, Iwaki, Japan.
http://www.igc.co.jp

Jackson Laboratory, Bar Harbor (ME),
USA.
http://www.jax.org/

JRH Biosciences, Inc., Lenexa (KS), USA
http://www.jrhbio.com/

Kaken Pharmaceutical Co. Ltd, Tokyo,
Japan.
http://www.kaken.co.jp/

Kord Products Inc., Ontario, Canada.
http://www.kord.ca/

Lamb, see **Raymond A Lamb LLC**.

Leica Microsystems AG, Wetzlar,
Germany.
http://www.leica-microsystems.com/

Leitz GmbH & Co KG, Stuttgart, Germany.
http://www.leitz.org/

Linco Research, Inc., St. Charles (MO), USA.
http://www.lincoresearch.com/

Marabuweke GmbH & Co. KG, Tamm,
Germany.
http://www.marabu.de/

Matsunami Glass Industries, Osaka Japan.
http://www.matsunami-glass.co.jp/

MatTek Corporation, Ashland (MA), USA.
http://www.mattek.com/

Mediatech, Inc., Herndon (VA), USA.
http://www.cellgro.com/

Menzel-Gläser, see **Gerhard Menzel,
Glasbearbeitungswerk GmbH & Co. KG**,
Braunschweig, Germany.
http://www.menzel.de/

Mercodia AB, Uppsala, Sweden.
http://www.mercodia.se/

Microm UK Ltd, Bicester, (Oxon), UK.
http://www.microm.co.uk/

Molecular Probes™, see **Invitrogen**.
http://probes.invitrogen.com/

Multi Channel Systems MCS GmbH,
Reutlingen, Germany
http://www.multichannelsystems.com/

Muto Pure Chemicals Co., Ltd,
Tokyo, Japan.
http://www.mutochemical.co.jp

Nalge Nunc International, Rochester
(NY), USA.
http://www.nalgenunc.com/

Nova Biomedical Corporation,
Waltham (MA), USA.
http://www.novabiomedical.com/

Oxoid Ltd, Basingstoke (Hants), UK
http://www.oxoid.com/

PAA Laboratories GmbH, Linz, Austria.
http://www.paa.at/

Paul Marienfeld GmbH & Co. KG,
Lauda-Königshofen, Germany.
http://www.marienfeld-superior.com

PeproTech Inc., Rocky Hill (NJ), USA.
http://www.peprotech.com/

Perkin Elmer, Inc., Boston (MA), USA.
http://www.perkinelmer.com/

Pharmacia Biotech AB, Uppsala, Sweden.
http://www.pharmacia.com/ *and see*
http://www.pfizer.com/

LIST OF SUPPLIERS

Plano GmbH, Wetzlar, Germany.
http://www.plano-em.de/

Prime Tech Ltd, Ibaraki, Japan.
http://www.primetech-jp.com/

Promega Corporation, Madison (WI), USA.
http://www.promega.com/

QIAGEN, Inc., Valencia (CA), USA.
http://www1.qiagen.com/

R&D Systems, Inc., Minneapolis (MN), USA.
http://www.rndsystems.com/

Raymond A Lamb LLC, Apex (NC), USA.
http://www.ralamb.net/

Research Systems, Inc., Boulder (CO), USA.
http://www.rsinc.com/

Riken Cell Bank, Ibaraki, Japan.
http://www.brc.riken.go.jp/

Roche Applied Science, and **Roche Diagnostics**, part of **F. Hoffmann-La Roche Ltd**, Basel, Switzerland.
http://www.roche-applied-science.com/

Sankyo LifeTech, Tokyo, Japan.
http://www.sankyo-lifetech.co.jp/

Santa Cruz Biotechnology Inc., Santa Cruz (CA), USA.
http://www.scbt.com/

Schütt Labortechnik GmbH, Göttingen, Germany.
http://www.schuett-labortechnik.de/

Serotec Ltd, Oxford, UK.
http://www.serotec.com/

SERVA Electrophoresis GmbH, Heidelberg, Germany.
http://www.serva.de/

Sigma-Aldrich Inc., St. Louis (MO), USA.
http://www.sigmaaldrich.com/

Specialty Media, part of **Cell & Molecular Technologies Inc.**, Phillipsburg (NJ), USA.
http://www.specialtymedia.com/

StemCell Technologies, London, UK.
http://www.stemcell.com/

Stoelting Co., Wood Dale (IL), USA.
http://www.stoeltingco.com/

Stratagene Corporation, La Jolla (CA), USA.
http://www.stratagene.com/

Strathmann Biotec AG, Hamburg, Germany.
http://www.biotec-ag.de/

Sutter Instrument Company, Novato (CA), USA.
http://www.sutter.com/

SYNTHECON™ Inc., Houston (TX), USA.
http://www.synthecon.com/

Takara Bio Inc., Shiga, Japan.
http://www.takara-bio.com/

Teikoku Zoki Co., Tokyo, Japan.
http://www.teikoku-hormone.co.jp/

Teva Medical Co., Ashedod, Israel.
http://www.teva-medical.co.il/

Thermo Shandon Inc., Pittsburgh (PA), USA.
http://www.shandon.com/

Upstate Biotechnologies, Inc., Waltham (MA), USA.
http://www.upstatebiotech.com/

Vector Laboratories, Inc., Burlingame (CA), USA.
http://www.vectorlabs.com/

ViaGene Biotech, Inc., Los Angeles (CA), USA.
http://viagenbiotech.com/

VWR International, West Chester (PA), USA.
http://www.vwrsp.com/

Wako Pure Chemical Industries Ltd,
Osaka, Japan.
http://www.wako-chem.co.jp/

Worthington Biochemical Co.,
Lakewood (NJ), USA.
http://www.worthington-biochem.com/

Yellow Springs Monitoring Systems (YSI) Inc., Yellow Springs (OH), USA
http://www.ysi.com/

Zymed Laboratories, see **Invitrogen**
http://www.zymed.com/

Index

action potential recordings 164
Activin 115
adult (progenitor) stem cells 289
   transplantation 234
albumin
   ELISA 233-4
   synthesis 229
angiogenesis 265, 267, 277, 280-1
animal transplantation models 294-5, 318
   hepatic 235-6
   neural progenitors 203-11, 310
      in utero transplantation 206-11
      organotypic slice cultures 203-6, 207-8
   pancreatic 234-5
aphidicoline 56
assisted reproduction 106, 106

bFGF (*equivalent to* FGF2) 261, 299, 300, 314
   and see FGF2
β-mercaptoethanol 114, 119
bioenergetic threshold 76
BMP
   haematopoietic differentiation 174
   hES-cell differentiation 261
   mesoderm differentiation 114-16, 119
Brachyury 115
brain synaptosomes 74
   isolation 80, 88-90
BrdU 87, 96, 97
BSA 114, 116, 117
   replacement with PVA 117

CAP^R 73
   cell lines 73
   marker 77
   mtDNA 73, 79, 96

CDM 114
   for mesoderm and haematopoietic differentiation 124, 125
   for neuroectoderm differentiation 124, 125
   preparation 119, 120
   recommended mES cell lines 118
   and see defined culture systems
cES cells 294-319
   cell-cycle time 296
   cellular heterogeneity 308
   characterization 301-2, 304
      antigenic markers 296, 301-2
      karyotype 302, 304
   cloning efficiency 296, 297, 307, 310
   feeder cells 296
   LIF 296
   morphology 296
   pluripotency 294, 296
      retention 301, 303
      teratomas 307
   therapeutic potential 316, 318
cES-cell derivation 296-301
   efficiency 297
   feeder cells 296, 297-8
   ICM isolation 296, 298-9
      anti-monkey antiserum 298
      colony dissociation 299-300
      embryos 298
      immunosurgery 298-9
cES-cell derivatives
   haematopoietic cells 317
   lens tissue 309, 314
   mesenchyme-like cells 310
   neural and neuronal cells 312-14
      immunochemical analysis 311, 314-16
   pigmented retinal epithelium 309, 310, 314, 318

INDEX

cES-cell differentiation in monolayers
  cES/OP9-cell coculture system 316
  haematopoietic differentiation 308-9, 316-17
    erythrocyte differentiation 316
  lens tissue 309
  mesenchyme-like cells 310
  neural and neuronal differentiation 309-16
    immunochemical analysis 311, 314-16
    method 312
    and see SDIA-based neural differentiation systems
  pigmented retinal epithelium 309-10
cES-cell differentiation via EBs 302, 305, 308
  eryrthrocyte differentiation 308
  heart muscle differentiation 308
  neuronal differentiation 308
cES-cell maintenance 301-3
  bFGF 299, 300
  cryopreservation 301, 303-4
  enzymatic dissociation 297, 302, 303
  FCS *vs.* serum replacement 297
  mechanical dissociation 297
  spontaneous differentiation 297, 301, 302, 305, 310
  zoonosis 297
cES/OP9-cell coculture system 316
cell viability 133
CFACs 181
CFU 277, 317
CFU-OP9 171, 181
  colony assay 182-3
CFU-E and CFU-GM 171
  colony assays 182-3
chorionic gonadotrophin 264
collagen (type) I 227, 280, 312
collagen (type) IV 279
confocal imaging 165-6
controlled differentiation (of mES cells) 112-29, 189, 218-19
  hepatic 226-7
  neural 189-217
  pancreatic 225-6
  and see guided differentiation (of hES cells)
C-peptide 229
cybrid transfer technique 83
  and see mES cybrids
cynomolgus monkey 295
cynomolgus monkey ES cells, *see* cES cells
cytoplast 54, 73
  isolation 85-90

db-cAMP 134, 136, 141
dedifferentiation 4-5, 6

defined culture systems for hES cells 261
defined culture systems for mES cells 112
  culture requirements 113
  for cardiogenesis 134
  for neural differentiation 191
  in vivo relevance 128
  rationale 116
  and see CDM, serum-free medium
definitive erythroid-cell differentiation 174, 176-86, 316-17
  differentiation-inducing factors 176, 317
  erythroid-myeloid lineages 177
  FACS® analysis 177, 317
  lineage-specific cell-surface antigens 177
  morphology 175, 176
definitive haematopoiesis 169-88
  definition 169
  induction 178-81, 316-17
  and see definitive erythroid-cell differentiation
differentiated state 5-6
  reversibility 6
DiI-acetylated low-density lipoprotein 280
directed differentiation (for hES cells) 316
DMSO (as morphogen)
  in muscle-cell differentiation 134
dopamine assay 313

EC cells 4, 73, 113, 239
ECs 269, 279, 280
ECM 279, 280
  ECM-associated factors 219
  laminin 261
  Matrigel™ 261, 280
EBs
  differentiation system 4, 190
  ISH on cryosections 159-60
  plating 119, 222-3
    critical factors 222, 131
  timecourse of EB formation 268
  whole-mount ISH 157-9
EGFP *see* GFP
ELISA 227, 228
  for albumin 233-4
  for insulin 232-3
embryos
  cynomolgus monkey 298
  human 240
embryo collection (mouse) 24-25, 47-8
embryo-culture medium (mouse) 43
embryo transfer (mouse) 43
embryogenesis
  endodermal lineages 218
  haematopoiesis 169-86
  liver development 218

muscle lineages 130
neural precursors 190
pancreas development 218
endothelial progenitor cells 279
epiblast
   as source of ES cells 9
   culture 32
   microsurgical isolation 9, 28-34
Epo 174, 317
erythrocyte differentiation 173-81, 316
   eryD 169, 317
      ALAS-E 176
   eryP 169, 317
      εγ-globin 175
      morphology 175
   May-Grünwald/Giemsa staining 176-8, 184-5, 317
   RT-PCR analysis 175, 317
erythroid-myeloid differentiation 115, 177
ES cells
   cell-cycle synchronisation 56-8
   criteria 295-6
   differentiation 1-2
   inter-species comparison 239, 253, 255, 258, 294, 295
   inter-species differences 239, 255, 294, 296
   in vivo relevance 260, 265
   nature and origin 2
   pluripotency 2, 41
   properties 2
   publications 1
   therapeutic potential 2, 189, 260, 316, 318
   and see cES, hES and mES cells
ES-cell derivation
   contaminants 3
   critical reagents 3
   immunosurgery 3
ESGPs 192
   differentiation 195
ESNPs 192
   differentiation 195
EtBr 83
extraembryonic tissues 260, 264

FCS
   batch-testing 13, 15, 18, 19-20, 114, 173, 219
   for haematopoietic differentiation 114, 173
   for hepatic differentiation 219, 223
   for mesoderm differentiation 114
   for muscle differentiation 131
   for pancreatic differentiation 219
   inherent factors 115
   masking effects on differentiation 114
   replacement 117
feeder cells for cES cell culture 296
   MEFs 297, 299

preparation 297-8
   STO fibroblasts 297, 299
feeder cells for hES cell culture 239
   cryopreservation 241
   feeder cell-conditioned medium 261, 262-3
   MEFs 241
   MEFs *vs.* STO fibroblasts 241
   preparation 242
feeder cells for mES cell culture 8, 11, 93, 171
   cryopreservation 17
   MEFs *vs.* STO fibroblasts 11, 219
   preparation 13, 16
   SNL76/7 fibroblasts 93
female mES cell lines 35, 81, 93
female blastocysts (mouse)
   GFP-based selection 99-102
FGF2 (*equivalent to* bFGF) 243
   and see bFGF

gelatin-coating of culture dishes 11, 133, 171
genetic clock 104
genetically modified cES cells 304-5
   electroporation 306-7
   transfection methods 305
   transformation efficiency 305
genetically modified hES cells 289-90
genetically modified mES cells 1
   for cardiomyocytes differentiation 139
   for haematopoietic differentiation 170
   for hepatic disease models 235
   for liver reconstitution models 235
   for neural differentiation 196-8
      neuronal selection 196-203
   for pancreatic differentiation 219, 225
GFP (and EGFP) 191, 196-200, 203, 206, 305
   X-linked 99-102
   visualisation 68
GPI isozymes 46, 68
guided differentiation 277
   and see controlled differentiation
GUP medium 83, 96

haemangioblast 169
haematopoietic chimerism 290
haematopoietic differentiation 266-7, 277, 278, 316, 317
haematopoietic progenitors 170, 171, 178, 181, 182-3, 184, 186, 317
   morphology 175, 181, 317
HAT medium 87, 94-6, 97
HCM 223, 226
hepatic differentiation 226-7
   characterisation 228
   and see mES-cell differentiation via EBs
hepatic progenitor cells 224
hepatocyte-like cells 224

INDEX

hES cells 1–6, 106, 211, 215, 238–93
  antigenic markers 249–50
  cellular heterogeneity 249, 255
  characterisation 244, 249–59
  criteria 244–9
  current challenges 289
  gene expression 253, 255
  genetic modification 289–90
  immunochemical analysis 249–51
    antigenic markers 249–50
    FACS analysis 252–3, 255
    immunofluorescence staining 250–1
  immunological intolerance 290
  in vivo relevance 260, 265, 268–9, 289
  karyotyping 244
  morphology 245
  pluripotency 238, 241, 244, 249, 253, 258, 260–90
  therapeutic potential 260, 289–90, 294
hES-cell derivation 239
  anti-human antiserum 240
  ethical considerations 2, 239
  feeder cells 239–41
  human embryos 239–40
  ICM isolation 240–1
  immunosurgery 240–1
  media 240, 241
    serum-free 241
    serum-supplemented 240
  methodology 238
hES-cell derivatives
  antigenic markers 215, 275
  immunofluorescence analysis 269, 271–2, 275–8, 280
  ISH analysis 212–14
  microarray analysis 255, 268
  RT-PCR analysis 255, 256–7, 273–4
hES-cell differentiation 260–93
  angiogenesis 265, 267, 277, 280–1
  haematopoietic differentiation 266–7, 277, 278
  lineage specific differentiation 260
  mesodermal differentiation 265, 267, 277, 279, 282
    mesodermal progenitor cells 279–83
  neural differentiation 211, 215
    neuronal differentiation 267
  spontaneous differentiation 241, 245, 249
  vascular differentiation 277, 279, 282–3
    in vivo relevance 280–1
  and see hES-cell pluripotency
hES-cell differentiation in monolayers 277, 279–81
  angiogenesis 280
  endothelial progenitor cells 279
  mesodermal differentiation 277, 279, 282

  mesodermal progenitor cells 279–83
  neural differentiation 309
  vascular differentiation 279, 282–3
    assays for vasculogenesis 283–4
hES-cell differentiation via EBs 267
  cardiomyocyte differentiation 272, 285
    antigenic markers 275
    contractile EBs 272, 275, 276
    electrophysiological properties 272, 275
    MEA systems 272, 276
  dynamic culture systems 284–8
    advantages 288
    RCCS^{TM} 285
    STLV^{TM} 287, 286–7
  EB agglomeration 285
  hanging-drop method 267, 284
  in serum-free medium 267
  in serum-supplemented medium 267
  in suspension 267, 270, 284
  in three-dimensional matrices 288–9
    cardiomyocyte differentiation 288
    haematopoietic differentiation 288
    vascular differentiation 288
  in vivo relevance 268–9, 289
  timecourse of EB formation 268
  vascular differentiation 269, 275
    antigenic markers 275
    ECs 269
    immunofluorescence analysis 271–2
    v-SMC 269
hES-cell maintenance 243–7, 261–4
  aneuploidy 244
  bFGF 261, *and see* FGF2
  cellular heterogeneity 249, 255
  colony dissociation
    enzymatic 241, 245, 246–7
    mechanical 241, 243, 245
  critical factors 244
  cryopreservation 244
    thawing and recovery 247–9
    vitrification 244, 247–9
  defined culture systems 261
  feeder cells 239–41, 261
  feeder cell-conditioned medium 261, 262–3
  feeder-free culture 261
  immunomagnetic isolation 253, 254–5
  karyotyping 244
  media 243, 243–4, 246–7
    serum-free conditions 246–7
    serum-replacement *vs.* serum-supplemented 243–4
  MEF-conditioned medium 262–3
  scale up 281
  spontaneous differentiation 241, 245, 249
  subculture 243, 245

zoonosis 261
hES-cell pluripotency 260-90
   differentiation into extraembryonic tissues 260, 264
   in vivo teratomas
      analysis 258-9
      in limb muscle 265-7
      in testis capsule 257-8
   spontaneous differentiation 241, 245, 249, 263, 264-5
   trophoblast differentiation 264
   and see hES-cell differentiation
hyperthermic treatment of mouse embryos 52-3
   embryolethality 52
HPRT
   enzyme 87, 94, 96
   gene 115
HSC 116, 125, 181

insulin
   ELISA 232-3
   synthesis 229
islet-like cells 223, 224
IL-3 174
immunofluorescence analysis 228-31
   hepatic derivatives 228
   immunostaining 231
   microscopic imaging 234
   pancreatic derivatives 228
   sample fixation 230
immunological intolerance 290, 294, 305
immunosurgery
   cynomolgus monkey embryos 298-9
   human embryos 240-1
implantation delay 3, 8, 32
ISH 149, 151-3, 155-6, 157-60, 212-14
ITSFn 191

karyoplast 41, 73
karyotyping
   cEs cells 302, 304
   hEs cells 244
   mEs cells 33, 36-8

lacZ staining 68
laminin 223, 225
LIF
   in cES cell culture 296
   in mES cell culture 8, 118
lineage restriction 4-5
lineage selection 144-8, 218-19
   antibiotic selection 144-8, 196-8
   cardiomyocyte selection 139
   hepatic 223-4
   neural lineage selection 190, 196-203

astrocytes 192, 196, 209, 214, 215
neuronal 199-200
oligodendrocytes 190, 191, 192, 196, 203, 204, 214
oligodendroglial 202-3, 204
using FACS® 199-200
using immunopanning 201-2
pancreatic 223, 225-6
lineage-specific differentiation 295

Macaca fascicularis 295
Matrigel™ 263
MEA systems 165, 272, 276
mES cells
   epigenetic instability 42, 54, 69, 131
   genetic imprinting 69
mES-cell characterization
   karyotype 33, 36-8
   pluripotency 33-4
mES-cell derivation 7-40
   from intact blastocysts 22, 26
   from isolated epiblast 23, 28-32
   LIF 8
   media and supplements 11, 12
      batch-tested FCS 15, 18, 19, 219
   mouse strains and efficiencies 7-10
   transgenic mice 9, 10
mES cell-derived mice 41-71
   developmental abnormalities 41, 69-70
   mES-cell contribution 68
   perinatal mortality 41, 43, 55, 69-70
mES-cell derivatives
   haematopoietic 169-88
      erythrocytes 173-81, 316, 317, 316
         eryD 169, 317
         eryP 169, 317
   hepatic 218-37
   immunofluorescence analysis 228-31
   mesodermal 178-9
   mucle130-68
      cardiomyocytes 130, 134, 136, 139
      skeletal myocytes 135-6, 138, 140
      SMC 130, 141-4
      VSM 134, 141-2
   neural 189-217
      astrocytes 192, 196, 209, 214, 215
      neuronal 199-200
      oligodendroglial 202-3, 204
      oligodendrocytes 190, 191, 192, 196, 203, 204, 214
   pancreatic 218-37
   and see animal transplantation models
mES-cell differentiation 112-237
   cardiomyocyte selection 145-8
   haematopoietic differentiation 169-88
   hepatic differentiation 218-37

INDEX

mES-cell differentiation (cont.)
  lineage selection 144-5
  mesoderm differentiation 178-9
  muscle differentiation 130-68
    cardiomyocytes 130, 134, 136, 139
    electrophysiological measurements 164-6
    functional evaluation 163-6
    in vivo relevance 128, 130-1, 136, 141
    mES cell lines 131
    mES-cell maintenance
      requirements 131-2
      FCS 131
    muscle-cell isolation 142
    rate measurements 163-4
    skeletal myocytes 135-6, 138, 140
    SMC 130, 141-4
    VSM 134, 141-2
  neural differentiation 189-217
    mES cell lines 193
    neural differentiation strategies 189-90
    neural precursors 189
    generation 191-5
    selection 196-203
    neuronal lineage selection 198-203
    oligodendroglial lineage selection 202-4
  pancreatic differentiation 218-37
mES-cell differentiation in monolayers
  haematopoietic differentiation 169-88
    and see mES/OP9-cell coculture system
  neural differentiation 190
mES-cell differentiation via EBs 4, 113, 114, 116, 130, 194, 219
  EB differentiation 4, 113, 115
  EB disaggregation 142, 195, 223
  EB formation 4
  EB plating 222-3
    critical factors 222
  erythroid-myeloid differentiation 115
  haematopoiesis within EBs 169, 170
  haematopoietic differentiation 124, 125
  hepatic differentiation 226-7
  in CDM, hanging-drop culture 122-4
  in CDM, suspension culture 122, 123-4
  in hanging drops for
    cardiomyocyte differentiation 137-8, 139
    hepatic and pancreatic differentiation 221-2
    muscle differentiation 135
    SMC differentiation 140
    VSM differentiation 141-2
  in suspension culture for
    neural differentiation 194
  in vivo relevance 128
  mass culture for
    cardiomyocyte differentiation 138
    hepatic and pancreatic differentiation 222

mesoderm differentiation 124, 125
muscle differentiation 130-48 *and see* mES-cell differentiation
neural differentiation 190-203
neuroectoderm differentiation 124, 125
pancreatic differentiation 225-6
RNA isolation 125, 126-7
mES-cell maintenance 16-20, 132, 219
  caveats for haematopoietic differentiation 171
  caveats for muscle differentiation 131-2
  conditioned medium 8
  critical factors and euploidy 42, 43
  feeder-free culture 118
  LIF 8, 118
  nucleoside supplement 44-5
  other supplements 117
mES cell-NT 53-68
  by direct injection 54, 64-6
  by electrofusion 54, 64
  by Sendai virus-induced fusion 54, 62-63
  critical factors
    confluence of mES cells 56
    epigenetic status of mES cells 54
    recommended mouse strains 54-6, 69
  and see NT (mouse) embryos
mES cybrids 73, 81, 94
  culture and selection 94-7
  generation 90-3
  genotyping 98-9
  morphology 98
mES/OP9-cell coculture system 170-1, 173-6, 178-81
  analysis of derivatives 183-6
    FACS® analysis 177, 184-6
    May-Grünwald/Giemsa staining 177, 183, 185
  cytokine and growth factors 174
  evaluation of system 186-7
  key steps 174
  mesoderm differentiation 179
  role of FCS 173, 181
  time course of haematopoietic differentiation 178
mesodermal differentiation 115, 277, 279, 282
  influence of EB plating density 119
  mesodermal progenitor cells 279-83
microarrays 255, 268
mitochondrial
  disease 72, 78, 105, 106
  genetics 74
  oxidative stress 105
  pathogenic mutations 75
  proteome 74
monkey ES cells, *see* cES cells
mtDNA-deficient cells 83
  and see ρ^0 cells

mtDNA mutations 78
  age-related diseases 104, 105
  aging clock 104, 105
  capture 74, 78, 80, 81, 90-3
    by electrofusion 90-3
  characterisation 81-2
  complementation *in trans* 77
  heteroplasmy 73, 76, 94, 106
  homoplasmy 73, 76
  maternal inheritance 73, 75, 93, 106
  pathology 78, 104
  recombination 106
  segregation 76, 77, 79
mtDNA polymerase γ 72, 80, 105
mtDNA polymorphisms and sequence variation
  mouse strains 79
  mouse subspecies 78
MTG 114, 118, 119, 133
muscle-cell differentiation from mES cells 130-68
  isolation of cardiomyocytes 142-4
  isolation of skeletal muscle 142-4
  isolation of SMC 142-4
  muscle-specific gene expression 148-51
  muscle-specific proteins 156, 160-2
muscle-cell differentiation from hES cells
  cardiomyocytes 272, 285, 288
  v-SMC 269
mycoplasma contamination 20, 43
  detection 21, 22

neoplasia 265
neural differentiation 189-217
  and see mES-cell differentiation via EBs
neural progenitors 189, 191-5, 310
  characterisation and detection 211-15
    electrophysiological recordings 205, 206
    immunohistochemical 211, 214-15
    ISH 212-14
  functional assays 203-15
    transplantation models 203-11, 310
      in utero transplantation 206-11
      organotypic slice cultures 203-6, 207-8
NGF 118
nocodazole 56
NOD mouse 234-5
NT (mouse) embryos
  chemical and electrical activation 66-8
  embryo transfer 66
  and see mES-cell NT
nuclear reprogramming 6, 54, 69-70
nuclear transfer 5, 41, 43, 44, 53-70
NZB mice 79, 82, 106

oocyte collection (mouse) 59
oocyte enucleation 58-60

ooplasm transfer 106
OP9 cells 116, 170, 308, 316
  origin 170
  maintenance 171-3
OXPHOS 74, 78, 106

PA6 cells 190, 308-10
  origin 309
  maintenance 310-11
pancreatic differentiation 225-6
  characterisation 228
  and see mES-cell differentiation via EBs
pancreatic progenitor cells 223
  morphology 224
PAS reaction 139
Pax4 219, 225
PDM 223, 225
PNA 80
poly-L-ornithine 223, 225
primate ES cells 295
  cell-cycle time 296
  morphology 296
  role of LIF 296
  and see hES cells *and* cES cells
primitive erythroid-cell differentiation 174-6, 178-81, 316-17
  differentiation-inducing factors 175
  FACS® analysis 317
  morphology 175
primitive haematopoiesis, 169-88
  definition 169
  induction 178-81, 316-17
primordial germ cells 2
progenitor/stem cells
  hepatic 218, 224-6
  pancreatic 218, 234-6

Q-PCR 149, 154-5
qRT-PCR 149, 151, 153

ρ^0 cells 77, 80, 83, 84, 90, 96
RA (as a morphogen) 308
  in muscle differentiation 134, 140
  in neural differentiation 189-90, 314
reverse transcription 151-2
R6G 73, 77, 83, 94, 95-6
RNA isolation 148
ROS 80, 104
RT-PCR analysis 255-7, 270, 273-4, 276
  mES-cell derivatives
    hepatic derivatives 227-8
    muscle derivatives 148-53
    pancreatic derivatives 228
  hES cells and derivatives 255-7, 270, 273-4, 276

INDEX

SCF 174
SDIA-based neural differentiation systems 309
   critical factors 310
   for cES cells 309-16
   for mES cells 309
   neural precursors 310
   system refinements 309-10
serum, *see* FCS
serum-free medium,
   for cardiogenesis 134
   for neural differentiation 189, 191
   and see CDM *and* PDM
SMC 130, 136, 141-4
sperm mitochondria 75
spontaneous differentiation
   cES cell 297, 301, 302, 305, 310
   hES cells 241, 245, 249
      role of BMP 261
SRM 264
SSM 244
stem cell
   adult 234
   niche 5
   transplantation 234
STZ 235
suicide genes 290
superovulation (mice) 47
synaptosome isolation 88-90

T 115
tetraploid (mouse) embryo 41
   in vitro development 48
   in vivo development 45
   production by electrofusion 45, 47-8
tetraploid-morula aggregation 50-51
   lectin (PHA)-assisted morula aggregation 48-50
   mES cell lines 46
therapeutic cloning 105, 106
TG selection medium 87, 97
   6TG 87
$TGF\beta_1$ 134, 136, 141
tissue-restricted promoters 144
   muscle 145, 148

TK 87, 94, 96
transmitochondrial cybrids 76
   generation by electrofusion 90-3
transmitochondrial mice
   genotyping 103
   mitochondrial disease model 105-7
   phenotypic analysis 102-3
   production 73-4, 81
teratomas
   from hES cells 257-8, 265-7
   from cES cells 307
trophectoderm 240
trophoblast 260, 264, 299

universal cell 290

vasculogenesis 265
   and see hES-cell differentiation in monolayers
      and via EBs
VEGF 115
   in haematopoietic differentiation 174, 277, 279-84, 317
   vascular cell marker 274
voltage clamp recordings 164
von Willebrand factor 280
VSM 134, 141-2
   induction by db-cAMP and $TGF\beta_1$ 134, 136, 141
v-SMC 269, 277, 279-80, 282-3

wild mice 77, 78

xeno-mitochondrial cybrids 77, 83

zona pellucida (cynomolgus monkey embryo)
   removal 299
zona pellucida (mouse embryo)
   cutting 51, 61
   removal 49
zona pellucida (human embryo)
   removal 239, 240
zoonosis 261, 297